AF631397

AND THE BAMBOO FLOWERS IN THE INDIAN FORESTS

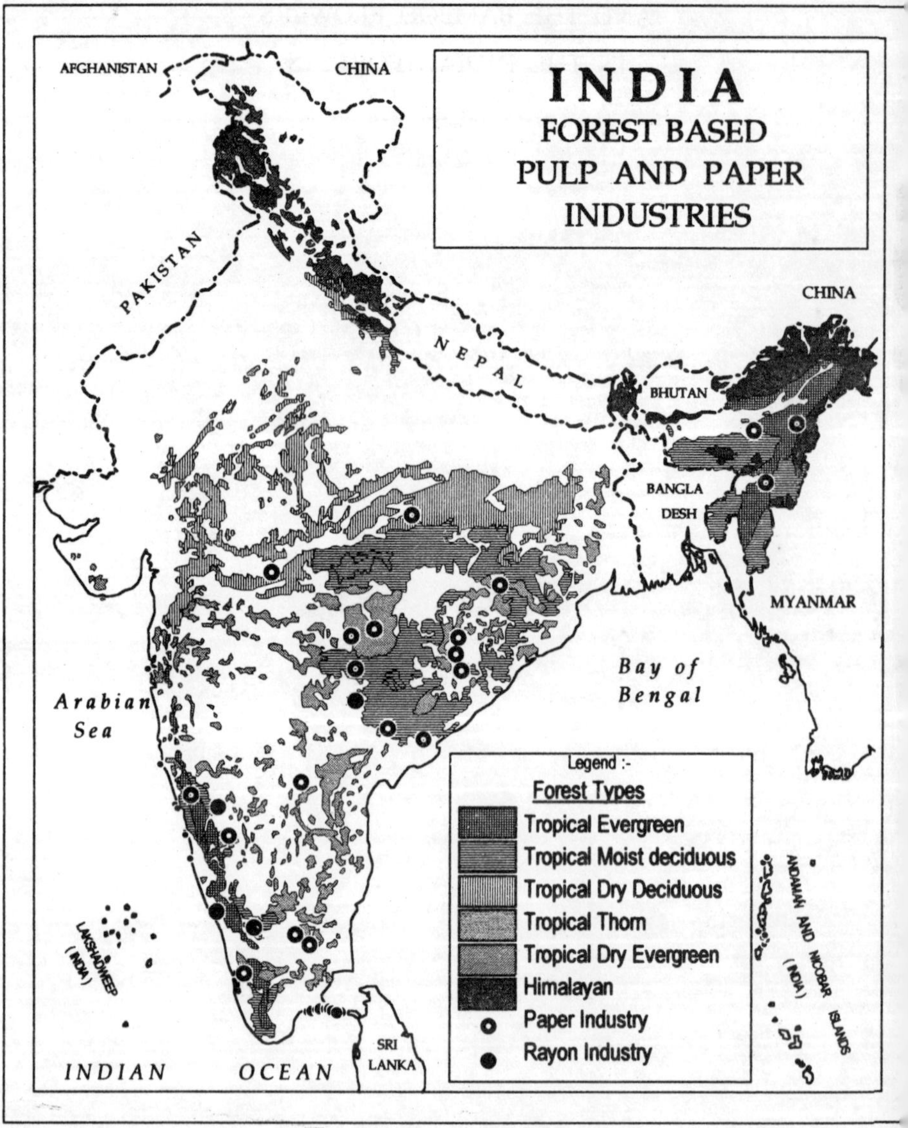

INDIA
FOREST BASED
PULP AND PAPER
INDUSTRIES
AFGHANISTAN
CHINA
PAKISTAN
NEPAL
CHINA
BHUTAN
BANGLA
DESH
MYANMAR
Bay of
Bengal
Arabian
Sea
LAKSHADWEEP
(INDIA)
ANDAMAN AND NICOBAR ISLANDS
(INDIA)
SRI
LANKA
INDIAN
OCEAN
Legend :-
Forest Types
Tropical Evergreen
Tropical Moist deciduous
Tropical Dry Deciduous
Tropical Thorn
Tropical Dry Evergreen
Himalayan
Paper Industry
Rayon Industry

And the Bamboo Flowers in the Indian Forests

What Did the Pulp and Paper Industry Do?

Volume I

MANORAMA SAVUR

IDPAD

MANOHAR
2003

First published 2003

© Manorama Savur, 2003

All rights reserved. No part of this publication may be reproduced or transmitted, in any form or by any means, without prior permission of the author and the publisher

ISBN 81-7304-413-9 (Set)
ISBN 81-7304-414-7 (Volume I)

Published by

Ajay Kumar Jain for
Manohar Publishers & Distributors
4753/23, Ansari Road, Daryaganj
Delhi 110 002

Typeset at

Kumud Print Service
Boileauganj, Shimla 171 005

Printed at

Rajkamal Electric Press
Delhi 110 033

For my beloved parents

RAMNATH AND RATNABAI SUJIR

who gifted me with sight, and
with whom I would have dearly wished
to share this experience

Contents

Figures

Maps

Tables

Preface

This study deals with the consequential effects that the growth of the forest-based pulp and paper industry has left on the tropical forests in India which are spread over nine states.

The pulp and paper industry is owned by the most powerful section of the Indian business class. This class has enough clout to manipulate the State structure and is responsible for both the denudation of bamboo forests and apparently for imposition of untried exotics on a massive scale on the natural forests. But eucalyptus as a pulpable species was introduced by FAO. It is important to know why. Being a state subject, the forest departments' reactions to such intrusions vary. There are also interesting differences with which state forest departments cope, or fail to cope in relation to the growing demand on their forests. These are often related to local social history and to the local environment movements. Forests have faced a variety of other pressures, starting with the pressure initiated by the colonial government. Here again there are interesting state-wise differences. We therefore feel that the wealth of information and the complex network of relationships should be analysed and presented state-wise. Only then will the policy implications of this study become clear and useful.

There were, of course, broad similarities which became the bases to test out the hypothesis that emerged through the course of the study and on the basis of the methodology employed. The similarities were also useful to arrive at some generalizations.

The project is funded by the Indo-Dutch Programme on Alternatives in Development (IDPAD) in collaboration with the Indian Council of Social Science Research (ICSSR), New Delhi, and the Institute for Social Science Research in Developing Countries (IMWOO), The Hague. IDPAD, however, does not take responsibility for the facts stated, opinions expressed and conclusion reached in this report. Recommendations are made as a part of IDPAD's mandate.

This study was made possible by the space given by the Western Regional Centre of the ICSSR Bombay and the unstinted cooperation and help by its Deputy Director Mr. Mohan R. Prabhu.

It is mandatory to list the participants in the project and indicate their precise contributions to the preparation of the project report.

The contributions of some are intangible. The tangible ones are:

1. T.S. Ramnarayan, M.Phil. (1 July 1991–30 March 1992)

   Collected basic source material on PPI and on BILT and NEPA. He suggested PPI should be analysed in terms of its pulping technology, types of paper it produces and the new raw material it uses.

2. P.V. Sunderashwar, M.Sc. (1 July 1991–30 April 1992)

   Visited forests of Chanda and M.P. Evaluated and wrote on the catchment areas of BILT (Ch. 4), NEPA and Orient (A) (Ch. 3).

3. P. Dongre, M.A. (1 July 1991–14 December 1992)

   Prepared Maps: Forest Types and Industrial Location (Frontispiece), Sources of raw material for BILT (Ch. 4). Volume of raw material supplied by FD and FDC to PPI (Ch. 10), Base maps (Chs. 2-4). Suggested: Marginal and Locational Maps for Ch. 1.

4. Shilpa Mahimakar, M.A. (1 July 1992–30 June 1994)

   Visited PPI units in Andhra Pradesh, Karnataka, Kerala, Tamil Nadu and wrote draft reports on them (Part 3 in Chs. 5-9).

5. B.M.H. Ahmed, Ph.D. (24 June 1992–April 1993)

   Collected data on forest raw material from Orissa. Visited catchment areas of PPI in Andhra Pradesh.

6. Anita Varghese, M.Sc. (1 June 1993–15 December 1993)

   Wrote on the natural and social history of the forests of Tamil Nadu. Visited forests of Karnataka, Kerala, Tamil Nadu, Assam and Nagaland and prepared draft report on the status of their catchment areas. Wrote on the misuse of effluents in Tamil Nadu

7. Manesha Divekar, M.A. (1 August 1993–6 April 1994)

   Prepared maps and Spatial and Temporal locations of PPI units (Ch. 1), Base maps for Chs. 5-9.

8. Anagha Gore, M.A. (1 August 1993–30 June 1994)

   Prepared maps on Spatial and Temporal Locations of PPI units (Ch. 1). Sources of raw material for BILT and PPI units in Andhra (Chs. 4-5), Movement of Birla, Singhania and Thapar as PPI Giants (Chs. 2 and 4), Base maps (Chs. 5-9).

9. Vijay Rao M.Sc. (August 2001–September 2002)

   Prepared the entire set of maps again as the original ones had deteriorated.

10. Indra Munshi Saldhana Ph.D. (1 July 1991–30 June 1994)

    Read the Report. Made useful corrections.

11. Swapna Banerjee Guha Ph.D. (1 July 1991–30 June 1994)

    Read the Report. Made useful suggestions.

12. Manorama Savur Ph.D. (1 July 1991–30 August 1994)

    Visited all the sites. Evaluated and wrote on:

    1. The catchment areas of PPI units in Orissa, AP, NE, Kerala and made a second evaluation of Chanda.
    2. PPI units in Orissa, MP, Chanda and Kerala.
    3. The natural and social history of all the regions except TN.
    4. The political economy of forest, paper and newsprint.
    5. Ecological question in Karnataka (Ch. 6) and on the social question of priority for bamboos in Kerala (Ch. 6).
    6. Introduction and the verses.
    7. Also counterchecked information at FAO, Rome in October 1995.
    8. Summary and conclusions and recommendations.

This project was a grand learning process for the entire team.

Mumbai MANORAMA SAVUR
15 August 1996

Acknowledgements

A study of this nature spanning so vast an area could not have been carried out without the support and cooperation of a large number of people and I am grateful to every one of them. They range from chief executives of powerful paper units to knowledgeable jeep drivers who unlocked the mysteries of the forests and the young researchers who shared their exciting findings with us. Not everyone can be formally acknowledged. All this was possible because the chief executives of state forest departments and the chief executives of the paper mills and UNFAO gave not only their time for discussions but set their teams to help us. I am most grateful for their kindness and their gracious hospitality. They are:

1. In Maharashtra—Dr. A.R. Maslekar, PCCF Maharashtra; A.C. Taneja, Chief G.M. BILT.
2. In Orissa—Dr. G.S. Padhi, PCCF, Orissa; Dr. A.N. Ahmad, G.M. OFDC; V.D. Bajaj, M.D., J.K. Industries: P.B. Gupta, M.D., Orient Paper Mills (B).
3. In Madhya Pradesh—the PCCF; V.C. Arzare, G.M., NEPA.
4. In Andhra—Dr. K. Kesav Reddy, PCCF, AP; H.K. Desai, M.D., APFDC; K.M. Banthia, G.M. SPM; M.P. Malliwal, M.D., ITCB; N.V.G. Rao, M.D. and L.M. Satyanarayana, G.M., APPM.
5. In Karnataka—S. Parmeshwarappa, PCCF, Karnataka; J. Lamba, M.D., KFDC; Vijay Gore, M.D. and C.H. Devraj, G.M. MPM; G.D. Maheshwari, M.D., WPM from N.K. Somani.
6. In Kerala—G. Mukundan, PCCF, Kerala; James Varghese, M.D., FDC; Surja Singh, M.D. and S. Bhargava, G.M., HNL.
7. In Tamil Nadu—M. Hari Krishna, PCCF, TN; P. Subramaniyam, IFS, TAFCON; N. Gopalaratnam, M.D., SPM; N.K. Tushu, G.M., TNPL.
8. In Assam—H.K. Chaudhari, PCCF, Assam; P.V. Rao, HPC; S.N. Rehman, NPM; R.S.D. Pandey, CPM; R.S. Tripathi, NPPC.

9. In Nagaland—K. Ao, PCCF, Nagaland.
10. In Rome—K.H. Schmincke, Director, Forest Products Division, FAO, Rome; J.P. Lanly, Director, Forestry Division, FAO, Rome.

My very special thanks go to three scholar foresters—Dr. A.N. Yellappa Reddy and Dr. S.N. Rai, Senior Conservators, Karnataka and to M.G. Gogte, CF Poona. To A.K. Mukherjee IG and Narayan Singh, Asstt. IG of Forests, New Delhi, for the time they gave for discussions on the methodology of the study, and of course, Dr. S.C. Mudgaul for his sympathetic understanding of the issues that were being investigated. Disagreement with FAO policy apart, I am most grateful for the gracious help offered by FAO officials of both forestry and paper division.

Many insights on the State Forests were given by retired foresters. I would particularly like to thank

U.N. Sarangi, retd. PCCF Orissa
Pushp Kumar, retd. PCCF and C.H. Kamalapati retd. CCF of AP
S. Shyam Sundar, retd. PCCF Karnataka
T.P. Vishwanathan, retd. PCCF Kerala
John Joseph, retd. PCCF Tamil Nadu
Dr. P.C. Gaswami, retd. PCCF Assam.

I am also grateful to Karl Jacques Lahussis, Wood Branch, Forest Product Division, FAO Rome and Dr. B. Kurkland, retd. Forest Product Division, FAO, Rome.

Among the scientists I would particularly like to thank Prof. Madhav Gadgil, ISC Bangalore; Dr. S. Shankar who put me in touch with his colleagues at KFRI, Peechi; Prof. V.M. Maher Homji of the French Institute Pondicherry; Prof. G. Obliswamy TN Agricultural University; Prof. Vinaya Rai, Mettapalayam; Prof. A.S. Juwarkar, NEERI Nagpur; Prof. S.B. Chapekar, Poona University; Dr. K.V.S. Rao, University of Bombay and Prof. R.S. Rao of Sambalpur University. I would also express my gratitude to Mr. Prasad Gogte for help with the maps.

Insight on industry-forest relations came from two young scholars. Arvind Sardana (Hoshangabad—M.P.) and B.K. Sudhakar Reddy (Hyderabad). Both had embarked on similar projects for their doctoral dissertations, and our understanding on the newsprint question came from Narayana Murthi of the India Newsprint Manufacturers Association, New Delhi.

I am exceedingly grateful for the generous grants given by Swiss

Aid, and to J.J. Bhabha who made money available from Dorabji Tata Trust.

I am also grateful for the gift of a free ticket by Air India, grants by Naval Tata Trust, N.M. Wadia Trust, Lotus Trust, and Pirozshah Godrej Trust, Thakurdas Trust, and to Kisanbhai Mehta. I also thank P. Anthony who so willingly helped me and V. Sathyanarayanan for typing this report and Janardhan Burkul for his gracious help.

The project involved constant reference work. I would like to place on record my appreciation for the help extended by Prof. Tikekar, Prof. Vijaya Rajhans and Shalini Rao of the University of Bombay Library and their team, to Issac Kimkar of Bombay Natural History Society, Bombay, the librarians of the forest departments at Bangalore, Trivandrum, Peechi, Hyderabad and Poona, FRI, Dehradun and the librarians of Centre of Development Studies, Trivandrum and Centre for Monitoring Indian Economy, Bombay; the State Bank of India, Bombay, IDBI and ICICI, Bombay and the librarians of FAO, Rome.

I am most grateful to Prof. D.N. Dhanagare, the former Member-Secretary of the ICSSR for smoothening out administrative problems, and to Dr. Shreekant Khandewale. Words cannot quiet express my gratitude to Prof. Ritu Dewan who as usual came to my rescue, and without whose unstinted help the draft report could not have been submitted and the information from FAO at Rome collected; and to Usha Hemmadi for improving the study by her editing.

Despite so much help, my spirits began to flag reading and re-reading the proofs till Abha Choudhuri, a geographer, happily took on the task of reading large sections of the typescript and for excellent editorial revisions in the final stage. Her willing and happy temperament revived me.

I am tremendously grateful to my husband Gopalrao Savur for bearing most patiently with me and my constant disappearances into the forests, leaving him to fend for himself. I also owe him gratitude for helping me with the graphs, tables and many textual matters.

MANORAMA SAVUR

Abbreviations

APFDC	Andhra Pradesh Forest Development Corporation, Hyderabad.
APPM	Andhra Pradesh Paper Mills, Rajahmundry.
APR	Andhra Pradesh Rayons, Kamalapuram.
ARC	Agriculture Refinance Corporation, Bombay.
BIFR	Bureau of Industrial Finance and Reconstruction.
BILT	Ballarpur Paper Industries Ltd., Ballarpur.
BOD	Biological Oxygen Demand.
BNHS	Bombay Natural History Society.
BPB	Bhadrachalam Paper Boards—(ITCB), Bhadrachalam.
CCF	Chief Conservator of Forests.
CF	Conservator of Forests.
CIDA	Canadian International Development Agency, Dehradun.
CMIE	Centre for Monitoring Indian Economy.
COD	Chemical Oxygen Demand.
CPPRI	Central Pulp and Paper Research Institute, Saharanpur.
CSIR	Council for Scientific and Industrial Research.
FAO	Food and Agricultural Organization of the United Nations.
FD	Forest Department.
FDC	Forest Development Corporation.
FRI	Forest Research Institute, Dehradun.
GRASIM	Gwalior Rayon Industries, Mavallor.
HNL	Hindustan Newsprint Ltd., Veloor.
HPC	Hindustan Paper Corporation, Calcutta.
HPF	Harihar Polyfibre, Harihar.
IFGTB	Institute of Forest Genetics and Tree Breeding, Coimbatore.
ITCB	Indian Tobacco Company Bhadrachalam, Bhadrachalam, formerly Bhadrachalam Paper Mills.
JK	JK Paper Mills.
KFDC	Karnataka Forest Development Corporation, Bangalore
KFDC	Kerala Forest Development Corporation.
KFRI	Kerala Forest Research Institute, Peechi.
MFDC	Maharashtra Forest Development Corporation, Nagpur.
MPM	Mysore Paper Mills, Bhadravati.

MPM	Mandya Paper Mills, Mandya.
MPRVVNL	(Nigam of MPFDC) Madhya Pradesh Rajya Vana Vikas Nigam.
NFP	National Forestry Programme.
NFAP	National Forestry Action Programme—India.
NABARD	National Bank for Agriculture and Rural Development.
NI	Newspaper Industries.
NP	Newsprint.
NPM	Nagaland Paper Mills, Tuli.
NPM	Nowgong Paper Mills, Nowgong.
NEPA	Formerly National Newsprint Ltd., E. Khandesh.
ODA	Overseas Development Agency, Britain.
OFDC	Orissa Forest Development Corporation, Bhubaneswar.
OPM(A)	Orient Paper Mills, Amlai.
OPM(B)	Orient Paper Mills, Brajrajnagar.
PAPRI	Pulp and Paper Research Institute, Raygada.
PCCF	Principal Chief Conservator of Forests.
PPI	Pulp and Paper Industry.
RBI	Reserve Bank of India.
RPM	Rayalseema Paper Mills.
RNI	Registrar of Newspaper Industry.
SFRI	State Forest Research Institute.
SEBI	Securities and Exchange Board of India.
SICOM	Small Industries Corporation of Maharashtra.
SIV	South Indian Viscose.
SIDA	Swedish International Development Agency.
SEOD	Stock Exchange Official Directory.
SPB(SPM)	Seshasayee Paper Boards (Mills).
SPM	Sirpur Paper Mills.
STC	State Trading Corporation.
TAFCORN	Tamil Nadu Forest Corporation.
tpa	tonnes per annum
TNPL	Tamil Nadu Newsprint and Paper Ltd.
USAID	United States Agency for Development.
WB	World Bank.
WCPM	West Coast Paper Mills

CHAPTER 1

Introduction

And out of the ashes rose the bamboo
To soothe the bruised earth
And succour man and beast alike.
But what did some men do
To make the bamboo sad
And burst into flower?

M.S.

The face of the Indian forests keeps changing. Forests are dynamic and evolve through the ages. Abundant sunshine and copious rainfall perhaps contributed to the fullest development of vegetation in India. The complexity of the Indo-Malaysian tropical rain forests had surpassed that of the Amazonia and Congo tropical rain forests (Whitemore 1975). However, the rapid changes in the last two centuries have not made for greater maturity or greater bio-diversity.

Forests: Reasons for Complexity

Most of the once dense forests in the 'paper belt', the area of this study, are found on mountain ranges and hill slopes of the Western and the Eastern ghats, the Vindhya and the Satpura ranges, the Assam hills and the Eastern Himalayas (Chs. 2-9). Elevation gives respite from the intense heat of the tropical sun to give rise to montane forests (Ch. 9) as well as complex *sholas* in sheltered depressions (Ch. 8). At lower elevations a combination of dense foliage evolved through the ages and the tree trunks which cut light and wind velocity helped to reduce evaporation. The microclimatic zones thus created, harbour moisture-loving, shade-tolerant species, eventually establishing symbiotic relations. The result is a series of complex but fragile multi-storeyed forests in the torrid tropics enjoying heavy seasonal rainfall. Geological movements contributed to the spatial drift of the species

(Ch. 5), as well as to the changes in the mineral content of the sites. Minerals are useful although many are generally absorbed only as trace elements; they have nevertheless acted on the specificity of the species. The richness of the soil being created by the forests themselves has become, along with the mineral content, the controlling factor in the distribution of plants. The rich fauna, even the browsing ones, have not destroyed forests. On the contrary they may have contributed in controlling the density of a plant species to make room for diversity. The first assumption of this study, therefore, is that forests are complex ecosystems and trees by themselves do not make a forest. Incidentally, the broad natural history of the forest belt also gives a spatial dimension to this study.

The second assumption is that forests are life-support systems. Their biological function is a recent discovery. But cultural historians have shown that forests were the most useful environment for the social and cultural evolution of humans (Gordon Chide 1951). People used products from nature for their development and this is the subject matter of our study. It is believed that there are no inherent contradictions between human beings and nature. Contradictions appeared very late in history,[1] when, instead of using nature's products wisely for social development, humans outstripped and exploited nature with impunity. Intruders, ignorant of well-established local people-forest relationships, were responsible for the breakdown of the forests ecosystem (Section 1, Chs. 2-9). In certain areas, for instance in western Orissa, the forests were so dense and the terrain so forbidding that as late as the eighteenth century, no conquerors had dared to enter them. Adventurers from neighbouring states, however, had no such inhibitions, and integrated themselves with the local culture, living in a condition of mutual tolerance, while the local people continued to live in harmony with nature.

This is history both natural and social, the continuum of which, when intercepted, provides answers about when and why the face of Indian forests began to change, and the regressive trend began. But the colonial foresters merely classified the original and the retrogressed forests into such categories as wet evergreen, moist deciduous, dry deciduous and scrub forests, in that order. Other disciplines categorized forests without explaining that the present state of the forests was the result of a retrogressive process, a rapid reversal in natural history.

If other disciplines have correlated the prevalent forest types with climate, relief, soil, drainage, etc., it is because of the attempts by the wet evergreen forests to adapt and rehabilitate themselves to a lower evolutionary stage within the given environmental conditions, as colonial and post-colonial pressures on the forests mounted. But it must be emphasized that there were no contradictions between the natural and the social history of the local people whether they were the *adivasis*, the original inhabitants of the forest, or the more advanced agricultural communities (Chs. 2, 3, 4 and 6). The contradictions arose only when the pace of change was accelerated by outsiders, a process which the vulnerable insiders could not rectify.

Role of State

The forests were vast and no group could destroy them without the backing of a powerful institution. The most powerful institution is the state. Viewed historically, the establishment of the nation-state is one of the landmarks in the social history of humankind. A state is all powerful. But that power is abstract and to a large degree controlled by the class in power. This complex relationship can be utilized by a powerful class to manipulate the state machinery even beyond its territorial boundary in order to extend control over extra-territorial areas. It first began in the late fifteenth century when the rich merchant classes in Europe utilized their state power to trade with the opulent Orient. The line between trade and plunder being faint, territorial control was the next logical step to acquire finished goods and, later, raw material for its own forward movement. The impact of the social history of the mercantile class on the natural history of Indian forests soon began to manifest itself. The concern of this study is with forest raw material.

The European mercantile class was the first to deal a death-blow to the fragile forest ecosystem, rich in timber trees. The nutrients accumulated in the biomass of the extraordinarily tall timber trees for a century or more, were lost to the system when the trees were felled. The nutrient cycle was broken and the nutrient content of the ecosystem was drained out and depleted. The 'enlightened' self-interest of the British 'foresters' towards reviving timber (Ch. 8) and other plantations on hill slopes was the second grievous onslaught (Ch. 6). The same factors, heat, rain and elevation, which had contributed to the richness and complexity of the Indian forest turned negative when large forest areas were cleared or when the canopy was opened (Ch. 2). The fierce tropical sun burnt the shade-and moisture-loving species as well as the micro-organisms in the soil. Rainfall, particularly concentrated in the monsoon season, often resulted in the soil and the humus leaching out, with no root system left behind to protect and hold them back. In other words, tropical forests are fragile ecosystems, which is the third assumption. Exploitable areas were exploited to the hilt and only unapproachable areas were left untouched. The German foresters, the experts called in to train the British, were themselves ignorant of the complex tropical forests. They struggled to understand, but their expertise in coping with forests enjoying much simpler temperate climates was of little worth and quite irrelevant to India. Their experiments went awry (Chs. 2 and 8). One of the objectives of this study is to provide an insight into the policies which led to the denudation of the forests.

The white man was much too arrogant to learn from the local population whose empirical knowledge of felling had been perfected through centuries of trial and error (Ch. 9) and preserved through folklore (Ch. 6). The colonial foresters failed to notice that the early methods of acquisition of knowledge were the same as those of modern science. Also slash-and-burn agriculture (Chs. 2-4, and 9), had not destroyed the forests. Soon compelled to cope with social reality, the British administrators realized that traditional methods were not harmful and recorded their observations accordingly (Ch. 6). Nevertheless, the forest dwellers were declared ignorant destroyers of forests by the government. This occurred when they resisted the interlopers' attempt to 'enclose' their homelands as Reserve Forests, reserved for timber exclusively for colonial, military and commercial purposes. The colonial foresters devised their own rules for felling.

Since the transfer of power, the Indian government and a sizeable section of its foresters have continued to echo this alien anti-people sentiment. They too fail to see that the people-forest ratio declined only with enclosures. The Planning Commission through its Five-Year Plans offered no alternate space, no alternate methods for the development of the forest dwellers. This is a tragic but inevitable result of the fact that the state is pressurized by one of the most powerful sections of the ruling classes which needs forests for its own growth and because of which it cannot tolerate even the original inhabitants (Ch. 2). How this class acquired so much power is discussed in the section, 'Antecedents to the Growth of PPI in India'.

Expansion of household cattle into herds and their pressure on the forests is again a product of British interference and misrule (Ch. 4).

distarbed of land and with no alternate sources of survival, the impoverished peasants began to rear cattle to be sold in times of acute crisis. If large herds of herbivores did not destroy undisturbed natural forests, and indeed played a positive role in their evolution, then in theory domestic herbivores cannot be used as a ruse to explain the exhaustion of forests (Ch. 4). But since the forests were disturbed, the area available to the people and their few animals diminished. Over a period of time common village lands were apportioned or put to a variety of other uses.

By the 1970s, non-browsable but pulpable eucalyptus was planted as fuel, fodder and food species, in common village lands and wherever space was available, and christened 'Social Forestry' (SF). The state progressively turned anti-social, this time under the advice of an international body, the forestry wing of the FAO of the UN, and was actively supported by other international agencies such as the WB, USAID, SIDA, etc. Even earlier, negative processes had been set in motion, and either deliberately or slavishly the 1952 Forest Act followed the British Act of 1894. The British had made 'Reserve Forest' synonymous with their national and industrial interests (see the 1894 Act). With such a definition, not only swollen herds but forest dwellers too become 'pressures' on the forests. And by the same definition they are pushed outside the context of national interests. SF is technically outside the scope of this study but is mentioned to show that more and more space would be occupied by exotic pulpwood trees.

This study examines the specific question of the impact of the pulp and paper industry on the bamboo forests in India which is its major objective. In fact, this study started with the assumption that production of cultural paper and newsprint are essential for a rounded development of India.

Bamboos are an understory, and bamboo forests are secondary forests due to anthropogenic interference but nevertheless an outgrowth which protects the ecosystem. The British foresters put it down to slash-and-burn, but clearing forests for plantations and opening out the canopy have had identical effects. The British foresters were annoyed with the proliferation of bamboo but later used it as a raw material for paper-making. There is no reason to believe that the British-owned paper industry was kinder to the bamboo than its Indian counterpart; perhaps it was, but so too was the Indian PPI; this, however was only till the late fifties when an alternate potential raw material was suggested by the

FAO. Having exhausted their own forests, the British turned to India for material for paper-making.

History of Paper Industry

The use of paper[2] in India started late. *Jadpatra* (palm leaf), *tamrapatra* (copper plates), *lohapatra* (iron sheets), *bhuajpatra* (beech), etc., were used for preserving the Vedas and other philosophical and literary writings during the ancient and classical period (Poddar 1979: 4).

The art of paper-making is believed to have reached India through the Arabs who initially learnt it from Chinese prisoners when they raided parts of China. The Chinese cultivated paper mulberry trees for cellulosic material. A less popular version is that some Indians who visited Mongolia learnt it directly from the Chinese. Whatever the origin of paper-making, the technique was kept a closely guarded secret by some Indian families, known as *Kagzis*, who flourished under the patronage of the Mughals. Initially the raw material used was rags, later it was mixed with birch bark or jute using *sajjikar*. Hindus did not like rags (Poddar 1979: 4-5).

William Carey in 1812, with the help of local *Kagzis* at Serampore, made efforts to mechanize the Indian paper industry. In 1820, came the first steam engine, and in 1832 the first Fourdinaire machine; but the mill failed due to government's apathy. Production began in 1870, with *sabai* grass and rags, at Bally Paper Mills, Hooghly, followed by the mills at Titaghur, Lucknow and Punalur (Poddar 1979: 5).

The technique was modernized in 1820, first with a steam engine, later by a Fourdinaire machine (MF). In 1905, the process of pulping bamboo was initiated by Pearson, while Raitt conducted the revolutionary studies. However, bamboo was not utilized as a pulpable material for paper-making till 1925, thanks to World War I and its after-effects. The number of mills then rose from six to nine between 1925 and 1937. Till 1947, these mills existed mostly in and around West Bengal while the raw material came mainly from East Bengal. There is no available information on the effects of the enormous out-turn of bamboo on the East Bengal forests. After Partition the mills turned to Orissa which, once rated as an inexhaustible source of bamboo, is now almost totally exhausted. There is no report to indicate the precise quantum drawn from East Bengal or the area exhausted by these mills. After Indian independence the British left India and their mills died a quiet death. Wealthy Indian merchants, who aspired to become an industrial bourgeoisie, entered the paper scenario directly in 1937 (Chs. 2, 4, 5), this being no easy task.

The Indian merchants' entry into this narrow industrial space, which had been a tightly-held British monopoly, was indeed historic. Their power over the state structure also needs to be explained.

Antecedents to the Growth of Indian PPI

India had a variety of industries. Its system of production including paper was familial and caste-based. The *Kagzis* for instance, produced paper and were patronized by the Mughals. European powers were attracted by the Indian commodities, especially by the variety of exquisite fabrics. Every one of these powers tried to get a foothold in India, but finally it was the British East India Company which established itself a political power in the void left by the decay of the Mughal empire.

It is also important to remember that the British East India Company broke India's basic economic structure by introducing unfamiliar land revenue system: the Permanent Settlement in 1793, and in some regions the *ryotwari* system. The revenue system spawned wealthy disinterested landowners and oppressive intermediaries, which resulted in a drastic fall in food grain production while killer famines became a regular feature.

Meanwhile, with the invention of the Spinning Jenny, the British chose to break Indian industry too and reduce the country to a hinterland, in order to draw its raw material. Ricardo and Adam Smith provided the rationale to justify trade and exchange, maintaining that these would be beneficial to both the countries entering into the deal (Kurien 1995: 29). But the newly independent USA rejected such theories and so did Germany. Both these countries raised strong tariff walls to protect their domestic industries and forged forward to become powerful industrial nations. India could not take these steps and thus suffered the consequences of the British colonial policy.

The earliest measure, taken by the colonial government in India was to introduce plantations, initially indigo—a dye for its cotton industry—and, later plantations of coffee, tea, spices, and of course timber trees in the fragile forest region which caused considerable damage to the forest ecosystem.

In the countryside, commercial crops, especially cotton, began to replace the food crops when the American sources began to dry up after their War of Independence and the Civil War. Incidentally, increased trading activity brought in more wealth to the trading communities in Gujarat and Kutch; the American Civil War, which created a crisis for the British cotton industry, gave an opportunity to some of these ambitious business families to set up cotton mills in Ahmedabad and in Bombay, though the British Parliament tried to block this attempt by administrative action as well as by its tariff policy in 1877 (Palme Dutt 1948: 148). But the even richer Marwaris from Rajasthan who used to control internal trade and finance and were to be found in all the nerve centres of India (Nehru 1948: 276), missed the nineteenth century opportunity to industrialize via textiles and had to bide their time. The Bombay Stock Exchange had closed its doors and the Rajasthan desert was sterile.

British colonial ambitions moved much beyond India. China had to be tamed and opium was a means of doing so. Opium grew in Rajasthan and it provided an opportunity to the wealthy financiers-cum-traders of the region to enter into a trading partnership with the British. They then moved eastwards mainly to Calcutta—the British commercial centre in the last decade of the nineteenth century-where small opportunities would open out and later larger ones would be seized, before their entry into the arena of paper production (see Map 1.1.).

This study had earlier indicated that the earlier British land revenue policy had led to repeated famines and numerous revolts. The revolt of 1857 is much publicized and it is definitely the spark that led to politicization of even the intelligentsia.

The Famine Commission Report (FCR) of 1880 and again of 1891 foresaw the danger and suggested the industrialization of India to stem the tide of growing discontent and political movements this generated. The British parliament protested against the FCR's suggestion of industrialization. But the colonial administrators in India, sensitive to the ground realities, encouraged it, mainly the opium trading agents—the potential PPI giants, its allies and trading partners to set up rice milling, cotton ginning and oil pressing units. Later in 1905, Lord Curzon set up a Department of Industries and Commerce. Nothing much happened, except that from this study's point of view research to evaluate cellulosic content of bamboo was initiated at the Imperial Forest Research Institute at Dehradun with the intention of strengthening of the paper industry in India. Incidentally, even so sagacious a person as Dorabji Tata had to wait till 1907 to set up his iron and steel

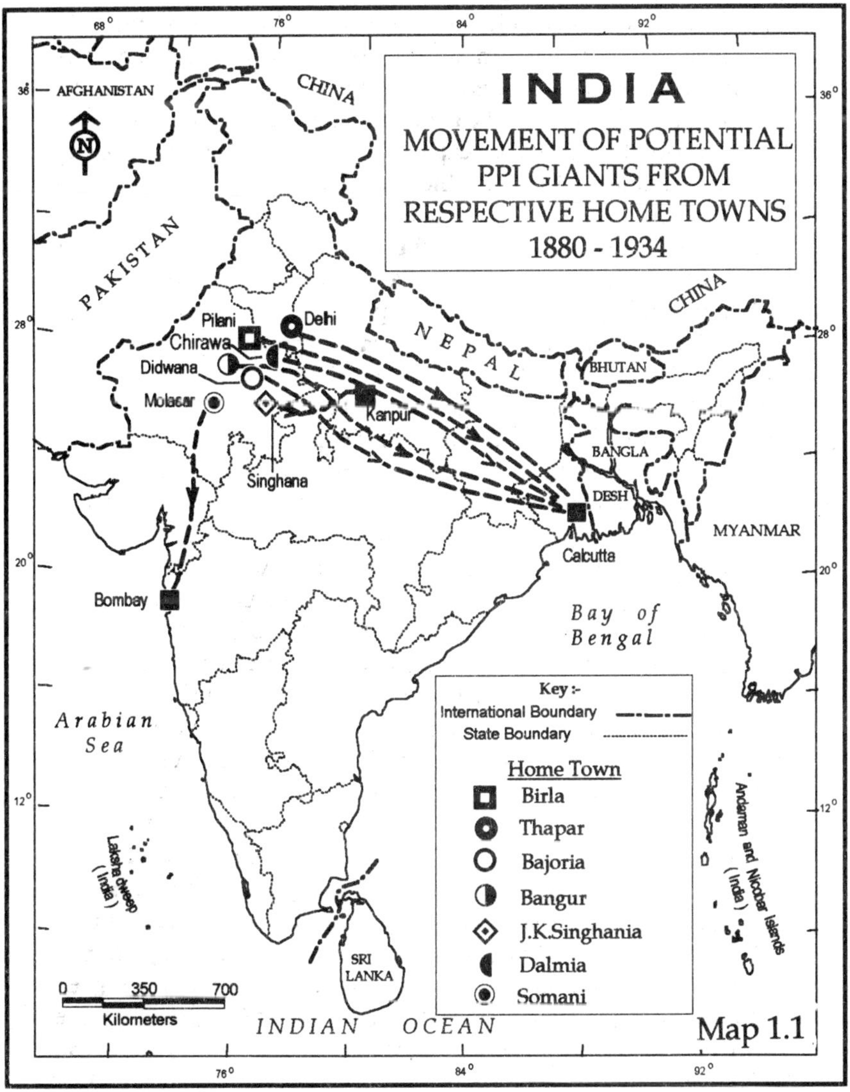
INDIA
MOVEMENT OF POTENTIAL PPI GIANTS FROM RESPECTIVE HOME TOWNS 1880 - 1934
AFGHANISTAN
CHINA
PAKISTAN
NEPAL
BHUTAN
BANGLA DESH
MYANMAR
SRI LANKA
Pilani
Chirawa
Didwana
Molasar
Delhi
Kanpur
Singhana
Calcutta
Bombay
Arabian Sea
Bay of Bengal
INDIAN OCEAN
Lakshadweep (India)
Andaman and Nicobar Islands (India)
Key :-
International Boundary
State Boundary
Home Town
Birla
Thapar
Bajoria
Bangur
J.K.Singhania
Dalmia
Somani
0 350 700
Kilometers
Map 1.1

works and was forced to raise the necessary equity capital from a small group of rich Indian financiers in Bombay, as the London Stock Market turned a cold shoulder to his project in 1882 (Elvin 1958: 36). Till, 1907, British investment in India remained as low as 11 per cent of its total foreign investment. It was largely confined to extractive industries, mining and plantations for the benefit of England. Palme Dutt (1948: 133) points out that the British capital invested in India, even for laying the infrastructure, was raised in India itself through plunder.

Industrial development was very slow till 1914. There was a complete reversal in the colonial policy after the First World War broke out. Industrialization was set as the aim in the economic field and responsible government in the political field as is evident in Lord Hardinge's dispatch to the Indian Secretary of State, dated 26 November 1915. Following this, an industrial commission was appointed in 1916 and Birla, the future paper giant, set up his first cotton mill in the Gwalior State in 1916. The political stranglehold was loosened by the Montague-Chelmsford Report on Indian Constitutional Reform, 1918. Palme Dutt (1948: 50-1) puts down the remarkable reversal in colonial policy to three factors: (a) for military strategic reasons; (b) to stem the tide of foreign competitors who might turn India into a dumping ground for their manufactured goods (protective tariffs were a means to prevent it, which in the long run would help the emerging Indian industrialists, particularly those who entered into production of pulp and paper); and (c) to secure the support of the Indian businessmen during political disturbances of the War and the post-War periods. The strategy was to give a few economic and political concessions and promise more to this class. By the Act of 1919, industry was made a Provincial subject. That year J.K. Singhania set up his first cotton mill.

A fiscal commission was appointed in June 1921, and a Tariff Board set up in July 1923, to work out discriminatory protection for British industry. The hopes of the Indian capitalist class were raised so high that its party, the Swaraj Party, defeated the non-cooperation politics of the Gandhian leadership in 1923. In 1924, fiscal autonomy was given to India and Singhania entered the field of jute and engineering. The British investment in India rose to 20 per cent of all its foreign investment and it would further rise to 72 per cent by 1947. During the inter-War years the industrial base broadened from cotton textiles and sugar to cement, iron, steel, chemicals, ferro-alloys, diesel engines, machine tools and paper. The Indian industrialists faced tough competition from the British, and the discriminatory policies of the British towards the newly emerging Indian industrialists frustrated them. They could not enter into the arena of paper industry. At least one senior member of these families quietly began to forge links with Gandhi (see Ch. 2). The Indian National Congress had the support of the people and the popular movements. The economic depression of the late twenties and the gathering war clouds in the mid-thirties sapped British power till the potential paper giants set up their units. It is a historical fact that this expediency occurred when the British power had weakened further and the setting up of popular governments was imminent. Sites were chosen in politically safe areas, as the British still ruled India (see Map 1.2).

Meanwhile, Birla, the most prominent among the emerging industrialists and leader of the PPI, is known to have maintained cordial relations with the colonial government and had played the role of a mediator. Singhania was knighted. Jayaprakash Narayan did not trust the industrialists who had military contracts with the British (Chopra 1994: iii). Popular movements under the leadership of the Congress were gaining ground and the British started investigating the Congress funding. G.M. Tyson, Chairman of the Public Relations Committee, Government of Bengal, noted that 'Birla was not only helping the Congress movement but was also trying to persuade the British businessmen to join hands with Indian businessmen in the demand for National Government' which would ensure British business interests after independence (Chopra 1994: iii). Among the PPI owners Birla's involvement with the Congress figures in the *British Secret Documents* in 1931 whereas Dalmia, Singhania and Karamchand Thapar, as funders of the Congress, figure in these documents only in 1942. (Chopra 1994: 59, 60, 81, 145). The industrialists continued to play an important role, sometimes behind the scenes, during the crucial period of the interim government (Chopra 1994: iv).

From small beginnings, the Indian capitalist class began to develop along with the overall growth of modern industries. Most of the industries grew in the inter-War period. World War II was an opportunity to grow further, but it was thwarted by the British as much of industrial production was only for the War efforts. Nevertheless, after independence, they moved rapidly forward leaving behind the small scale sector. Two of the most powerful entrepreneurs, viz., Birla and Dalmia, were owners of paper industries. Singhania and the Thapars forged ahead to the front ranks. Investment in PPI is heavy. At no point of time were there more than seven houses in control.

The growth of big business houses after the transfer of power was possible only because they opted for a huge public sector. There was a large component of this class in the Congress National Planning Commission (Nehru 1948: 423) and in the group of industrialists who formulated the *Bombay Plan* or *Tata-Birla Plan* towards the end of the World War II. India's Industrial Policy Resolution of 1948, as well as that of 1956, show that the State had agreed to step in only where private enterprises could not do it alone. State intervention was meant for rapid development, as also to prevent excessive concentration of economic power which would ruin small and middle business. This remained a theoretical proposition; concentration did take place, and big industrial houses took the centre stage as the Monopoly Commission reported (MCER 1965: 121-2). Big business now had the capacity to manipulate the State to its own advantage (Chs. 2-9) resulting in uneven development on the one hand and paper famine on the other, where the PPI was involved (Ch. 4).

World War II had created a scarcity of paper; prices soared. Some of the planners were already experienced owners in the paper industry. These planners had realized that paper was necessary for a newly independent country to catch up and move forward. Paper was necessary for vital sectors like education, banking and communications; therefore it was listed as one of the 17 basic industries of importance. It was an era of 'socialist planning' and inexpensive paper was necessary for free and compulsory education for children and the printing of text books for them. Retrospectively, it appears as an ideological prop to justify the close industry-State relations; whereas the actual growth of PPI was facilitated by numerous subsidies (see Maps 1.2-1.7).

This study had earlier indicated that by and large all the emerging industrialists in the paper sector hailed from Rajasthan; barring of course the two erstwhile native states which had progressive Dewans. The first six entrepreneurs to set up paper units were Birla, Bagoria,[3] Dalmia,[4] J.K. Singhania and the Mysore and Hyderabad States. Around 1946, M.C. Chidambaram Chettiar set up his rayon unit in Kerala at the invitation of C.P. Ramaswamy, the Dewan of Travancore-Cochin. The other most powerful business houses entered the paper industry in the fifties. These were Thapar[5] and Somani with financial assistance from Bangur. Bangur ousted Somani first from Andhra and then from the Karnataka unit. Not wanting to be pushed out of the core industry the Andhra and Tamil Nadu governments selected local entrepreneurs and strengthened them both financially and infrastructurally. Two international organizations took advantage

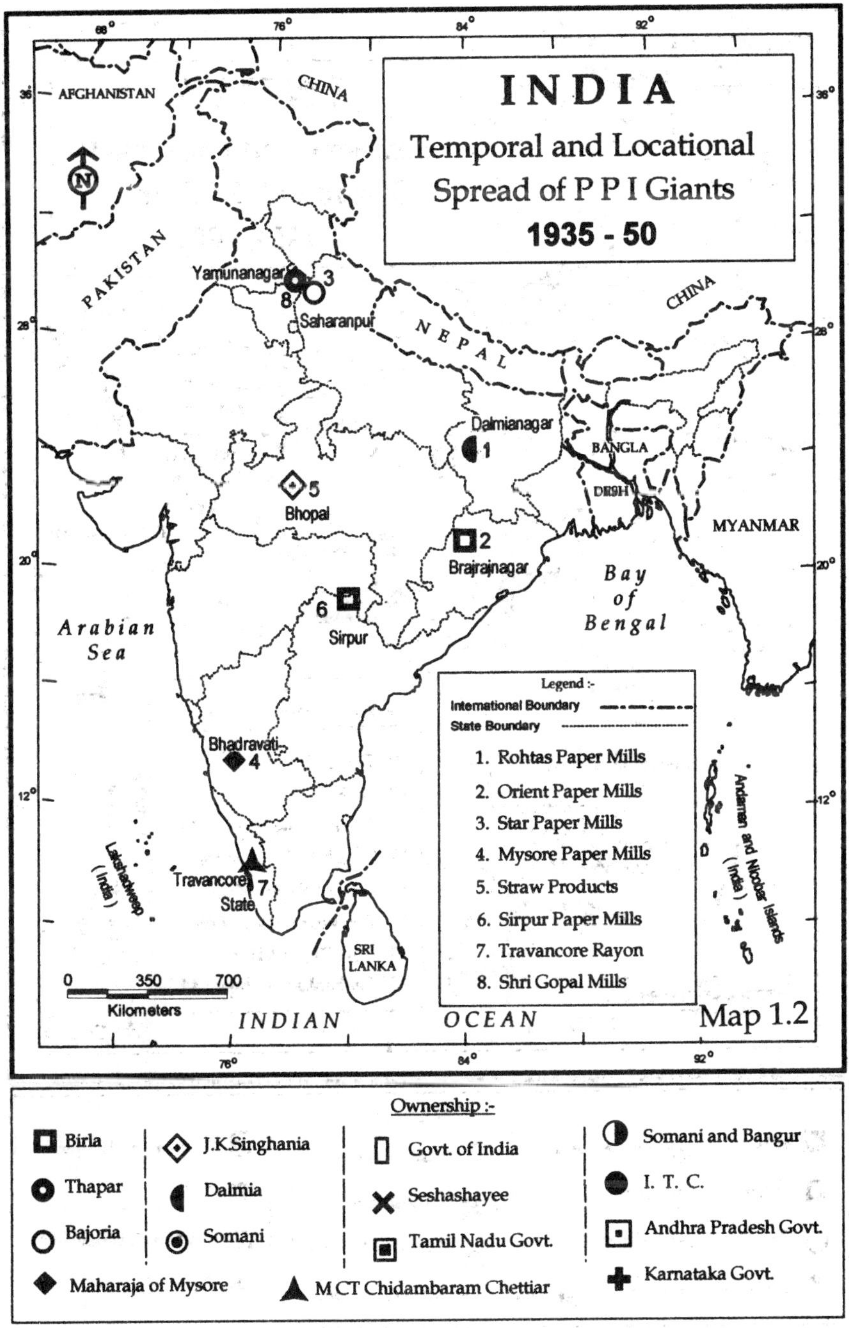
INDIA
Temporal and Locational Spread of P P I Giants
1935 - 50
AFGHANISTAN
CHINA
PAKISTAN
NEPAL
BANGLA DESH
MYANMAR
SRI LANKA
Arabian Sea
Bay of Bengal
INDIAN OCEAN
Lakshadweep (India)
Andaman and Nicobar Islands (India)
Yamunanagar
Saharanpur
Dalmianagar
Bhopal
Brajrajnagar
Sirpur
Bhadravati
Travancore State
Legend :-
International Boundary
State Boundary
1. Rohtas Paper Mills
2. Orient Paper Mills
3. Star Paper Mills
4. Mysore Paper Mills
5. Straw Products
6. Sirpur Paper Mills
7. Travancore Rayon
8. Shri Gopal Mills
0 350 700
Kilometers
Map 1.2
Ownership :-
Birla
Thapar
Bajoria
J.K.Singhania
Dalmia
Somani
Govt. of India
Seshashayee
Tamil Nadu Govt.
Somani and Bangur
I. T. C.
Andhra Pradesh Govt.
Karnataka Govt.
Maharaja of Mysore
M CT Chidambaram Chettiar

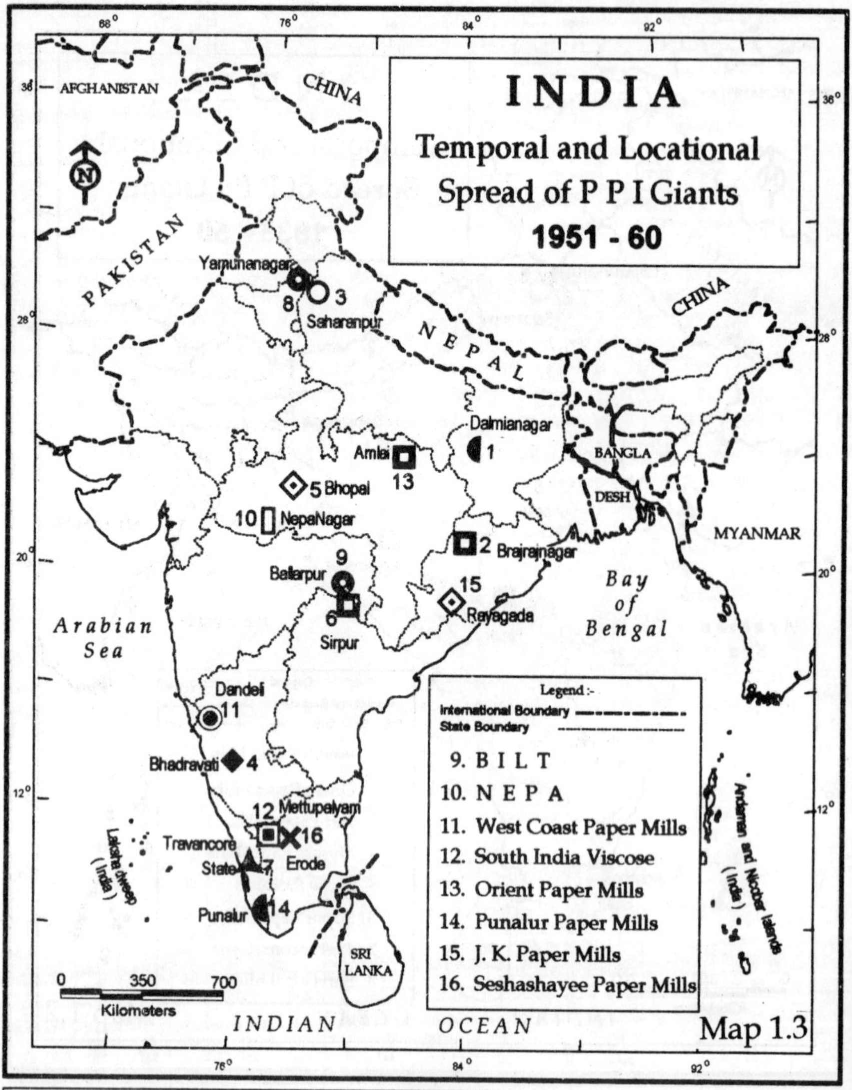

INDIA
Temporal and Locational Spread of P P I Giants
1951 - 60
68°
76°
84°
92°
36°
28°
20°
12°
AFGHANISTAN
CHINA
PAKISTAN
NEPAL
CHINA
BANGLA
DESH
MYANMAR
Arabian Sea
Bay of Bengal
INDIAN
OCEAN
SRI LANKA
Lakshadweep (India)
Andaman and Nicobar Islands (India)
Yamunanagar
8
3
Saharanpur
Dalmianagar
1
Amlai
13
5 Bhopal
10 NepaNagar
2 Brajrajnagar
9
Ballarpur
6
Sirpur
15
Rayagada
Dandeli
11
Bhadravati
4
12 Mettupalyam
16
Travancore State
7
Erode
Punalur
14
0
350
700
Kilometers
Legend :-
International Boundary
State Boundary
9. B I L T
10. N E P A
11. West Coast Paper Mills
12. South India Viscose
13. Orient Paper Mills
14. Punalur Paper Mills
15. J. K. Paper Mills
16. Seshashayee Paper Mills
Map 1.3
Ownership :-
Birla
Thapar
Bajoria
Maharaja of Mysore
J.K.Singhania
Dalmia
Somani
M CT Chidambaram Chettiar
Govt. of India
Seshashayee
Tamil Nadu Govt.
Somani and Bangur
I. T. C.
Andhra Pradesh Govt.
Karnataka Govt.

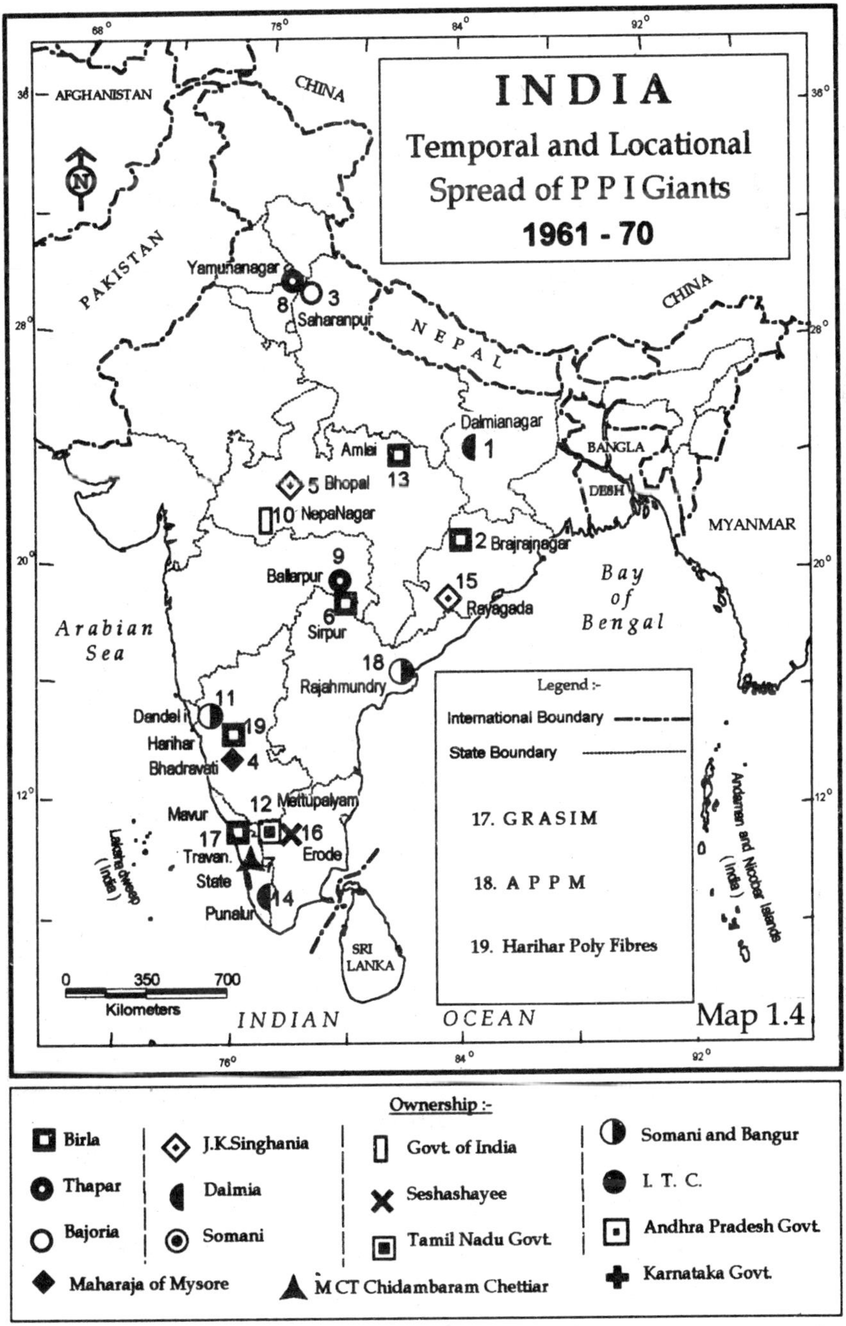
INDIA
Temporal and Locational Spread of P P I Giants
1961 - 70
AFGHANISTAN
CHINA
PAKISTAN
NEPAL
BANGLA DESH
MYANMAR
SRI LANKA
Arabian Sea
Bay of Bengal
INDIAN OCEAN
Andaman and Nicobar Islands (India)
Lakshadweep (India)
Yamunanagar 8
3 Saharanpur
Dalmianagar 1
Amlei 13
5 Bhopal
10 NepaNagar
2 Brajrajnagar
9 Ballarpur
6 Sirpur
15 Rayagada
18 Rajahmundry
11 Dandeli
19 Harihar
Bhadravati 4
12 Mettupalyam
Mavur 17
16 Erode
Travan. State 7
14 Punalur
Legend :-
International Boundary
State Boundary
17. GRASIM
18. A P P M
19. Harihar Poly Fibres
0 350 700
Kilometers
Map 1.4
Ownership :-
Birla
Thapar
Bajoria
Maharaja of Mysore
J.K.Singhania
Dalmia
Somani
Govt. of India
Seshashayee
Tamil Nadu Govt.
M CT Chidambaram Chettiar
Somani and Bangur
I. T. C.
Andhra Pradesh Govt.
Karnataka Govt.

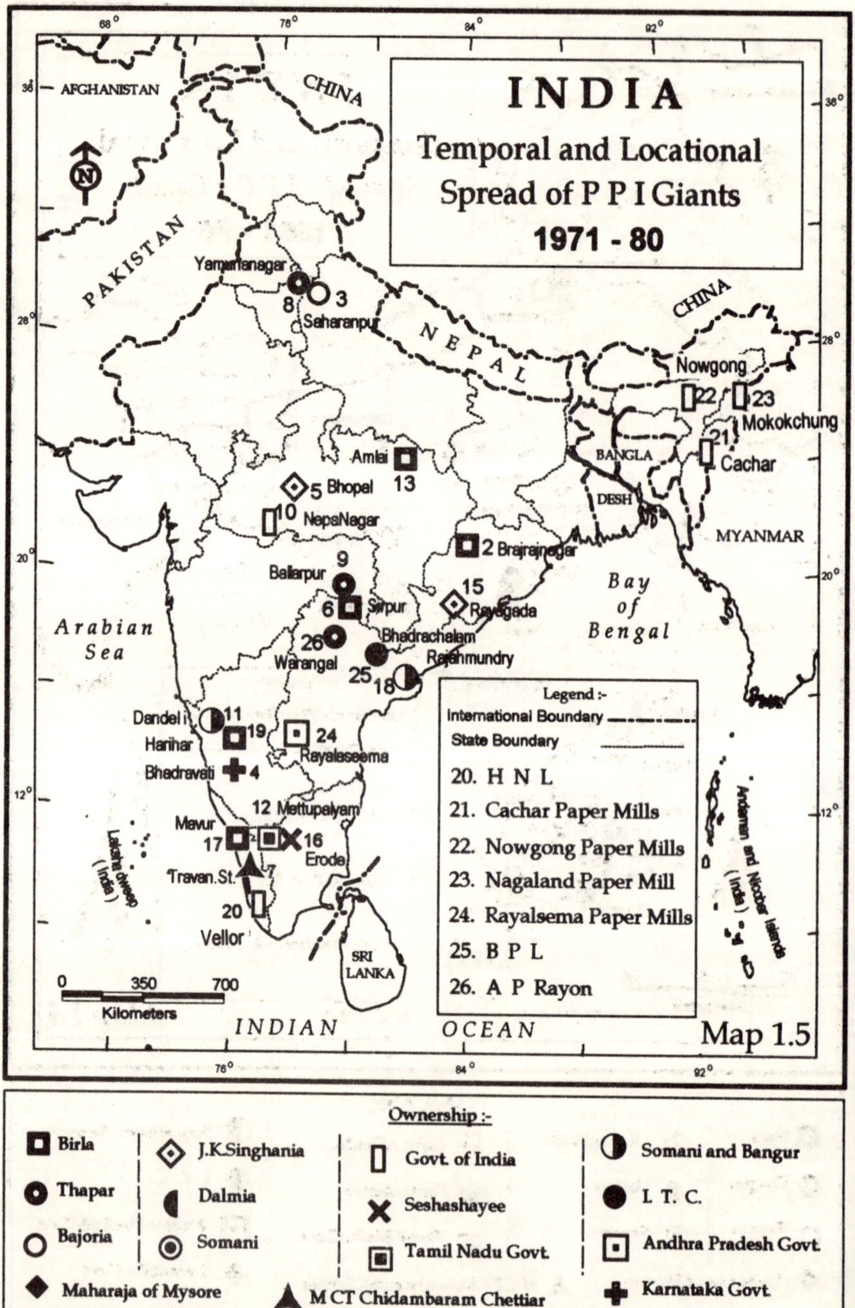
INDIA
Temporal and Locational Spread of P P I Giants
1971 - 80
AFGHANISTAN
CHINA
PAKISTAN
NEPAL
BANGLA DESH
MYANMAR
Arabian Sea
Bay of Bengal
INDIAN OCEAN
SRI LANKA
Lakshadweep (India)
Andaman and Nicobar Islands (India)
Yamunanagar
8
3
Saharanpur
Nowgong
22
23
Mokokchung
21
Cachar
Amlai
13
5 Bhopal
10 NepaNagar
2 Brajrajnagar
9
Ballarpur
6 Sirpur
15
Rayagada
26
Warangal
Bhadrachalam
Rajahmundry
25
18
Dandeli
11
Harihar
19
24
Rayalaseema
Bhadravati
4
12 Mettupalayam
Mavur
17
16
Erode
Travan.St.
7
20
Vellor
Legend :-
International Boundary
State Boundary
20. H N L
21. Cachar Paper Mills
22. Nowgong Paper Mills
23. Nagaland Paper Mill
24. Rayalsema Paper Mills
25. B P L
26. A P Rayon
0 350 700
Kilometers
Map 1.5
Ownership :-
Birla
Thapar
Bajoria
Maharaja of Mysore
J.K.Singhania
Dalmia
Somani
M CT Chidambaram Chettiar
Govt. of India
Seshashayee
Tamil Nadu Govt.
Somani and Bangur
I. T. C.
Andhra Pradesh Govt.
Karnataka Govt.

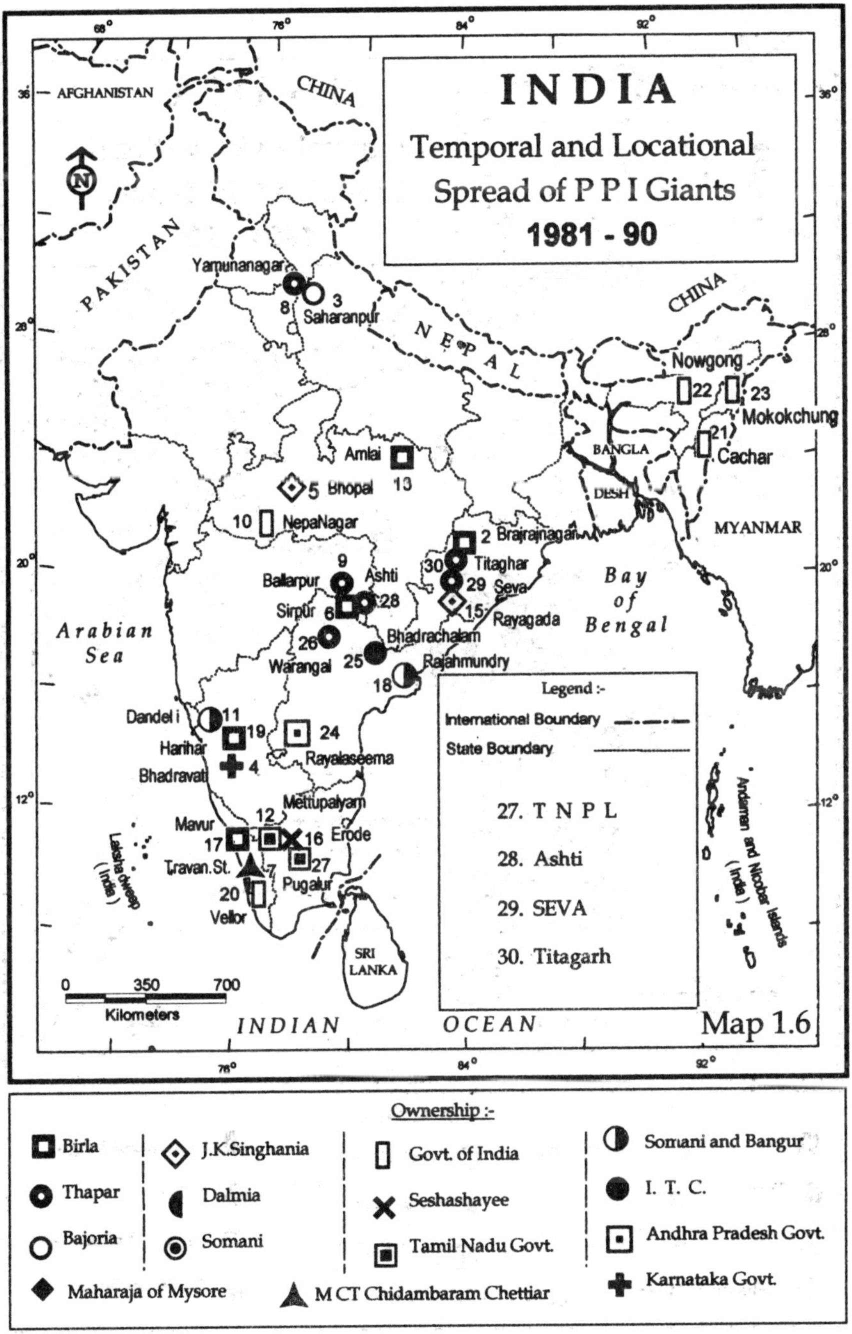
INDIA
Temporal and Locational Spread of P P I Giants
1981 - 90
AFGHANISTAN
CHINA
PAKISTAN
NEPAL
BANGLA DESH
MYANMAR
Arabian Sea
Bay of Bengal
INDIAN OCEAN
SRI LANKA
Lakshadweep (India)
Andaman and Nicobar Islands (India)
Yamunanagar
8
3
Saharanpur
Nowgong
22
23
Mokokchung
21
Cachar
Amlai
13
5 Bhopal
10 NepaNagar
2 Brajrajnagar
30 Titaghar
29 Seva
15 Rayagada
9
Ballarpur
Ashti
Sirpur
6
28
26
Warangal
25
Bhadrachalam
Rajahmundry
18
Dandeli
11
19
Harihar
4
Bhadravati
24
Rayalaseema
Mettupalyam
12
Mavur
17
16
Erode
Travan.St.
7
27
Pugalur
20
Vellor
Legend :-
International Boundary
State Boundary
27. T N P L
28. Ashti
29. SEVA
30. Titagarh
0 350 700
Kilometers
Map 1.6
Ownership :-
Birla
Thapar
Bajoria
Maharaja of Mysore
J.K.Singhania
Dalmia
Somani
M CT Chidambaram Chettiar
Govt. of India
Seshashayee
Tamil Nadu Govt.
Somani and Bangur
I. T. C.
Andhra Pradesh Govt.
Karnataka Govt.

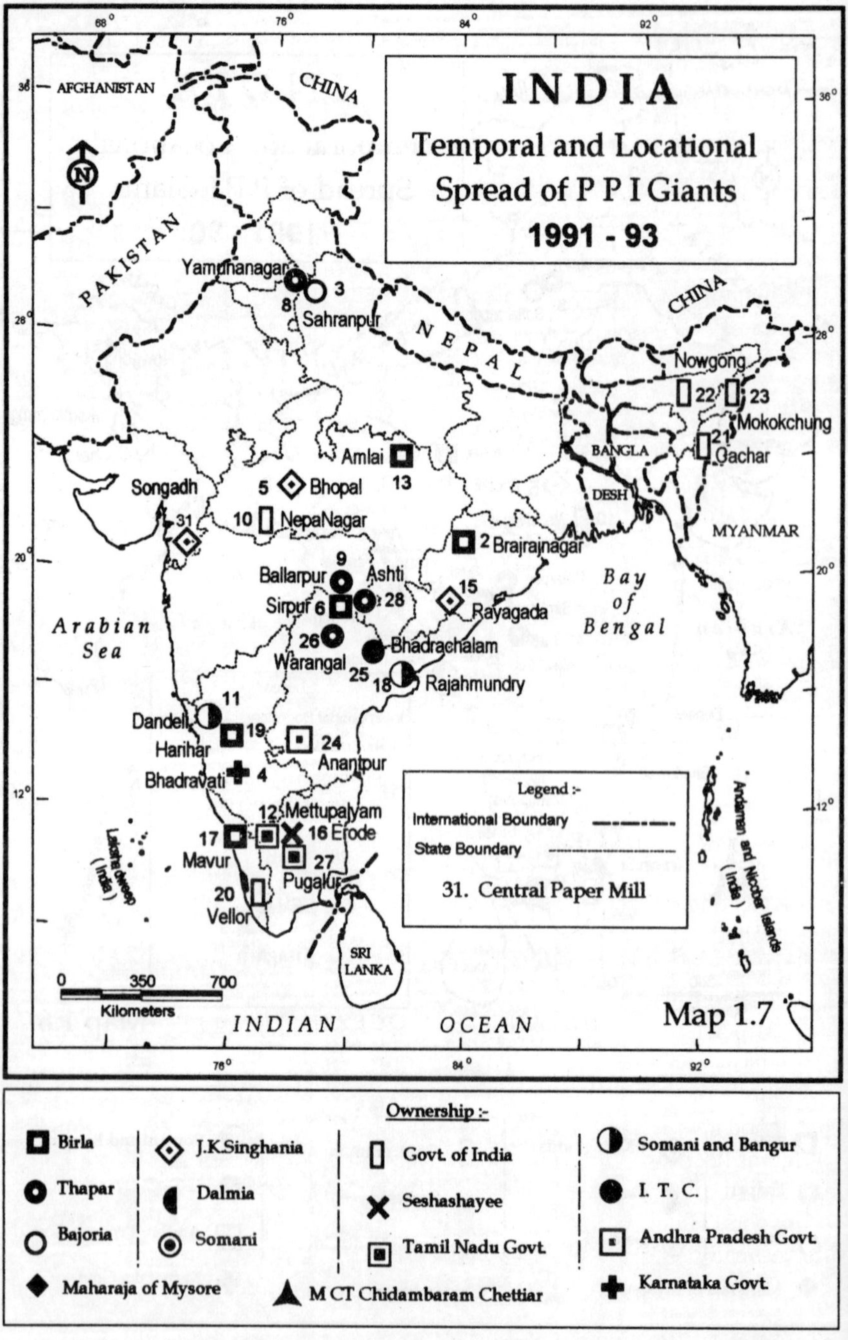

INDIA
Temporal and Locational Spread of P P I Giants
1991 - 93
AFGHANISTAN
CHINA
PAKISTAN
NEPAL
BANGLA DESH
MYANMAR
SRI LANKA
Arabian Sea
Bay of Bengal
INDIAN OCEAN
Yamunanagar 8
3 Sahranpur
Amlai 13
5 Bhopal
10 NepaNagar
Songadh 31
2 Brajrajnagar
9 Ballarpur
Ashti 28
Sirpur 6
15 Rayagada
26 Warangal
25 Bhadrachalam
18 Rajahmundry
11 Dandeli
19 Harihar
24 Anantpur
Bhadravati 4
12 Mettupalyam
17 Mavur
16 Erode
27 Pugalur
20 Vellor
Nowgong 22
23 Mokokchung
21 Cachar
Andaman and Nicobar Islands (India)
Laksha dweep (India)
Legend :-
International Boundary
State Boundary
31. Central Paper Mill
0 350 700
Kilometers
Map 1.7
Ownership :-
Birla
J.K.Singhania
Govt. of India
Somani and Bangur
Thapar
Dalmia
Seshashayee
I. T. C.
Bajoria
Somani
Tamil Nadu Govt.
Andhra Pradesh Govt.
Maharaja of Mysore
M CT Chidambaram Chettiar
Karnataka Govt.

of the situation and entered into the scene, one as a joint sector unit and the other utilizing state resources for strengthening itself (Ch. 5). Obviously therefore the political economy of the paper industry becomes an important component of this report.

Anthropologically viewed, Birla, Bagoria, Dalmia, J.K. Singhania, Thapar and Somani hail from the north-west. Thapar is from Delhi and the rest are from Rajasthan. Except for Singhania, their movement was via Calcutta. Somani, though based at Bombay, sought finance from Calcutta-based Bangur. Singhania found his route via Kanpur. None of these PPI giants, however, as is evident in the analysis of the paper industry, is fully the focus of this study.

Today, the growth of paper industry, however skewed, is threatened. There are several reasons for this, the immediate one being the so-called liberalization of the economy and the relentless push of international capital to which the government has succumbed. The big paper corporations are gearing up to face this situation by reverting to production of less profitable printing and writing paper for which there would be no competition from American or European paper giants. Secondly, they are saving on fuel and using stringent chemical recovery processes. This implies that the private sector misused the concessions given by the government. The concessions were in terms of very low tariff on power, and reduction in excise duty on critical chemicals (First Five-Year Plan: 171-7). The largest concessions were in terms of low royalty on bamboo, a mere rupee for a tonne (Ch. 6). Throw-away prices of forest raw material resulted in squandering what was believed to be a renewable resource. The central government instead of curbing the industry, first brought the Forest Department, which was a state subject, under its control by putting it on the concurrent list. Secondly, it called a meeting of the state agricultural ministers in October 1985 to ask their respective Forest Departments to maintain a regular supply of wood to PPI.

In fact, the concern started early. The FDs were also asked to carry out forest surveys to evaluate the available resources (Second Five-Year Plan). The government's move is understandable. In the Second Plan period it was in a hurry to speed up industrial development. It offered to grant licenses to import machines for production of specialty papers and in anticipation put a ban on the import of expensive specialty papers. The PPI did not immediately grab this opportunity; it was not yet ready and possibly because of the price control order. This order was finally lifted in 1968 when the industry as a whole deliberately created a paper famine in order to increase prices. The industry flouted its obligation to produce levy paper, steadily moving towards production of high value-added specialty papers.

Polite requests by the government and its several Price and Production Control Acts in the seventies had little effect. The threat of the Emergency helped for a short while. The government again took two defensive measures. On the one hand it invited small entrepreneurs, this time offering several attractive concessions to them to set up small scale paper mills, the only stipulation being that they use non-conventional raw material. The forest raw material was reserved for its favoured big units. The small entrepreneurs accepted the challenge and delivered the goods, only to be sacrificed in 1987 at the suggestion of FAO (Kurkland: personal communication). Application of MODVAT was all that was needed for this sector to collapse (Ch. 4). At the same time, unsure of the ability of the small entrepreneurs, the government defensively set up its own mills in a geographical area rejected by the PPI as totally unsuitable (Ch. 9). When the public sector units began to produce a sufficiently large quantity of printing and writing paper, the central government lifted the Price and Paper Control Orders and absolved the large PPI units of their social obligation. Their accumulated stock of cream wove-laid paper was, furthermore, also permitted to be sold to the newspaper industry. The shrewder of the PPI units recovered and soared. But the move boomeranged on the government's own newsprint units (Ch. 3).

Every one of the Newsprint (NP) units is in the public sector. The central and the state governments set up the NP units as the private sector was unwilling to do so. The investment in NP is much heavier than in a paper unit while the returns are considerably lower. The post-1980 units are all automated and equipped with more advanced pulping technology that saves scarce forest raw material. The new pulping technology utilizes as much as 80-85 per cent of the cellulosic raw material, in sharp contrast to 30-35 per cent used by the private sector. Unlike the private sector, which uses the obsolete and dangerous chlorine dioxide process to bleach its pulp, the newsprint units use the modern but expensive hydrogen peroxide (Ch. 7). Despite this, the public sector units, except perhaps the one backed by the World Bank (Ch. 8), were under threat of closure in manoeuvers of the newspaper industry. The process which began sometime in the 1950s and was initiated by newsprint dealers was deliberately ignored by the Ministry of Information and Broadcasting (Ch. 7).

Forest-based wood is pulped not only for paper and newsprint but also for viscose rayon. This is not only a fall-out of but is an aberration in the history of the pulp and paper industry in India, especially when the forest raw material is not easily renewable. While no attempt has been made to understand the growth of this industry, Map 10.1 (Ch. 10) shows the drain on the forests often exceeds that of a socially useful paper and newsprint unit within the same state. The paper and newsprint industry is therefore considered only in terms of its relations with the forests. One of the rayon units does not draw raw material from the forests but resorts to farm forestry. The types of exotic species cultivated for the purpose are often mismatched to the site and hence detrimental to the fertility of the soil.

Research in the use of bamboo for production of rayon was initiated by Birla in 1954. The first plant was set up in Nagda in MP, then the industry moved on to Kerala and then to the neighbouring state of Karnataka when the raw material was getting exhausted. Eucalyptus plantations were raised to feed the mills (Chs. 3, 6 and 7). Rayon is therefore rejected by this team as socially and ecologically wasteful. Another rayon unit is given preferential treatment by the state over paper and newsprint (NP), and fed with questionable species grown in ecologically vulnerable high forests (Ch. 8).

Two of these rayon units belong to Birlas who continue to rate as one of the most powerful business houses (Ch. 7). Now the Thapars have also claimed this status (Ch. 4).

Birla is known to have blatantly used the State structure for his own growth and to acquire preferential treatment (Chs. 2, 3, 6 and 7) much to the annoyance of other PPI units (Chs. 6 and 7). Rayon figures peripherally in this report because we regard it as a drain on forests, and no more.

The large industry continued to ravage the bamboo and almost wiped it off the face of the Indian forests. An alternate raw material was promised, viz., eucalyptus under the advice of the FAO which will be dealt with later. One of the objectives of this study is to understand the ecological consequences of setting up plantations of exotic species in forests and on farm lands.

FAO and Eucalyptus Plantations in Indian Forests

This raises the question of why the Food and Agriculture Organization (FAO) and to some extent the World Bank (WB) are deeply involved in both forestry plantations and with India's PPI. Viewed ideologically their agenda is much broader, viz., post-War reconstruction under the leadership of the USA.

FAO was the first UN organization, set up in October 1945, followed by the WB in December 1945. In India, WB works in tandem with FAO by providing loans for social and farm forestry with pulp wood species. It also periodically reviews the functioning of some of the State Forest Departments. WB's formal function is to provide long-term capital requirements (loans) to national governments.

The antecedent to the setting up of FAO was the US President Roosevelt's request to convene a meeting on Food and Agriculture [*Encyclopaedia Britannica* (*EB*) 1992: vol. XXIX, no. 149] in 1943. FAO was transferred to Rome in 1951 where the International Institute of Agriculture (IIA) was based since 1905; IIA's function was to protect European farmers from slumps and gluts. FAO, however, was meant for the third world.

The decision to include a Forestry and Forestry Product wing in an organization meant for increasing food and agriculture to relieve hunger in the third world, battered by war and colonial pressures, was brought up during the Second World Food Congress, at the end of three years of debate between 1943-5 in the Congress (FAO 1968: 1). The arguments varied from the beneficial effects of forests on agriculture through its action on hydrology and soil protection, which in turn would raise agricultural production, to provision of fuel wood and benefits of forest-based industries to the host country in which FAO functions. FAO saw milling, plywood, particle boards, pulp and paper as forest-based industries to improve the living standards of the people and states that it is working towards it (FAO 1968: 35, 42). It argues that the rich and varied forest resources of the tropics are sources for export since neither the US nor Europe has timber comparable to the tropics. Its main argument is that extraction and sale of these products will improve the balance of payments which is imperative for countries that cannot produce sufficient food for themselves (FAO 1968: 54, 57 and 59). Extra budgetary funds became available from 1950 onwards for FAO's operation in Burma, Ceylon and India with the creation of UN Expanded Programmes of Technical Assistance. These countries had efficient British trained forest services, unlike in the vast forests of Latin America and Africa; and the first assistance was for logging and extraction of timber (*Journal of Forestry* 1951: vol. XLIX, no. 8) and only later for PPI. Such a policy had negative effects on the Indian forests (Chs. 8 and 9). Our main concern is with the FAO's involvement in PPI and pulpwood plantations which it initiated to feed the PPI in India. Incidentally, FAO's net is flung very wide over all the tropical forests located in former colonial countries.

In 1947, UNESCO, another UN body, was concerned with a need for adequate supply of paper and newsprint, and FAO under its charter had to ensure the output of forests both in quality and quantity to fulfil that need (FAO 1954: 1). The direction and scope of its energetic programme was derived from the Resolution adopted at the FAO conference at its sixth session in 1951 (FAO 1954: 2). It may not be out of place to mention that the US did not have sufficient raw material for its PPI. Till the mid-1950s it had to import pulp from Europe as its own pine forests were in inaccessible regions and Canada was not under its shadow at that time. Even more interesting is the fact that the US had started research in tropical wood (broad leaf) for pulp in the 1930s and its efforts were intensified during World War II (FAO 1954: 61). FAO's two-pronged strategy was to find raw material for the paper industry and to start PPI in countries that did not have unlimited foreign exchange to buy paper, the demand for which was rising (FAO 1954: 6).

Did FAO take its mandate to identify pulpable species of tropical hard wood and produce adequate quantities of it overzealously? Did the Indian PPI in anticipation of the new raw material become careless with bamboo felling and destroy the forests? These were questions that came to the fore as our research proceeded.

In 1951 UNESCO asked FAO to undertake investigation to determine the possibility of expanding manufacture of pulp and paper (PP) and special attention was to be given to identify tropical hardwoods besides other raw materials (FAO 1953: 1). In April that year the UN Secretary General sent out an appeal to member nations to invite technical research on substitute material for PP making; 15 countries responded and India was one of them. In 1952, the Secretary General sent out letters to member governments asking them to invite FAO to dispatch to their countries expert missions to study the possibility of creating and developing PPI, and again the same 15 countries responded. In 1951 UNESCO had collected information on the production, trade consumption and prices of paper and newsprint (FAO 1954: 1). It found India was self-sufficient in paper and board and a newsprint mill was being set up with *Boswellia serrata* as the raw material which was found close to the mill site. The report (FAO 1954: 33-4) also states that FAO's preliminary investigation had revealed that there was plentiful bamboo in the country to last 20 years but in the long run it would have to turn to eucalyptus plantations and to Himalayan conifers. It held Sweeden's man-made forestry as a model without taking note of the sharp differences in the natural and social history of the two countries. This issue is discussed later (Ch. 10). The main question is: why did FAO think in terms of eucalyptus as early as 1952 when it was impressed with the excellent pulpable qualities of bamboo (FAO 1953: 110)?

In 1951 FAO missions carried out preliminary surveys of forests of 25 countries (FAO 1951: vol. XLIX, no. 8). The report (FAO 1953: 84-5) states some tropical trees produce pulp equal or even superior to temperate broad leaf species but confesses tropical forests contain mixed species and the pulpable qualities of the species had not yet been determined; it would be difficult to ascertain whether an area of 5,000 ha would yield pulp wood to manufacture 200 tonne per day which is an economic minimum quantity for a PPI unit. FAO estimated that tropical forests occupy 800 million ha in Africa and 600 million ha in South America where the species were not classified and neither continent had infrastructure, this being a vital consideration for movement of both raw material and finished goods.

We do not know if Australia was one of the 25 countries FAO had surveyed. We presume it was. One of the representatives of the 34-member team to Australia in 1952, Sukam Thirawat (1955: 224), expressed very serious doubts about the regenerative power of eucalyptus species in Asia, which at that point of time was specifically planned to be introduced in Thailand, and perhaps even in India. He even mentioned the case of *E. marginalia*, a splendid species which grows only in western Australia and explained its exclusive location on the basis of the unique type of soil in the region, found nowhere else in Australia. In other words, he is stressing the importance of matching the species to the site which is of course of fundamental importance that both FAO and the WB ignored. Thirawat (1956: 663) quotes R.S. Troupe's classic study of 1921 on Indian flora wherein 74 species of eucalyptus are enumerated, and *E. globulus* is mentioned as the only successful species cultivated in India on the Nilgiri Hills. Thirawat also mentions the work of forest botanist R.N. Parker and his two monographs published in 1925 on eucalyptus which give several reasons why eucalyptus is bound to fail in India, particularly in the plains (Ch. 8). Nevertheless *E. tereticornis* was introduced in Madhya Pradesh in 1956 (Ch. 3) on a massive scale and failed, and *E. grandis* in 1958 in Kerala (Ch. 7) where it was almost wiped out by pink disease and could never reach the size that it does in Australia. In both areas they were introduced without trial and where there was no shortage of bamboo. Why?

This study asks the question why for two reasons: (a) was FAO aware of the danger of monoculture and susceptibility of such plantations to disease which was not understood at that time, and (b) was it also aware of erosion of soil in the long run (FAO 1954: 61-5). But it warns there can be no doubt that man-made forests are going to play an ever-increasing role to supply pulp need (FAO 1954: 61-3). A later report (1968: 57) says the demand in the US is rising and it may not be able to meet its own needs by the end of the century. Incidentally, FAO-WB collaboration to raise eucalyptus plantations for pulp began first in 1954 somewhere in Latin America. This study is confined to India.

There is much literature on bamboo in *Indian Forester*, a prestigious journal of the Forest Research Institute, Dehradun. None of the articles before the late 1950s mentions gregarious flowering which does coincide with the species' normal physiological cycle. Contribution to this journal is by foresters and by scientists concerned with cellulosic raw material for paper. The physiological cycle of bamboo varies with species; it ranges from 20 to 60 years. But bamboo began to flower gregariously in the catchment areas of all the paper mills within two years, especially in those set-up in the second half of the fifties. This was about the time when FAO initiated eucalyptus plantations in India, the first perhaps in Madhya Pradesh in 1956.

Bambusa arundinacea, the tall elegant bamboo, grows in wet evergreen forests. Its 'branches' are 'thorny' and the culms tend to get tangled at the top. It is time-consuming to follow the elaborate felling rules prescribed by the working plans. 'Time is money' for an industrial organization. It is not clear whether the felling practices were ignored once the industry felt 'assured' of eucalyptus as an alternate raw material. Circumstantial evidence seems to point towards this. Several FAO officials repeatedly mentioned they were close to the Industry Ministry but not to the Forest Department which they reported as hostile (personal communication: October 1995, Rome). The link between the PPI and central government is also very close. There is no reason not to believe that the industry knew about the potential new raw material.

The government of India accepted and incorporated the FAO expert, Von Monroy's suggestions (1961:1) in the Third Plan document. He recommended 0.06 million ha or 1 per cent of the most productive forest zones to be selected for a ten-year period from 1965-75 to be clear-felled and planted with fast-growing exotic species at the rate of 1,50,000 acres per year and also for its intensive management to ensure 4.5 million tonnes of industrial raw material from 1975 onwards. He also recommended a change in the raw material consumption of PPI from predominantly bamboo to predominantly hardwoods, viz., eucalyptus (Ch. 3). Full assistance was given by the Centre to the State governments for this purpose; as a result 87,000 ha of eucalyptus plantations were raised. This annoyed the State Forest Departments (Ch. 4). Bamboo had been given away almost gratis to PPI, yielding no revenue. The only resource available for the State Forest Departments was the central government's grant to raise eucalyptus (Ch. 8). This shortsighted plastic surgery did not repair but only further distorted the face of the Indian forests.

The pace of distortion increased in the Fifth Plan. It recommended that 2 per cent of the potentially most productive forests be replaced with eucalyptus. The Planning Commission's position was strengthened by the setting up of the National Commission of Agriculture (NCA) which also recommended this plastic surgery. It was euphemistically termed as 'dynamic forestry'. It was presumably based on the FAO sponsored 'Indicative Plan' drawn up by the Ministry of Agriculture in 1965 (NCA 1976: vol. IX, 57). Letting in other international agencies like the World Bank, USAID, SIDA and ODA (Chs. 3, 7) has rendered the face of Indian forests grotesque. However, it is still far from clear why the World Bank is again seen to be so totally involved with Indian forestry in Maharashtra and Andhra Pradesh. 'Beggars cannot be choosers', said a much respected Principal Chief Conservator of Forests sadly (personal communication, Hyderabad), August 1992).

Exotic pulpwood, especially eucalyptus, had to be cultivated even beyond the forests, in common lands and all available space, and christened 'Social Forestry' (SF) to placate villagers deprived of grazing rights for their cattle and their right to faggots as firewood from the nearby woods. Social Forestry with eucalyptus was not very social; its leaves were not browsable and no grass or edible weed would grow under its shade. The wood was expensive and in great commercial demand as sturdy poles for scaffolding and also perhaps as pulpable material for PPI. The villagers could compete with neither the core industry nor the builders. They could only protest and put to fire the eucalyptus plantations. The WB and SIDA have participated very actively in social forestry (Ch. 8) and SF is now questioned by the foresters. By

1977 FAO began to have doubts about cultivating eucalyptus in arid tropics and as a renewable resource (FAO 1979). This was further confirmed in a species-wise analysis in 1980. The two-volume report also evaluated the paper-making qualities of the various species (FAO 1980, vols. I, II). It is not clear why India persisted in the folly, despite the negative features of exotics on alien soils that FAO confessed. It was aided by FAO's *Economic Analysis of Forestry Project* which was a practical guide for planners in government agencies grappling with the complex problem of raising pulpwood plantations with exotics (FAO 1979). The focus was on cost-benefit analysis. It noted that foresters ingrained with the notion of sustained yield were reluctant to introduce eucalyptus plantations by clear-felling forests (FAO 1979: 35). Perhaps to placate the villagers' growing resistance, FAO published *Forestry and Rural Development* in 1981 (FP: 26). The document recommends Farm Forestry, which is not the brain child of FAO—but Farm Forestry with exotics certainly is. There seems to be little concern for the damage it does to the ecosystem and to the survival of the poor peasantry whose holdings are used both by the richer peasants and the industry through unequal deals (Ch. 5).

India started with indicative planning. The Five-Year Plan documents are useful instruments with which to judge the government's stand. As the years roll by the Plan documents are progressively silent on the issues concerning forests and on the production of paper and rayon. This is possibly a fall-out of loud protests on the question of environment. With the setting up of the powerful Department of Environment and passing of the Forest Conservation Act 1980, the Sixth Plan stressed the need to intensify efforts to restore the eco-balance. And so does the Seventh Plan. Meanwhile, Wasteland Development Projects (WDP) have fallen into disrepute. PPI was looking forward to WDP as a renewed source of pulpable wood but the head of that organization disregarded this and had to go, according to an FAO official (October 1995). Farm Forestry is attempted to be rectified by adding another term 'Sustainable Development' (FAO 1994), a term that is a fall-out of the Paris Declaration of 1991. NGOs began to figure as agents of change in FAO's Forestry Papers from 1990 onwards.

Only with the establishment of the Indian Council of Forest Research and Education at Dehradun, a document routinely listing the negative features of eucalyptus has been published since 1993, although the FAO knew about them by the mid-1970s and the government of India by 1980. The reasons for such damaging silence are far from clear.

METHODOLOGY

The problem to be investigated is complex. The major research questions, the objectives as well as the assumptions of this study have been clarified above to give a total perspective of the continuously changing scenario.

At the risk of over-simplification, it may be said that the two empirical realities this study sets out to analyse are (a) the reasons for the death of the bamboo forests, when bamboo itself was a protective device as a succession, and (b) the reasons for the persistent attempts made, without trial, to set up the plantations of exotic species which often refused to respond.

Pure silvicultural analysis cannot indicate the plausible causative factors of these two issues. Historical data points out that proliferation as well as destruction of bamboo are the result of two different forms of anthropogenic interference, in the nineteenth and the twentieth century respectively, and so too the entry of the exotics. Decimation of bamboo can be reduced by modernizing pulping technology but a pure technologist or an industrial or a chemical engineer without historical insight cannot link over-consumption, gregarious flowering and the death of bamboo forests. In other words, only a methodology which can view forests as complex ecosystems, where flora, fauna and human beings can live in harmony, would be useful. Such a methodology would look at production of paper and newsprint as vital for all-round social development without regarding this as an inherent contradiction to the growth of bamboo forests. The historical method fulfils this need.

Natural history tells us about the growth and complexity of the natural forests, i.e. when and how the face of the forests began to be transformed. The social history of the various forest regions gives an added focus to the situation and the industrial history with its heroic struggles and murky deals in the colonial setting that took place along with the development process. In other words, the power struggle gives a realistic dimension to history and provides the context for empirical assessment of the metamorphosis of the Indian forests. Political economy brings out the contradictions and tensions between the two. And entry of FAO and other international organizations in the forestry sector highlights the integration of India in the world economy on an unequal basis.

The study observed and analysed every single forest-based pulp and paper unit that was under production between June 1991 and June 1994, viz., the project period of active research. These include 12 pulp and paper units and four pulp and newsprint units. Only the consumption pattern of three pulp and rayon units was studied, not the growth of that industry as it is an aberration in the history of PPI. It is an aberration since forests are not easily renewable and alternate sources are available for production of fabrics.

In order to get a total time perspective the study did dip back very briefly in this introduction to the Mughal period and also pointed out that setting up of five industrial units in 1936-7 was not a coincidence but was history. Neither was the central government's decision, taken in the mid-1970s, to encourage and support the small sector to manufacture paper and at the same time to initiate public sector PPI and newsprint units with the latest raw material-saving technology a mere coincidence.

All the PPI units except one are located in the forest areas from which they draw their raw material. Whether the areas still harbour bamboo forests or not is another matter, and if they do not, it is a mute testimony of the complex relationship between the industry and the forests as well as between the industry and the central government (Ch. 3).

The geographical dimension of the study is spread over nine states. The forest belt stretches right from the deep south in Kerala along the Western Ghats, northwards into Karnataka especially on its eastern flank, and eastwards into Tamil Nadu and its Eastern Ghats.

The forest on the Satmala range in the west stretches through the central India plateau to join the Eastern Ghats. On the southern slopes is the Telengana forest belt which became the catchment area of three paper units of Andhra Pradesh and on the northern slopes are the famed Chandrapur forests of Maharashtra which provide ample bamboo for one PPI unit.

To the north are the highlands of Madhya Pradesh, its once rich forests stretch across the Satpura and Vindhya ranges. On the eastern flank of the Vindhyas are again the once unpenetrable forests of Orissa. The Eastern Himalayas provide bamboo for PPI units in Assam and Nagaland.

This state-wise analysis of forests is made for two reasons. Forests have been a state subject since 1927, and only in 1977 were brought under the control of the central government as well. Its negative impact is felt in pulpwood plantation programmes. Federal states in India are also linguistic states each possessing a cultural identity of its own, which is useful in understanding the social history of the forest region. Forests are also a state subject. The findings of the study are therefore chapterized state-wise. Viewed historically, Orissa is the first state where Indian business invested in prestigious paper industry. Almost simultaneously units were set up in MP, AP and Karnataka and Bihar. Soon after independence a unit was set up in Maharashtra as well. The last of the units was set up in Assam. A combination of history and continuity is used for state-wise chapterization. Each chapter almost stands on its own; cross references are sometimes given to avoid repetition of information.

Since this study investigates the impact of PPI on the Indian forests, which now extend beyond bamboo as the raw material, it examines the availability of the alternate raw material and the reasons for rather wasteful consumption of forest-based raw material, bamboo as well as eucalyptus. In other words, pulping technology and the comparative yield of cellulosic raw material is also taken note of. The percentage yield varies from 35 to 95. Incidentally, all the private sector units, without exception, use obsolete wasteful technology, while the public and joint sector units use modern technology which yields as much as 85-90 per cent. The question is: how did the yield get reduced to as low as 5 to 10 per cent below the expected international level? The next question is the type of paper the industry manufactures. If more time and space than what is strictly necessary is devoted to this question, it is primarily to highlight the deviation in the third world production pattern from socially useful writing and printing paper to speciality papers that yield only corporate profits. Such a production pattern is sustained at the expense of the forests; it also generates enormous pollution. Incidentally, the finer the paper is, the larger is the quantum of chemicals used to produce it, all of which flow out into the rivers, the very lifetime of the villages downstream (Ch. 3). The first environmental protest movement began on this issue in the seventies. Following closely is the evil of the clear felling of natural forest for planting exotic pulpwood species, and at the same time denying forest-dwellers the traditional rights to forest produce—mainly grass and faggots as head-loads of firewood.

The central issue is evaluation of the destruction of forests by PPI. This is done primarily by evaluating the felling practices of PPI in its lease areas in the forests—mainly bamboo forests. The industrial unit is expected to follow the guidelines laid down for selection felling in the working plan (WP) drawn up by the State Forest Department. The coupes in the felling circles are worked rotationally. The WP also lays down rules for felling different species. Either the FD or the industry pointed out the place for evaluation, and they were under the felling series. A diagrammatic representation of a felling series is given in Chapter 4, where the first fieldwork area of this study is discussed. Other chapters contain only relevant information and analysis but not details, as there was no need to repeat such information. Sites where plantation forestry is carried out were also evaluated. They are established and tended either by the Forest Development Corporations (Ch. 8) or by the industry itself (Ch. 5), or by both (Ch. 6). Here the growth pattern of the exotics was noted and the impact on the ecosystem evaluated, in consultation with soil scientists and plant physiologists (Ch. 6). The growing trend is to utilize partially-treated effluents to cultivate both sugarcane and eucalyptus. The ecological consequences are discussed in Chapters 4 and 8.

To get back to the question of the felling of bamboo, in every instance the felling practice by the industry was erroneous. Industry data indicated that after 1959 the bamboo in the PPI's catchment area has reacted negatively and vanished. Most often it has flowered gregariously within a couple of years after being felled for the newly set up paper and pulp units. This cannot be just a historic coincidence. The hypotheses that emerged in the course of our observation and as a result of our methodology is that flowering of bamboo is a defensive mechanism, a last ditch attempt by a species to survive through seeding and dispersal of the seed. Out hypotheses can be fully worked out only in the last chapter. So too the hypotheses as to why eucalyptus does not grow in India and why it is allelopathic to so many Indian flora and fauna, unlike in native Australia.

The cellulosic content of *E. tereticornis*, the most prevalent type, is also suspect. Why then is the central government promoting it, will also have to be discussed in the last chapter.

The struggle to grow eucalyptus, the resistance to forest plantations by the concerned foresters (Ch. 5) of various states and tensions generated with the Centre on this issue is discussed substantively in the chapters.

If bamboo protests and vanishes and eucalyptus refuses to grow, what are the other alternatives? Does bamboo really vanish or does it go into hibernation? Why does bamboo continue to flourish in the north-east where it is used as timber to build homes, as a saleable commodity to paper mills, and even as firewood? In other words, bamboo forests need not get exhausted. Besides, there are other raw materials that PPI can use for its growth. PPI's own R and D discovered these raw materials when it faced a raw material famine. These materials are pollutant-free in sharp contrast to the enormous quantities of black liquor that, for instance, eucalyptus generates, nor is the lignin from bamboo biodegradable. All of these issues are discussed in this study.

In a study of this nature, destruction of forests by the PPI should be mapped especially since it was the satellite images of 1970 that first made India aware of the rapid pace of deforestation. But bamboo, by and large, grows as an understory which neither can satellite images capture nor can Forest Survey of India Maps selectively portray. It is the timber industry which is the known major destroyer but PPI also uses hardwoods from the forests and this is not quantified. This study therefore presents only flow maps which indicate progressive outturn of bamboo or eucalyptus wherever division-wise data is available. The flow maps also indicate a movement into more distant areas (often new and virgin forests) as the yield in the old ones begins to dry up. 'Poaching' on other mill area is also shown. Thematic maps show growing control of some PP corporations ranging from proximate to far-flung forests. If the actual area that was devastated could not be mapped, physiography of every forest region is mapped to indicate the prime factors which contributed to the evolution of the world's once most complex forest ecosystem.

NOTES

1. Under Chandragupta (327 BC) there was a well-established forest administration with a superintendent of forests. Later, during the reign of Ashoka, the administration was responsible for raising trees and protecting forests. A number of restrictions were placed on felling of trees and hunting of wild animals. Under Akbar a number of trees were proclaimed as royal trees, and felling was restricted and offences were punishable (S.S. Negi, 1994).
2. Etymologically the word 'paper' comes from the word 'papyrus', an Egyptian agave which was converted to writing material. The general consensus is that paper-making was first discovered in China in the second century using the fibre of 'paper' mulberry, hemp and bark, and was kept a secret till the eighth century. Arabs, and then Morocco and later the Moors took it to Spain. Later it moved to France, England, Germany and then to Sweden in 1532. Lack of forests in the Middle-East brought about a change in the raw material, viz., from mulberry and softwood to rags (Ahuja: 19).
3. Bagoria Star Paper Mill is outside the scope of this study as it is situated in Saharanpur (UP) and used to draw its raw material, pines and spruce, from the sub-Himalayan forests of western UP, and *sabai* grass from the local forests. It now uses eucalyptus.

4. Dalmia's Rohtas (in Bihar) and Punalur Paper Mills have ceased to function. Punalur figures peripherally in Ch. 8.
5. Thapar's first mill is in Haryana. It does not draw its raw material from tropical forests but uses eucalyptus cultivated as farm forestry within the state. This is therefore outside the scope of this study.
6. This is not to absolve all the foresters. Corruption is said to be rampant in the FD, perhaps because of alienation. Recent rapid degradation of Kodugu forests is reported to be a result of the connivance of officials in the departments of forests, law, revenue and finance with the powerful timber lobby. The investigation was carried out by a sensitive forester A.N. Yellapa Reddy, IFS, Secretary, Environment and Ecology, Bangalore (Alladi Jayashree 1994).
7. A part of these partially treated effluents is used for irrigation, the excess runs downstream into the water. Water from upstream is drawn by the industry for its own use.

REFERENCES

Ahuja, S.P. (n.d.), *Paper Industry in India*, The Institute of Market Research, New Delhi.

Alladi Jayashree 1994, 'Trees Felled in Karnataka Despite a Ban', *Times of India*, 22 June 1994.

Chattopadhyaya, Paresh 1971, 'Modernization of Economic Life', in A.R. Desai (ed.) *Modernization of Underdeveloped Countries* (ed.), A.R. Desai, Thacker, Bombay.

Chopra, P.N. 1986, *Quit India Movement, British Secret Documents*, vol. I, Interprint, New Delhi.

———, 1991, *Quit India Movement, Role of Indian Big Business, British Secret Documents*, Interprint, New Delhi.

Elwin Verrier 1958, *The Story of Tata Steel*, Commercial Printing Press, Bombay.

Encyclopaedia Britannica 1988, FAO, Chicago.

FAO 1953, *Raw Materials for More Paper* (FFSP no. 6), Rome.

———, 1954, *World Pulp and Paper Resources and Prospect*, N.Y.

———, 1979, *Economic Analysis of Forestry Project*, Rome.

———, 1980, *Pulping and Paper Making Qualities of Fast Growing Plantation Species*, vols. 1 and 2, Rome.

——— 1981, *Forestry and Rural Development*, Rome.

——— 1992, *Economic Assessment of Forestry Project Impact*, Rome.

——— 1993, *Assessing Forestry Project Impacts Issues and Strategies*, Rome.

——— 1994, *Technical Papers on Sustainable Forest Management*, Rome.

——— 1976, *National Commission of Agriculture*, vol. IX, Ministry of Agriculture and Forests, New Delhi.

Kimmins, J.P. 1987, *Forest Ecology*, Macmillan, London.

Kurien, C.T. 1994, *Global Capitalism and the Indian Economy*, vol. IX, Orient Longmans, New Delhi.

Monopoly Enquiry Commission Report 1965, New Delhi.

Nehru, Jawaharlal 1948, *The Discovery of India*, Signet Press, Calcutta.

Negi, S.S. 1994, *Indian Forestry Through the Ages*, Dehra Dun.

Palme Dutt, R. 1948, *India Today*.

Planning Commission 1952, *First Five Year Plan*, vol. I, GOI, New Delhi.

——— 1952, *Programmes of Industrial Development, 1951-6*, GOI, New Delhi.

——— 1952, *The First Five Year Plan*, GOI, New Delhi.

——— 1956, *Programmes of Industrial Development, 1956-61*, GOI, New Delhi.

——— 1960, *Second Five Year Plan—Progress Report, 1958-9*, Industry & Minerals, GOI, New Delhi.

——— 1961, *Third Five Year Plan*, GOI, New Delhi.

——— 1963, *Programmes of Industrial Development, 1961-6*, GOI, New Delhi.

——— 1970, *Fourth Five Year Plan 1969-74—A Draft Outline*, GOI, New Delhi.

——— 1971, *The Fourth Plan—Mid-term Appraisal*, vol. I, GOI, New Delhi.

——— 1973, *Draft Fifth Year Plan 1974-79*, vols. I and II, Ch. 5, Industry and Minerals, pp. 131-59, GOI, New Delhi.

——— 1979, *Draft Sixth Year Plan 1978-83*, revised, GOI, New Delhi.

——— 1981, *Sixth Five Year Plan 1980-5*, GOI, New Delhi.

——— 1983, *Sixth Five Year Plan Mid-term Appraisal*, GOI, New Delhi.

——— 1985, *The Seventh Five Year Plan 1985-90*, vol. I, Perspective, Objectives, Strategy, Macro-Dimensions and Resources, GOI, New Delhi.

——— 1985, *The Seventh Five Year Plan 1985-90*, vol. II, Sectoral Programmes of Development, Ch. 7, Industry and Minerals, pp. 167-95, GOI, New Delhi.

——— 1988, *Seventh Five Year Plan 1985-90—Mid-term Appraisal*, GOI, New Delhi.

——— 1992, *Eighth Five Year Plan 1992-7*, vol. I, Objectives, Perspective, Macro-Dimensions, Policy Framework & Resources, GOI, New Delhi.

Poddar, V. 1979, *Paper Industry in India*, Oxford and IBH, New Delhi

Troup, R.S. 1921, *Silviculture of Indian Trees*, vol. III, International Book Distributor, Dehradun.

Von Monroy, J.A. 1961, *Report to Government of India on Integration of Forest and Forest Industries*, FAO.

Whitmore, T.C. 1975, *Tropical Rain Forests of the Far East*, Clarendon Press, Oxford.

CHAPTER 2

Orissa

Whose Country? Paper for Whom?

Where is the Wild Odra?
Where is the jungle fastness?
Where are the Khonds
  Who romanced with the bamboo?
Where are the Gonds
  Who built their little kingdoms?
Remember?
  The white hunters have cleared them.

Yet, out of the muck
  emerged the Lotus
And the Grizzly Bear
  soon sucked its honey;
While,
  the crafty Jackal smirked;
And
  the Tiger stalked its kill
Oh then,
  will the Lotus remain untainted?
And,
  the Odra bloom again?

M.S.

INTRODUCTION

The question of whose country Orissa is, will be settled by history. It is a historical fact that through time and usage, land and what grows on it belongs to the people who are born and have lived on it, from time immemorial in perfect harmony. On the other hand, legal rights are given or taken away by those who control the state structure. Thus, the question of whose land comes up, because, through a period of time the local people, the aboriginals, were quietly dubbed as 'primitive' tribes and deliberately marginalized by the British rulers. This was justified by nineteenth-century ethnography. But the more recent sensitive scholars have debunked this viewpoint as is evident in the writings of Headland and Read (1989: 1940-1) or H.M. Wobust (1978). It is important to note that the aboriginal people or tribes as they are referred to, vary from simple food gatherers to advanced agriculturists; the Gonds and Bhijals for instance have even developed state structures. These aboriginals were largely insulated by impenetrable forests and mountains[1] and lived in relative peace. This study raises the question of their rights to the produce of the lands in the first section, because their lands were unjustly bestowed by British rulers to the more amenable non-Oriyans. Much later, even the forest produce was practically given away by the Indian rulers to powerful industrial houses.

The concern of this study is with the gifting away of the forest produce to the powerful paper industries. This is dealt with it in the third section. The study shows how two representatives of the Indian bourgeoisie too, had to struggle against the British stranglehold and how they eventually entered into this safe forested area to produce highly profitable paper. But, unlike the aborigines, they had the blessings and the help of the post-colonial Indian government in return for which the industries had to produce a specified quantity of cheap levy paper for the poorer sections. The question is: Does the industry keep its promise to produce levy paper for them? The second section deals with the status of the forests and the soils, rivers, climate, the rainfall that nurtured them, and the topography which protected them over the centuries, lastly, the havoc created by Britich foresters, Indian industries and state policies.

SECTION 1

WHOSE ORISSA?

Ordra,[2] Ordra-desa is the aboriginal name which now survives as Orissa. The people themselves connect Ordra with the red jungle rose of the same name and 'revere that flower as one of the five blossoms of heaven (mandar)' (Hunter 1956: 50). Orissa was outside the pale of Aryan civilization and remained unknown to Vedic writers. The writers of epics referred to it as the wild outlying strip peopled by barbarians

(Hunter: 52). The strong prejudice against its people appears in Vedic hymns where they are called *rakshashas* (devils), *yaturdhanas* (goblins), *karvyadas* (eaters of uncooked meat) and even as *Asutripas* (cannibals). The *Vishnu Puranas* picture them as a dwarfish race with flat noses and skin the colour of charred slake. The *Bhagwat Purana* describes them as black as crows, with tawny hair, red eyes, projecting chins, flat noses, short arms and legs. According to the *Matsya Purana*, they were black as collyrium. To these features the *Padma Purana* adds a wide mouth, large ears and a protuberant belly (quoted in Carmichael 1869: 84-5). These are typical descriptions of the aboriginal people of Orissa, who were left alone in their wilderness.

The Khonds, one of the predominant aboriginal communities, firmaly believe they have existed in Orissa from the very beginning, having either sprung from the soil itself or having been created contemporaneously with it (Carmichael: 89).

The important fact is that there was little interaction between Orissa and the rest of India, perhaps due to the very rugged nature of the country and the impenetrable jungles which appeared fearful to all but its aboriginal people, who the ages became a part of its eco-system. It is known, for instance, that at the time of Bindusara's death, practically the entire subcontinent came under Mauryan suzerainty, but Kalinga[3] (Orissa) remained unconquered till Ashoka's dreadful campaign which left bitter memories for its people (Thapar 1966: 71, 72, 93).

Topography and Forests Insulate Orissa

Even in the fourteenth century, Orissa remained the only part of India unexplored in what Hunter loosely refers to as the Musalman period. The writings of this period refer to the long tracts of jungles which extended from the mouth of the Ganges to the Godavari river. When finally the 'Mahammadan' tardily penetrated the terra incognita, says Hunter (1956: 53), they found it so deeply cut by rivers and so strong in mountain and jungle fastnesses, that they asserted their authority only on the longest-settled and oldest part of the delta. The hills on the west and the sea board districts on the east were left as they were under the native chiefs.

The 'Mahammadan' conquerors gladly got rid of this impossibly inaccessible province as a bribe to 'Marhattas' in 1751 (Hunter 1956: 53).

The much respected historian W.W. Hunter who was also a British civil servant in Orissa in the early nineteenth century, wrote that it was deemed unwise to subject its inland jungles to regular British control (pp. 52, 185).

Enter the British East India Company

Much earlier in the seventeenth century the English merchants had, in fact, set up a factory[4] in Balasore well within range of the English cannon. Trade conducted under such conditions of fear and rivalry with other European merchants could yield but small profit (Hunter: 184). Only total control would ease the situation.

The main question that the British raised was not how to protect the weaving villages but how to collect rents. The region was covered with forts which, while they served as strongholds against the 'Marhattas', also provided defence against the land bailiffs. Civil government and tax collection in a country covered with fortresses simply resolved into a military occupation. By means of infinite harrying, English merchants were able to collect more than half the land tax in 1789, their total demand being Rs. 3,73,700. Political control over this region was not easy.

Colonial Rule: British Land Settlement Acts and Peasant Resistance

Only with the defeat of the Marathas in October 1803 did the province of Orissa, stretching over 23,907 sq. miles with a population of 3 million, come under British rule. The British immediately set about consolidating the land revenue administration. This resulted in the tribals losing control of land as well as forests, and making them protest all over western Orissa (Tripathy 1988: 228). Meanwhile the British started experiments with a series of land revenue settlements[5] starting with Regulation XII of 1805 which empowered the Board of Revenue at Fort William to superintend the revenue administration (Sahu 1956: 395).

Hunter (1956: 193-4) confessed:

> The real trouble began with the end of the conquest. The British held the land but could find no proprietors and no peasantry to till the soil. Many soothing proclamations eventually attracted the proprietors to settle the rents, but the difficulty was to find the peasantry to till the land. The jungles yielded

an unfailing refuge to the miserable tribal peasants. The land settlements were disastrous, all classes were seething with discontent.

In 1817, for instance, the Khonds of Ghumsur joined the militant Paiks,[6] the landed proprietors who gave a call for rebellion[7] when they found themselves deprived of lands which they had enjoyed from time immemorial.

The Paik rebellion brought a great change in the land revenue administration. In 1835 a settlement for thirty years was made; liberal terms were offered to the proprietors and the cultivators to increase the cultivated area. But when the Khonds revolted after the annexation of Khondmals in 1855 (Tripathi 1990: 28), no land settlements were made for the Khonds although they were peasants who also practised hunting in their densely forested province.

It is important at this point to remember that in an early stage of human evolution humans are, in what R.S. Rao calls an 'idyllic predicament' of not being able to question man-nature relationship except in the domain of divinity' (Rao 1990: 13). Having to contend a great deal of uncertainty regarding their agricultural output the Khonds occasionally practised human sacrifice as a fertility rite.[8] This provided the British with an excuse to oppress the rebellious Khonds,[9] who however carried on their struggle in Patna state in 1869 and in Kalahandi in 1878 (Tripathi 1990: 28). They were willing to give up the practice of human sacrifice only on the promise that they would not have to pay the land tax (*Orissa District Gazetteer*, Boudh Khondmals: 1).

Tribal Revolts

In reaction to the oppressive new governance the tribals continued to trouble the British administration throughout the first half of the nineteenth century. A large number of them flocked to the Khond leader Chakra Bisoi to carry out a sort of *jihad* against the British. The enlightened Raja of Angul who was suspected of complicity with Chakra Bisoi was deposed in 1847 and jailed in Hazaribagh. Chakra Bisoi, was chased from Ghumsur to Baudh, and then to Madanpur in Kalahandi district and further to Algoan in Patna state. Chakra Bisoi moved on to Parlekimedi to invigorate the Savaras—the tribal agriculturalists who had revolted against the British under the leadership of Bisoi Raja Krishna Dandasena. When this movement was suppressed in 1856, Chakra Bisoi went back to his stronghold on river

Tel to prepare for future operations. It was at this time that the first war of independence broke out, Chakra Bisoi plunged into this war and was never heard of again. Incidentally, Korapat, Phulbani (Boudh Khondmals) and Kalahandi which have been shown as regions of resistance, are also the areas from which the two paper mills of Orissa draw their raw material. Another important first catchment area for the Orient Paper Mills is Sambalpur, which is further north.

Sambalpur 'was called into existence' by Regulation XIII of 1833 after the suppression of the Kolth rebellion in 1831-2 (Senapati and Mahanti 1971: 3). Surendra Sai, a Chohan ruler of Sambalpur, whose life's ambition was to drive out the British, was assisted by the Gonds[10] and Bijhal zamindars, who found that their privileges were threatened and their land encroached upon by the Hindu favourites of the Rani whom the British had instituted in Sai's place (Sahu 1956: 398). He was defeated and jailed along with a large number of other rebels. They were finally freed during the first war of independence. Although the rebellion was crushed, Surendra Sai and others continued the struggle.[11]

As late as 1931, civilian F.C. King, ICS (p. 72), wrote: 'The aboriginal element is still strong among the old families holding a feudal tenure, and of the sixteen zamindari estates in the district, no less than twelve are held by Gonds and Bijhals. A large number of the latter are also managers of villages; but as a rule in the more backward tract, they are not proprietors, only lessees holding under zamindars.'King continues 'The majority of the aboriginals now cultivate small patches of land in the wilder parts of the zamindari land are generally poor. The holdings are, as a rule, insufficient to provide them with a full livelihood, and they eke out the cultivation of millets and sesum by collecting forest products . . . the power of the aboriginal owners of their soil was gradually being broken.'[12] They were being gradually replaced by non-tribals in western Orissa. They did not support the tribals in the struggle to get control of their lands.[13]

Social Consequences of Peasant Resistance

The continuous series of revolts, the disastrous famine of 1865-6, combined with the failure of the monsoon and a flood that devastated the country finally made the British rulers take note.[14] Famines became a regular feature. The East India Company took control of Orissa in 1803. Its continued lack of concern brought about a series of famines, the worst occurring in 1806, 1808, 1809, 1817, 1828, 1836, 1837 and 1842 (Sahu: 399). Famines had to be contained if revolts were to be stopped. Roads were opened up to remove the isolation of the region. The Bengal Nagpur Railway line was constructed through the coastal region and opened in 1899. This unfortunately brought in traders and money-lenders who ultimately came to hold the people in debt slavery or *Gothi*, through incorrect accounts of debts. *Gothi* was formally abolished in 1948, but it is by no means a thing of the past. Even today, cases of money-lenders fraudulently taking away the landed property of the tribals who then moved from court to court as defendants, are very common. The mass agitation launched by the Savaras in 1951-2 to get back their lands (Senapati and Sahu 1966: 201, 121) is a pointer to the British programmes of modernization and the legal system which brought slavery in its train. As late as 1980, western Orissa agitated against Marwari traders. The monopoly of trade, both wholesale and retail, was in the hands of Marwaris, Gujaratis and Telugu Komtis.

The most important task for the British was to improve agricultural production which had dropped. The East India Irrigation Company which was started in 1862 dug a series of canals as a post-famine measure; tank agriculture was known in many districts but had been allowed to fall into disuse. It also began to open out or clear forests for agriculture. This raises another important question: who is the main destroyer of forests?

The Role of the Forest Department

Orissa's social structure was disturbed when the British deprived the local population of land and therefore, livelihood. Land-hunger was growing and the peasants were restive. Their sporadic, as well as sustained revolts were crushed by the British with the help of local chiefs (Senapati and Sahu: 74). What the new rulers were looking for were large stretches of agricultural land to increase agricultural production. It was in response to hunger and revolt that the first Forest Policy in 1894 gave primacy to agriculture over forest conservation. F.C. King (1931: 2) reports that much of the forest lands in Bargarh subdivision of Sambalpur district was brought under cultivation.[15] Dewar who did the settlement of Sambalpur in 1902-6 lamented, 'The cutting out [*sic*] of the forests and the spread of rice and cane cultivation into all the valleys and up all the streams have of late years curtailed the grazing grounds of wild animals and cut off their water supplies'[16] (quoted in Senapati and Mohanti 1971: 30).

Besides clearing forests for agriculture, the working of the Forest Department was itself faulty and degrading. Although the question of forest conservation of the valuable timber forest was raised in 1886, nothing was done except demarcating and classifying the areas as state, *malguzari*, and *zamindari* in the next two years. In 1889 the forests were managed by revenue officials,[17] whose only duties were collection and detection of offences. The state forests were divided into Reserved and Protected. The Reserve forests were formally closed to all cuttings while bases were given to some companies during 1915-16. In 1916 a rough plan was drawn up by the forest officer, Cooper. He introduced 'Improvement felling' which meant cutting of dead, dying and unsound trees. In forestry, 'unsound' implies 'unsound growth' or inferior species which interfere with the growth of the better class trees. Unfortunately, the high percentage of 'unsound' trees led to heavy felling, the canopy was excessively opened out and large blanks were created. The trees were not coppiced or cut back but simply girdled; thus the scope of future regeneration was destroyed, especially that of *sal* which requires shade for protection from heat.[18] All the blocks not reserved in 1907 were thrown open to cess-payers and other purchasers who promptly and indiscriminately cut down trees. From 1928 onwards, coupes were opened out every year and leased to successful bidders on contract who worked under the somewhat doubtful supervision of the forest staff (Senapati and Mohanti 1973: 1, 14-16).

Thus the scenario shows that the tribes are made scapegoats, first by the ignorant British foresters, and by the Indian foresters, following in their footsteps, and finally by the recent editors of the *Gazetteers*. These latter hold that slash-and-burn agriculture, and the villagers currently in search of firewood are the main destroyers of standing forests. But the earlier writers of the *Gazetteers*, the British administrators in the nineteenth century who were confident of their control over the colony, felt no need to rationalize. In 1872 H.G. Turner reported:

When civilization pushed wilder members of the hill tribes into the unconquered jungle, they commenced upon felling and burning generally without ploughs. He grows his rice in terraces hollowed out of the water course on the slopes of the hills. The burning is very superficial and the stumps, or at least a large proportion of them at once begin to grow again. And when cultivation is abandoned, which is generally after two years, the forest soon begins to recover. The evergreen trees suffer more than the others and these are more or less absent at first. For some years rank grass, much thorn and coarse undergrowth hold sway and fires periodically sweep through, but it is not till the growth arrives at an age of twenty years or more that there is a chance of much humus being added to the surface soil. Fires are then excluded, seedlings have a chance and shortly afterwards, rattan and tree ferns appear. The evergreen trees increase in number, and the undergrowth greatly changes in character and a species of *Acanthaceous* (Strobilanthes) appear as in moist western Sholas.

This is a sketch of what occurs after the first virgin felling. 'When a tract is allowed forty-to-fifty years to recover, it appears to return to its pristine vigour and form. Unless the base of the older tree is observed, a forester may even be deceived and fancy that it is a virgin forest' (Francis 1907: 116-18). It is a moot question as to whether their administrators could have done without the enormous revenue.[19] This study has shown, how in Bolangir the ignorance of the administrators regarding tropical forests was revealed in their method of felling, their greed to sell timber, and their need for revenue to control such valuable property, all of which signalled the death of the forest.

It is also important to realize that the slash-and-burn method of agriculture became a menace only because the agricultural communities or tribes were deprived of their land as a punishment for their resistance to British control, and because simultaneously much of their forest land was enclosed as reserved, thereby bringing the cultivator-forest ratio to a dangerous level. It is also important to remember that not all the tribal communities were primitive agriculturists; they ranged from pure food-gatherers and hunters, to irrigating and plough cultivators. The intermediary type who were hunters-cum-simple agriculturists alone, did practise slash-and-burn agriculture. The forests were vast, and it was on the lower slopes of the valleys that these hunters practised their *podu* cultivation. They tended to sow more seeds on the upper sides of the slope, some of which were washed down to the lower areas resulting in a uniform density of crop over the whole area. They raised rice, millets, *ragi*, *junha kanga*, *ghantia*, *birhi*, *kandul suan*, *guliji*, *jada*, *alasi*, *til* (Padhi 1991: 89). Padhi, the present PCCF, holds the forest managers responsible for the persistence of *podu*. He argues that had mango, jackfruit, *mahua*, *tentul*, *tala khajuri*, etc., been planted in the poor and ravaged areas, the tribals would in their own interest have not burnt them down. To them, exotics like *Acacia auriculiformis* and eucalyptus that are planted are an eyesore (p. 88). They have lost the forests because the Government's revenue department over-exploits forests for revenue and fells valuable species like *sal*, *asan*, *bija*, *kurum*, *arjun* and *jamun* (p. 89). However, they continue to feel that the forests belong to them and have their own laws regarding management of forests and harvesting of forest products. They have their own folk taxonomy for the flora and fauna and follow a type of binary classification according to availability, utility, edibility, medicinal properties, etc. (p. 85). And as P. Sainath (Sainath 1994) puts it, the Koya's relationship with bamboo, far from being a conquest of nature, is a 'romance with nature'. From their very childhood they are taught to cut the bamboos in such a way that it grows again. Koyas live in Malkangiri and use bamboos in myriad ways. They also barter it for other essential items like salt (Sainath 1994).

The New Province of Orissa

The British acquired a total stranglehold on its Orissa by 1803. For administrative purposes Orissa was hitched to Bengal, its southern end to Madras where the British were more strongly entrenched, and Sambalpur to the Central Provinces. The notion of making Orissa a separate province emerged only after the catastrophic floods of 1865-6. The question of making Orissa a state in itself was first mooted by General Chasney, and by Sir Stafford Northford in 1874, advocated by Cook, the administrator, in 1895 and accepted, sympathetically by Lord Curzon. However, except for transferring Sambalpur to Orissa in 1895, nothing more was done. Finally, in 1912, the Province of Bihar/Orissa was constituted, Bihar getting a sea-board (Sahu 1956: 399-400).

The need to form a separate province was first enunciated by the Utkala Union Conference in 1903. Under the leadership of Madhusudan Das all sections of the people joined the movement with enthusiasm. The political situation changed with the rise of Mahatma Gandhi, and Gopabandu Das came forward to join the All-India Congress movement for freedom. As a result, Orissa Congress and Utkala Union Conference parted company in 1920-1. The Congress movement achieved great success at the time of the Salt Satyagraha.

The British government following their policies of appeasement and divide-and-rule, in 1936 created a Muslim and a Hindu majority in the western and eastern regions of India respectively. Orissa was constituted as a separate province on 1 April 1936. The election for the Legislative Assembly was held on 1 April 1937 in which Congress won as many as thirty-six out of fifty-six elected seats; the first Congress government was formed in July 1937 remaining in power till 1976, when the emergency was declared.

Almost immediately after Congress came to power, Birla entered north-west Orissa and carefully selected Brajrajnagar in Sambalpur district to set up his paper mill. Thus a richly forested well-watered place, served with basic infrastructure, was chosen in order to set up his Orient Paper Mill.

Much later in 1959, Singhania chose to set up his new modern paper plant at Rayagadha in Koraput district also rich in bamboo forests and watered by the perennial river Nagaveli.

Despite four paper mills in the area, Orissa continues to be backward, while the rights of the local population are largely ignored, and the forest ravaged. Belatedly, the National Forest Policy of 1988 recognizes the symbiotic relationship between the tribal population and forests. This is evident in Dr. D.N. Tiwari's comment: 'In the context of growing unrest in the tribal belts of the country, the forestry sector must act as a nodal agency for alleviation of poverty. Only forestry programmes can generate productive and gainful employment and other opportunities to secure better food, clothing, shelter and health' (Tiwari 1991: 984).

SECTION 2

Forests of Orissa

Orissa was richly endowed with forests as the earlier section has indicated. A century later it classifies only 67,925 sq. miles of the land area as forests. This constitutes 43.59 per cent of the total geographic area of the state. The bamboo areas overlap other forests stretching over 10,533 sq. miles. At this point only the forest areas which are the regular bamboo catchment areas of the two paper mills Orient and JK are being considered. For this purpose, nineteenth century *District Gazetteers* are an excellent source, but often it has been necessary to fall back on later editions when the original ones were not available.

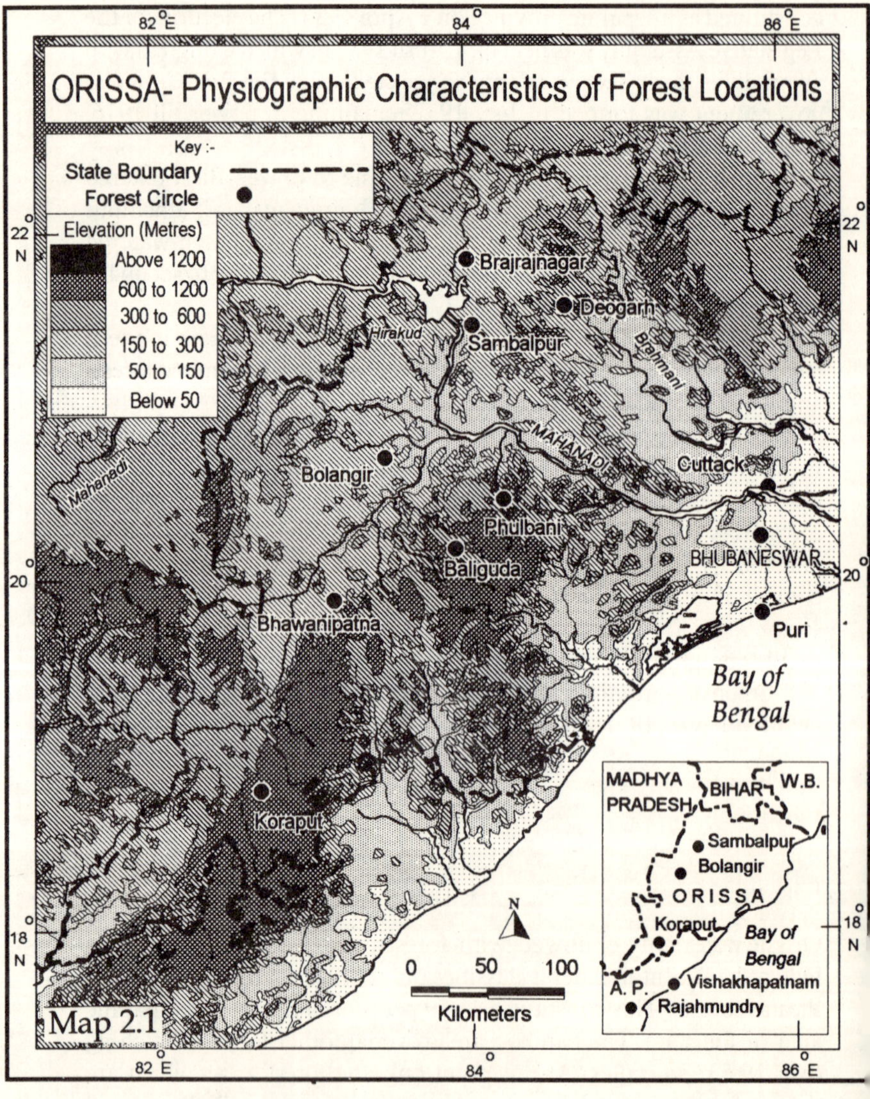
ORISSA- Physiographic Characteristics of Forest Locations
Key :-
State Boundary
Forest Circle
Elevation (Metres)
Above 1200
600 to 1200
300 to 600
150 to 300
50 to 150
Below 50
Brajrajnagar
Deogarh
Hirakud
Sambalpur
Brahmani
MAHANADI
Mahanadi
Bolangir
Cuttack
Phulbani
BHUBANESWAR
Baliguda
Bhawanipatna
Puri
Bay of Bengal
Koraput
MADHYA PRADESH
BIHAR
W.B.
ORISSA
A. P.
Vishakhapatnam
Rajahmundry
0 50 100
Kilometers
82° E
84°
86° E
22° N
20°
18° N
Map 2.1

Earlier Status of the Forests in the Catchment Area of the PPI

Several factors influence both the type and growth of forests. The most significant of these is latitude, in this case 17.49° N to 22.34° N, i.e. tropical, but modified by the altitude; then there are the prevailing winds and the rains they bring around 157.5 mm from the south-west monsoon, a little from the north-east when the south-west monsoon is retreating and convectional rainfall during March-May due to local depressions. Drainage and the soil types and their mineral content as well influence the species which constitute the forest types.

For administrative purposes, state forests are divided into Forest Divisions (FD). Such divisions are useful tools for classification of forest types.

Drainage

Broadly speaking, the Mahanadi with its tributaries is the main river system in the north-western sector. It enters the north-west of the Sambalpur FD and after crossing the Hirakud dam, takes a wide curve flowing south into Bolangir district. Its principal tributary in Sambalpur FD is the Ib.

In Bolangir FD, on the other hand, there are other rivers too, viz., Tel and Ang. The combined waters of the tributaries Jira and Data flow southwards to join the Mahanadi, while Salesinh-jor meets the Mahanadi to the south of its confluence with Jira.

These tributaries of the Mahanadi also drain the riverine plains of the Boudh, while the Tel drains the southern and western regions. The hill streams Bagh, Salki, Rushikalya and Dokarakani form the affluents of the Mahanadi and Tel. The Tel forms the boundary between Bolangir and Phulbani districts.

The northern part of the next FD Kalahandi, is the parting of the water shed of Mahanadi and Indravati. The plains in the south are drained by the Tel, Jonk and Nasunkunda rivers. In the open countryside the Tel, Sagada, Hutti, Ret and Utei are reduced to tiny streams in the lower reaches from February to June.

The southern-most forest division is Koraput. Here, the western port of Kashipur is drained by Golagada, the fertile Rayagada by Jhanjhawati and the undulating tableland by Indravati, Kolab and Muchkund.

To the north-east of Sambalpur is the Deogarh FD. It is drained by the Brahmani which runs from north to south as well as by numerous perennial hill springs; Pradhanput, Koradkot and Limura create waterfalls which are tapped for hydel power.

Soil

The soil merely forms a cap of varying thickness derived by alterations of the rocks below so that its nature depends on these rocks. Its mineral content in traces forms plant nutrients and is therefore important.

In Sambalpur and Deogarh forest divisions the unclassified granites which are so widespread generate a coarse sandy soil. Mica-schists occur in patches in the granitic area. The soil cap is agrillaceous coal and confined to the Ib river area. Other important minerals are lime-stone, dolomite, manganese, graphite, gold, lead and silver ore, clays, beryl, abrasives and bauxite.

The soil in Bolangir is light and sandy. A great variety of ores and minerals are to be found in this district, the important ones being graphite, galena (lead ore) manganese ore and bauxite.

Porous pitted laterite caps almost all the hills of Phulbani district, Kondalite being the less lateritized of all other rock types. It may range between the pleistocene age and those more recent in time. Residual soil varies in colour from light grey to brown and is a product of weath-ering. Important minerals are graphite, manganese, mica and quartz.

The lithosphere of Kalahandi region points to basement gneiss on the western part of the plateau between 250 and 1,000 m, composed of minerals like felspar, biotite, hornblende and pyroxene. The minerals are bauzite, graphite, manganese, sillimanite and galena. The bauxite in Kalahandi is usually ferroginous in character despite its silicon dioxide and titanium dioxide contents being low. There is graphite in the eastern ghats, manganese ore, magniferous horizon confined to quartz-sillimanite schistes and galena.

Koraput forest division lies on the section of eastern ghats. The minerals of the region are clay, gold, graphite, limestone, manganese and mica ochre.

Relief

Sambalpur and Bargarh Forest Divisions: Sambalpur and Bargarh (including Padampur) subdivisions together form an undulating upland varying in elevation from 146 m to 228.6 m. The general slope is from north to south, but much broken by the rugged range of hills and traversed in all directions by drainage channels. The principal range is the Jharghati whose highest point is 516 m. A noticeable feature of this hill system is the absence of flat-topped trap hills which are so common in the north and west.

To the north-east of Sambalpur is Deogarh.

Deogarh Forest Division: This is a hill system grouped under four ranges: (a) Khajuria range on the north, (b) Pradhanpat and Kaidanta ranges 743 m and 816 m respectively are to the north, just south of the Khajuria, leaving a vast plateau in between, running from west to east up to the bank of the river Brahmani, (c) On the eastern side of the river Brahmani, runs the Pawri range (Poudi) 678 m. It has reserved forests, and (d) The Ushakothi range extends into Kuchinda and Jamunkira towards the west. The hill ranges of the subdivisions are thickly wooded and their elevation is from 610 to 762 m above sea level.

To the south of Sambalpur lies Bolangir.

Bolangir Forest Division: The 10-mile long Gandhamardan range forms a natural boundary, a crest on the north-western side of the district. It consists of several hills, the highest being Gandhamardan (988.9 m). The hill range is topped by a cap of high level laterite. Rain water soaking through the laterite collects under it and then oozes out at Kapil Dhar springs on the northern and Hari Shankar on the southern slopes.

The other hill ranges are discontinuous and broken for several miles by plains. The main forest areas stretch along the western boundary bordering to Nawapara subdivision of Kalahandi district, and then turns to the east, running parallel to the Gandhamardan range. This forerst tract is broken by an occasional clearing and small settlements.

Phulbani Forest Divisions: Phulbani district is to the east of Bolangir. It is broadly divided into two distinct physical divisions—the fertile plain region of Boudh in the north, and the Khondmal highlands in the south about 518 m above sea level, girdled almost continuously by high ranges which cut it off from the surrounding country. On the north, east and west these ranges quite perceptibly rise abruptly from the plains of Boudh, while on the south they merge in the outliers of the eastern ghats in Baligurha subdivision. The high plateau lying within these ranges is broken up by numerous smaller ranges which form an endless series of valleys varying in size from 20 to 25 sq. km.

According to the *Gazetteer of Phulbani* (1977), primeval forest covered much of this tract and the villages lie in scattered clearings on the hill sides and in the valleys below, while some are in almost inaccessible places on the summits of the hills. The whole of this is, in fact, a network of hills and forests, interspersed here and there with the small hamlets of the Khonds. The western portion of the Khondmals subdivision and also its north-eastern corner form a compact block of rugged hills, range after range being visible for miles around. The area of cultivated land is small. The uplands and the slopes leading down from the foot of the hills are periodically cleared for raising dry crops and the low paddy lands have been permanently cleared and cultivated every year. The rest of the country is covered with thick forests.

Baligurha Forest Division: The entire subdivision of Baligurha is a plateau and lies at a height varying from 3,000 to 1,100 m above sea level. The eastern side consists of wide well-cultivated valleys separated by hills covered with sparse jungle. The western and southern portion comprising the Baligurha, Paringbarhi, Kotagarh and Belaghar are mountainous, covered with dense *sal* (*Shorea robusta*) forests inhabited by wild animals. This division is separated from the Ganjam plains by the well-known 11 km. long Kalinga ghat. The hills of the subdivision are a part of the eastern ghats.

Kalahandi Forest Division: This district is divided into two natural divisions (a) the thickly wooded hill tracts, Dongarula, covering an area of about 3,665 sq. km and lying on the main line of the eastern ghats and (b) the plain country to the south-west comprising the Nowpara plain.

The Sonabeda plateau extends over a 780 to 1,028 sq. m. Its height varies from 610 to 915 m. The forests of the Kariar nestle among the circular mountains.

Koraput Forest Division: The district lies on a section of the eastern ghats. It is divided into four natural divisions: (a) In the first division is Niamgiri, a rugged mass of hills on the border of Rayagada and Gunupur. This natural subdivision consists of two valleys, Nagavelli and Vamsadhara, (b) The second natural division is the 3,000 ft plateau which extends 110 miles from Kashipur tahsil to the border of east Godavari (AP), (c) The third natural division covers almost the entire extent of the Nowrangpur subdivision. It extends far west into Bastar in the north into Raipur district, and (d) In the south it falls by steep ghats into Malkangiri tehsil. In the north-east it slopes down to Kalahandi 1,000 ft below.

Malkangiri Forest subdivision: This is a thin strip along the eastern boundary including the ghats which fringe the 3,000 ft plateau and the valley of Muchkund. The area is sparsely inhabited by primitive tribes, the forest dwellers—the Bondas, Porojas and Didayis.

Forest Types

Sambalpur and Deogarh: This is the great *sal* belt. In the south-east, where the climate is somewhat moist, the forest is nearly evergreen. With the *Shorea robusta*, or *sal* are found certain of its usual associates, *Terminalia tomentosa* which yields the commonest of all building and tanning materials, *Pterocarpus marsupium* for making furniture, *Anogeissus latifolia* for building and invariably for cart axles, and *Terminalia arjuna* and *Eugenia jambolana*. In Sambalpur district *sal* forests do not reach their best as they are situated chiefly on dry hills or on flat lands where past anthropogenic interference has affected soil drainage, As a consequence, a drier type of mixed deciduous species occupies a large portion of the forest. This included *Cleistanthus collinus*, *Odllia wodier*, *Anogeissus latifolia* and *Kendu Diospyros melanokylon*. Just two decades ago this dry mixed formation, had in its turn to fight a stern battle with bamboo *Dendrocalamus strictus*, especially where the underlying rock is a granite or a gneiss (Senapati and Mohanti 1971: 2). In fact, bamboo, almost pure or mixed with the dry deciduous type of forest, approximate in extent to the more valuable and gregarious *sal*. This was true even in the seventies. Amongst other trees of economic importance found in the district are teak *Tectonea grandis* in small plantations, *Dalbergia latifolia*, *Gmelina arborea*, *Chloroxylon swietania*, *Terminalia chebula* and *Terminalia belerica* which yields inferior timber and medicinal fruit. Of trees growing in the open country the most important is the *Mahua* or *Bassia latifolia* and several varieties of ficus, tamarind, *kaith*, *bair*, jakfruit, *neem*, *bhelphal* and *jamun* which are planted for their fruit and oil.

There is a dearth of good grazing grasses, but the succulent *Cynodon dactylon* grows all over the district on sandy soil where there is some moisture. *Iseilema wightii* is confined to old fallows or to the boundary ridges of fields as it prefers clayey soil, and is rarely found in the forests. The commonest fodder is the *Pollinia argentea* found everywhere in the forests and in elevated places, while *Pollinia eriopada* is used for rope-making. *Heteropogon contorus* is very common in the forests (Senapati and Mohanti 1971).

Bolangir: The forests of the district are of a xerophilous nature due to several factors, chiefly, poorly distributed rainfall, geological formation which is resistant to weathering and shallow, and sandy and dry soil. By the 1970s, anthropogenic interference is said to have led to the depletion of the forests, which most likely had contained more *sal* earlier. It is even said to have resulted in the production of bamboo climax type of forests, and in some places to have retrogressed to even a savannah type of forest. The area of reserved forest is 453.32 sq. miles (724 sq. km.), and of the protected forests 257.72 sq. miles (412 sq. km.), the total being 20.80 per cent of the geographical area of the district. The *Bolangir District Gazetteer* (1967) has classified its forests under five heads according to vegetation: (a) *Sal* Forests, (b) Low Mixed Forests, (c) Alluvial Mixed Forests, (d) Teak Forests, and (e) Bamboo Forests. Only bamboo forests are dealt with in this study.

These are generally found in the valleys and on the lower slopes of the hills and occasionally on the higher slopes and hill tops. They grow gregariously so that there is a striking absence of all growth on the floor. Generally *Dendrocalamus strictus* is also found. Other bamboos such as *Bambusa arundinacea* are rare and confined to isolated patches. The common associates are *Acacia dynalolii*, *Gmelina arborea*, *Acacia catechu*, *Dalbergia latifolia*, *Chochlpermum grosypium*, *Sterculia urens*, *Bosewellia serrata* and *Launaea grandis*.

Phulbani: The Phulbani forest division extends over the two civil sub-divisions of Boudh and Khondmals. The vegetation of the two sections differs in character. The forests of Boudh are classified as (a) Northern Tropical Moist Desiduous, confined to the eastern part, and (b) Northern Tropical Dry Deciduous Forests on the western side of the sub-division. The forests of Khondmals subdivision are broadly classified as (a) Northern Tropical Moist Deciduous forests (moist *sal*) and (b) Northern Tropical Moist Deciduous Forests (dry *sal*). Mixed forests occur in pockets within the moist and dry *sal* forests. A large area of the forests int the northern part of Khondmals just along the Boudh border, from the Bagh in the west to Donga Bilabadi in the east, are covered with bamboo. Chiefly *Dendrocalamus strictus* is generally found on the slopes of the hills. Mixed forests are also found in pockets both in moist and dry *sal* forests. *Sal* occurs throughout the Khondmals either in pure form, or mixed with other species. The crops may therefore be divided into (a) *sal* with its usual associates and (b) Mixed forests. The reserved forests of Boudh subdivision in Phulbani forest division which comprise of about 965 sq. km. may be chiefly divided into the following four categories: (a) *sal* forest (moist peninsular low-level *sal*), (b) riverine forests, (c) teak forests (south tropical dry teak forests), and (d) mixed forests (Northern dry mixed deciduous forests (Sahu and Mohanti: 17). The *sal* forests, riverine forests and mixed forests do not exist in well-defined tracts. These types gradually merge with others and the intermediate types are easily recognizable.

The forests of Boudh may be broadly classified as Northern Tropical Moist Deciduous confined to the eastern part; and its northern Tropical Dry Deciduous on the western part.

Mixed forests, which occur extensively throughout the Boudh sub-division, tend to become moist in the eastern and dry in the western regions. They may be classified into two distinct categories. Valuable mixed forests which contain *Pterocarpas marcupium*, *Bridelia retusa*, *Adina cordifolia*, *Wodier ougemia*, *Diospyros melanoxylon* and *Bombax malabaricum*.

Four species of bamboo, *B. arundinaceae*, *D. strictus*, *B. tulda* and *B. nutans*, were said to grow profusely in the district in the seventies and were supplied as raw material for the paper industry. The species grows abundantly in the forests of Boudh, Phulbani and Baligurha.

Baligurha: The moist valleys of Baligurha harbour northern tropical semi-evergreen forests. The important species encountered are *Mangifera indica* (*mango*), *Diospyros embryopteris* (*makar kendu*), *Michelia champaca* (*champa*), *Dillenia pentagyna* (*rai*), *Macaranga peltata* (*manda*), *Mesua ferrea* (*nageswar*) and *Sarara indica* (*ashoka*), *Calamus* (canes) are found growing in shady moist places. Teak plantations have been raised with success and are being commercially exploited.

Four species of bamboo, *B. arundinasceae*, *strictus*, *B. Tulda* and *B. nutans*, were said to grow profusely in the district in the 1970s and were supplied as raw material for the paper industry.

Kalahandi: The flora of the valleys is evergreen where small patches of Reserve Forest occur, the crop inside is pure *sal* or plateau *sal*. Adjacent hillsides are covered with grass with no trace of *sal*. *Sal* is the relic of the area wherever it is found and extends only to about 18° 30″ N. The forest found on the outer slope of the plateau of 300 to

700 m. above sea level is essentially a dry mixed deciduous forest containing *D. strictus*. This type of mixed forest changes to *sal* on the red soil seen on the crest of Karlapat and Thuaml. The hill tops which range from 1,200-1,300 m. are for most part flat laterite plateaus. These are dry, except during the rains when they become grassy expanses. A few stunted trees are found along the water courses. In the villages, mango, tamarind, gulmohur and *sisoo* grow very well.

In the Khariar division 1,673 sq. km. of unclassed forests *sal* of a moderately moist type and its associates are found. In the dry mixed forest the number of species is too numerous to be mentioned. In the protected forest of Khariar Division (south) comprising 73 forest blocks, a typical mixed species of dry deciduous, teak and *D. strictus* is found. Just as *sal* shows a preference for more acid rock, the dry mixed forest follow the basic rocks on their hotter and drier aspect. Vegetation on the escarpment is often dry, mixed forest is shrubby, where hornblende schist is the common rock type, *sal* is invariably absent.

The only bamboo of importance is the *D. strictus*. It forms virtually pure patches as in the Gudia areas in the hills, but even there, it is commonly co-associated and surrounded by mixed forest and sometimes by *sal* where the underlying rock is Khondolite, or where the more basic parent rock (Charnockite or hornblende schist) has not been expressively exposed. The most extensive area under this type are found in Madanpur—Rampur and Lanjigarh, where areas are appreciable. Elsewhere, it occurs in small patches or as admixture with other classes of forest generally the dry mixed type (*District Gazetteer*, Kalahandi, 1976: 22).

It is important to note that bamboos were not leased out on a long-term basis although advised by the agency forest advisor in 1949-50. After the merger the area was brought under a scientific system of management on a sustained-yield basis. In 1954-5 a long-term, nine-year lease for bamboos was given to an Orissa Paper Mill.

The rich forests were almost intact till 1952 under the possession of Khariar zamindari who had sole proprietary rights for both conservancy and destruction.

Koraput: In the north of Gunupur tahsil and east of Bissam Cuttack there were once large tracts of evergreen forests containing some of the most valuable species of timber but largely degraded, where *sal* survives only above 2,500 ft. *D. strictus* is common in the entire Kashipur tahsil. Contiguous with the west and south-west of Rayagada is a wild region, a tangle of hills and valleys with a few plots of cultivable land.

In Kashipur tahsil, though drained by the Indravati, Kolab and Muchkund, the process of denudation has advanced so far that its hills are covered with low scrub or disfigured with bald patches.

In the extreme north-east is the region known as Pannabeda Mutha lying some 500 ft. below and separated by ghats and thick forest. Due to heavy 75 in. summer rainfall there is a fine growth of *sal* and other timber over the whole plateau.

There are hundred of square miles of continuous thick forests where *sal* is prominent. Due to its population and the fertility of the land it became the principle granary of Koraput.

Malkangiri Subdivision: Almost the whole subdivision is a vast jungle. Plantation schemes were reported to be underway in 1966. It is also rich in bamboo which used to be floated down the Sabri and Godavari rivers. During the rains it becomes swampy and impenetrable. The soil is poor. *D. strictus* (on basic rock) and *Oxytenanthera negrociliater* occur.

To sum up, Orissa is in the tropics and since it forms the meeting point of the southern and northern tropical forests, its striking feature is the merger of both teak and *sal*. In the north it is essentially *sal*, while bamboo occurs as an understorey of both species, though not in the forests of Mayurbhanj and Keonjhar districts. Yet both grow well in villages where they are planted. Reckless cutting of both teak and sandal *Santalum album* over the past decades has wiped out these two valuable species. *D. strictus* is more widespread than *B. arundinacea* and according to PCCF Padhi the two species of bamboo top the list and surpass even *sal* in their economic worth (Padhi 1991: 43-5).

The Changing Status of Forests of Orissa

As late as 1973 the IDBI, ICICI, IFC and AFC jointly carried out a survey to assess the industrial potential of Orissa. The survey noted that although three large paper mills—Orient, Titagarh no. III and JK (Seva not mentioned) drew as much as 5,90,000 tonnes of bamboo per annum, there were vast areas that were still unexplored (ICICI 1973: 66). Two of the units, Titagarh and Seva, had closed down in the 1980s but would officially go into production in 1993-4 under Thapar's management.

The original leased areas of the two currently working units, Orient and JK, were initially confined to Sambalpur and Koraput forest divisions respectively. The oldest Titagarh Paper Mill had been assigned the Angul division; the foresters credit this with treating the bamboo forests in its catchment areas with care and concern, evidence showing that the bamboo forests in Angul are still intact.

The 1980-1 Annual Administration Report of the Forest Department of Orissa puts down 3,298.44 ha or 3.98 sq. km. as disforested area, and lists five major causes. Diversion of forest land for, (a) irrigation, (b) rehabilitation of displaced persons, (c) transfer of land to the revenue department (d) raise plantation of cashew, etc., and (e) for industrial purposes (p.1). Although the main concern here is to asses whether the destruction of the bamboo was through over-exploitation by the paper industry, other issues too, particularly those listed in Chapters 2 and 3 are important. These diversions have resulted in bald hills, once thickly covered by forests. The real reason for this was explained by a retired Principal Chief Conservator of Forests, U.N. Sarangi. He put it down to a nexus between the bureaucracy and politics, which led to good forest lands on hills being gifted away; however, he added that the gradient of the area would have been much too steep for agriculture and rehabilitation of any deprived group. He points out that the *pattas* that were drawn up did not indicate the precise location of rehabilitation sites. The 'rehabilitated' individuals began to quietly extend their areas and get hold of forest wealth leading eventually to the balding of rows of hills all along the western chain of mountains running from north to south. The present PCCF, G.S. Padhi also blames the Revenue Department. He abserves: 'while the management of the protected forests is entrusted to the Forest Department the ownership of the land is vested in the Revenue Department. Large chunks of protected forests have been leased out in the past for agricultural purpose. It is exactly not known [*sic*] how much of these forest areas have been leased out for agriculture because of large scale encroachments'. However Padhi (1991), is certain of one thing: 'these forests are devoid of trees except in inaccessible areas'. Still another forester who sought anonymity, added another cause, viz., the veneer industry. Mangalam owned by Birla was granted forest leases for his unit. No further information is available as veneer and plywood are private limited companies.

Although records show that 43 per cent of the geographical area is under forest, the satellite imageries bring down the figure to as low as 22 per cent (Padhi 1991: 21, 22). The satellite also captures all the vegetation, and in Orissa the annual plantation programmes of all agencies totalled, works out to 30,000 ha or 300 sq. km. (p. 22). Plantation programmes began in the First Plan period.

Plantations do not make forests. Encroachments have to cease.

Any attempt by private individuals to bring these to the notice of the collector is penalized. R. Satyapathy (1994: 9) quotes the case of Radnath Pradhan who was jailed for his concern for the depletion of the Langigarh and Bhangpada in Kalahandi. While encroachments continue, the timber mafia in collaboration with the bureaucracy has its own way. The boundaries of Reserve Forests (RF) have been obliterated due to lack of maintenance and lack of funds. RF are not shown in settlement maps nor recorded in the Record of Rights. Sarangi therefore suggests that the boundaries of RF be properly redemarcated, surveyed and maps prepared by competent authorities such as the Survey of India or by the settlement organization of the state, but not by the forest staff who he thinks is not competent to undertake this task. It is also necessary to identify the boundaries of the protected forests in the field and mark them off with pillars, as well as to show them on the map. FD generally has no information on undemarcated protected forests, and therefore detailed information and *pattas*, Sarangi insists, should be issued to the FD. This would prevent some of the corrupt tahsildars/ revenue officials making any changes in the record of rights, from forests to wasteland or tenanted land. He insists that a separate organization be set up to maintain records, that forest officials need to be properly trained to investigate and apply relevant acts in courts of law, in forest cases (Sarangi: personal communication).

The 1982 Annual Administrative Report of FD stated that the major cause of deforestation in that year was because of provision of raw material for industrial needs. Actually in that year, 2,720 ha of reserve forests were converted into plantations of *A. auriculiformis*, *Casuarina* and pines at a cost of Rs. 19.14 lakh (p.12). This is slightly short of the previous year's record of conversion of natural 3,350 ha of forests for Rs. 36,44,610; which works out to Rs. 1,087 per ha of pulpable trees. Such monoculture is dangerous because it is vulnerable to destruction by pests. In 1981-2, for instance, young *casuarina* plantations in Balkunda, Konarak and Chilka divisions were attacked; the Malkangiri afforestation division was also attacked by grubs. Despite this very evident danger to monoculture, Rs. 5,180 worth of seedlings were distributed to farmers to raise farm forestry (*Annual Report*, Orissa FD 1980-1: 12). This was meant to feed the paper industry, but was found unworkable. The FD had exploited 1,20,405.46 ha of bamboo that year (1980-1: 12) of which 1,803.04 ha was pure bamboo area and 6,607.71 ha from overlapping circles (FD 1980-1: 10). The mills worked their own allotted areas. The royalty rate was low and the promised increase very gradual till 1988. A rebate of 30 per cent was promised.

Orissa was so rich in bamboo that it even allotted a four-year cycle to AP Paper Mills to collect bamboo from Malkangiri and Jaypur Division (G.O. no. 3368-6F-65/77 FFAH dated 2 December 1978). It collected Rs. 24,34,900 as royalty for 1980-1 (*AR* 1980-1: 31). Besides this the FD sold commercial bamboo both by auction and by calling tenders. Bamboo artisans were supplied commercial bamboo from

ROYALTY RATES FOR DIFFERENT TYPES OF BAMBOO

Year	Time period	*D. strictus* (in Rs.)	*B. arundinacea* (in Rs.)
1-10-1977	30-9-1980	2.90	5.80 per unit of 100 m.
1-10-1980	30-9-1984	3.83	9.58 per unit of 100 m.
1-10-1988	30-9-1989	5.75	11.50 per unit of 100 m.

Sources: Sanction letters, Government of Orissa Forest Fisheries, Animal Husbandry Department.
For Orient Paper Mill, letter no. 27426-6F-3/80 FFAH dated 11 November 1980.
For JKPM, letter no. 3739-6F-3/80 FFAH dated 11 February 1980.

CATCHMENT AREAS OF PPI UNITS, 1992

Name of the mill	Catchment area (Forest circle/Division)	Administrative districts
1. O.P. Paper Mills Brajrajnagar Sambalpur district	Sambalpur – Devgarh, Bolangir, Phulbani, Khariar, Ghumsur North, Bamra	Sambalpur, Bolangir, Phulbani, Kalahandi, Ganjam
2. J.K. Paper Mills Rayaghada Koraput district	Rayaghada Malkangiri Kalahandi, Phulbani, Balliguda, Ghumsur North, Ghumsur South Paralakimadi	Koraput, Kalahandi Phulbani, Ganjam

Source: *The Statistical Outline of Orissa Forests.*

unreserved forests on system at scheduled rates. As a minor forest product, revenue bamboo yielded Rs. 23,45,864 in 1979-80, rising to Rs. 28,28,058 in 1980-1 (*Annual Administrative Report*, Orissa Forest Department 1980-1: 31).

While bamboo of both species occurs as understorey though not in the forests of Mayurbhanj and Keonjhar districts, these grow well in villages where they are planted. Reckless cutting of both teak and sandal *Santalum albun* over the past decades has wiped out these two valuable species. *D. strictus* is more widespread than *B. arundinacea* and according to PCCF Padhi, the two species of bamboo top the list and surpass even *sal* in their economic worth (Padhi 1991: 43-5).

Assessment of the Felling Practices of Orient Paper Mills and JK Paper Mills

The original leased areas of the two working units of Orient (OPM) and JK Paper Mills (JK) were originally confined to the immediate vicinity of their factories, viz., in Sambalpur and Koraput circles respectively. The only available Annual Report of the Orissa Forest Department 1980-1 reports that while bamboo forests in Angul are still intact, large areas of forests in Koraput and practically all of Sambalpur are on the point of vanishing. The Orissa FD was alarmed at the state of affairs (OFDC *19th Annual Report*, 1982: 2), so the Government of Orissa nationalized the forest. Forests and minerals were the only source of wealth for the state. Industrial units with long leases with the Forest Department for raw materials were banned from felling their lease areas in 1986. The Orissa Forest Development Corporation (OFDC) has taken on the industries labour force to work the bamboo forests in the mills' erstwhile assigned areas. The labour force probably follows the old felling practices they used, while in the service of the two paper mills. Bamboo is the main raw material used by the mills, besides some quantity of hardwood. If proper silvicultural practice of felling is used, bamboo must flourish and continue to yield year after year through its vegetative system of reproduction until it flowers after long cycle of 40-60 years. If it flowers, it does so defensively to seed, and then dies. But the seeds sprout to eventually form new clumps. Seeding is a defensive reaction if bamboo is exhausted through over-felling, so that at this time they need to be carefully protected. But bamboo forests are vanishing. The question therefore is: Is this due to faulty felling? This study's evaluation will therefore be based on observation of felling practices in each of the catchment areas of the two respective industries.

Felling Practices of Orient Paper Mills in its Erstwhile Leased Area in Dhebreghad Wildlife Sanctuary (Sambalpur Forest Circle), January 1993

Dhebreghad is a two year-old wildlife sanctuary on a series of hills lying on the north-eastern edge of the huge Hirakud reservoir. Dhebreghad was established in 1930 by the then Deputy Commissioner of Sambalpur district (Senapati and Mohanti 1971: 34). The Dhebreghad felling series in the Batli range is in the old Baraphar (12 hill) range. The Hirakud dam water spreads into the forests and has had great influence on its ecology as well as on its wildlife. The sanctuary lies at a distance of 20 km. from Sambalpur town, its gates opening into a totally depleted forest which as one moves in begins to appear as semi-evergreen. But it then gives way to poor quality mixed deciduous forests with occasional bamboo clumps as undergrowth. Further uphill, 5 km. in the interior, bamboo *D. strictus* appears to be almost the main species. The other tree species that co-exist but are sparse and cannot be more than 5 years old with the girth never exceeding 15 in. just a little taller than the bamboo which ranges only between 12 to 15 ft in height.

The streams that now and again cut across the pathway were totally dry in January although the area is credited with an annual 527 mm of rain from the south-west monsoon and convectional rainfall during frequent depressions. The soil on the rough road meant for transport of forest raw material was very fine yellow earth. But on either side of the road the valley and the hill slope appear lateritic and leached, totally devoid of humus, with poor undergrowth,

This was a part of the OPM's leased area turned into a sanctuary, with OFDC workers felling bamboos in the sanctuary for OPM, according to the revised working plan 1990-1 to 1999-2000, prepared by A. Chitransi (pp. 257-72).

Dhebreghad is a part of the bamboo overlapping working circle of Sambalpur. The felling cycle is 4 years. Dhebreghad's bamboo area is 12,129 ha divided into 28 sub-compartments classified into A, B, C and D types. Under this working plan commercial/industrial felling is permitted only in A type. A, although well stocked, is degraded and damaged and C too sparse and scattered to permit harvesting of bamboo.

An area is typed as A and subdivided into I, II and III on the basis of height and the number of culms per clump. Broadly it is as follows:

A. Type I: height > 9 m and number of culms > 20 per clump
Type II: height 6-9 m and number of culms 15-19
Type III: height 6 m and number of culms 10-14

The prescribed silvicultural system is culm selection-cum-clump improvement and the working plans prescribe a definite sequence of felling. It also prescribes a minimum number of culms to be retained on the basis of the quality of the clump. Dhebreghad has only two areas marked as A type. The bamboo brakes according to Chitransi, is of the edapic type under Champion's classification.Chitransi complains, 'the regeneration of bamboo is going [*sic*] poorer day by day and in many areas (Sambalpur circle) bamboos are becoming extinct' (p. 258).

The study team observed that the clumps that were easiest to fell and closest to the road were hopelessly over-felled, the debris heaped but not cleared. OPM's official in charge of forest resources clarified that it is the general practice to first work the clumps close to the road and then work the more difficult undulating areas.

Silvicultural practices of working of bamboo are totally ignored. It is not unusual to find only six or seven current-year culms, sometimes even only two or three being left uncut in a whole clump. Now and again, only a couple of one-year old culms are left standing with the current year's growth. One knows that the danger of such a method of felling is that though each culm is tall and has reached its full height, it is still young and does not have the strength to withstand strong breeze. Orissa is prone to cyclonic storms. Silvicultural practices prescribe an equal number of mature culms to stand as supportive structure to vulnerable young ones. By and large mature culms are generally reaped, at other places, the central culms irrespective of their age are left intact in the clump as the peripheral one alone totally harvested. *D. strictus* is not thorny or bushy like *B. arundinecea*, so there seems no reason to commit this grave fault. It is the live, standing culms on the periphery, which encourage the outward spread of the rhizome. Again, the prescribed height, the need for a clean slash and an angular cut are totally ignored. The cuts are not always clean, the remaining bits of bamboo often splintered. Such injuries and straight cuts, not infrequently at the ground level, spell danger as rain water can stagnate instead of draining away, and encourage fungal attacks. This spells the death of the whole clump. Only an occasional high cut, in the current year bamboo culm was noticeable. The OPM's foresters blame bamboo artisans. But this runs contrary to the general information that artisans need mature culms while bamboo of any age can be pulped to produce paper. The number of such high cuts is too infrequent, perhaps one in thousand, to pose a threat to a bamboo forest.

Reactions of OFDC Officials

The reactions of OFDC officials were varied. A trained Indian Forest Service (IFS), a very senior official was taken aback by the observations of the study team, and angry at the mill's accusation that OFDC was responsible for wrong felling practices. He pointed out that, by and large, forest labour was trained by the industry and that it mechanically continued the wrong technique of felling. The same labour force is now on the payroll of the OFDC. Only one senior OFDC official conceded that his organization was responsible for faulty felling and the Forest Department should impose a fine on OFDC. Still another senior conservator, also now an OFDC official, was at a meeting with his field team when we met him. The group as a whole was upset and very concerned at such careless felling habits. They agreed that OFDC was new in this field, labour was untrained and so were the supervisors. But training lakhs of forest labourers, one of them argued, would imply heavy expenditure. The cost has to be borne by the organization which uses bamboo and needs a sustained yield, he continued, while the industry was totally unwilling to share the enhanced cost even if it meant a saving in the long run on the cost of raw material. Nevertheless the field officers expected that within a span of three or four years, the labour, if properly trained, would improve as it now works in happier conditions. The OFDC pays a minimum wage of Rs. 25 per day in contrast to Rs. 10 paid by OPM. Earlier such labour was controlled by labour contractors but it is not clear whether the contractor appropriated part of the wages earned by the labour.[20]

OFDC also hires labour through a contractor but with a difference—the labour contractor is called in only when there is a pressing need, say, of several lakh workers. The contractor is directly paid Rs.1.50 per labourer for the whole season, which works out to a substantial amount. He is not permitted to take a cut from the workers's daily wages.

Orient's Reaction to OFDC's Felling Practices

Orient now holds OFDC's faulty felling techniques totally responsible for the destruction of its catchment area and for creating a raw material famine. However, the fact is that OFDC took over the working of the leased bamboo areas in Orissa only after nationalization of the forests in 1988; nationalization itself was a defensive move by the Orissa government to save the rapidly dwindling forest under the onslaught by the paper industry (see OFDC *19th Annual Report*, 1982: 2). Orient's position that it carefully tended the bamboo forest in its own enlightened self-interest is not quite correct. Evidence also reveals that it had begun to feel the shortfall of bamboo as early as 1971 (*OPM Annual Report*, 1974 and *Stock Exchange Directory*). Its own Materials Manager confessed that by 1973 the mill began to mobilize all farmers within a radius of 200 km. into growing bamboo and other pulpwood species. It had started distributing seedlings to hamlets around Brajrajnagar and Jharsiguda. All along the way to Sambalpur from Brajrajnagar every hamlet has well-tended bamboo clumps, tall, handsome and massive clumps of *Bambusa neutons* and some even *D. strictus*. The latter were tending to shed their leaves (it was the third week of January 1993). There is, therefore, every reason to believe that the original bamboo forests in this region would have thrived had they not been over-exploited by OPM. Orient claims that it has its own bamboo-cum-*sal* plantation about 30 km. away from Sambalpur on the Sambalpur–Puri route, but that the bamboo cannot be harvested without permission from the Revenue Department.

Given below is the available tabulated data on bamboo made available to OPM by Sambalpur Forest Department and by OFDC respectively.

TABLE 2.1: BAMBOO SUPPLY FROM SAMBALPUR FOREST DEPARTMENT TO ORIENT PAPER MILLS LTD. (1985-6 TO 1987-8)

(*in million tonnes*)

Division	1985-6	1986-7	1987-8	Total
Sambalpur F.B.	6,070.9	5,253.9	5,325.4	16,650.2
Sambalpur H.B	2,96.3	2,68.2	350.0	914.5

Source: Forest Department Sambalpur.

Note: F.B.—Full Bundle, H.B.—Half Bundle.

TABLE 2.2: INDUSTRIAL BAMBOO SUPPLIED BY OFDC TO OPM (1988-9 TO 1992-3)

1988-9	1989-90	1990-1	1991-2	1992-3
1,04,805 S.U.	1,21,000 S.U.	1,10,000 S.U.	1,04,000 S.U.	1,04,000 S.U.

Source: OFDC.
Note: 1 S.U. = 2,400 mt; 1 S.U. = 1 notional metric tonne upto 1991-2.

JK Paper Mill and its Forest Raw Material

Preamble: A very senior JK official stated that the present site of the mill was deliberately selected because of its proximity to the abundant bamboo forests, despite the fact that the river Nagavalli did not carry sufficient water. Water is a critical component for manufacture, nevertheless JK gave priority to bamboo.

The first leased area for bamboo in 1957 was Rayagada in Koraput district (see Table 2.3) which was extended to contiguous forests in the neighbouring districts of Kalahandi, Phulbani and Ganjam, and more recently to distant Mayurbhunj with depletion of bamboo (the depletion is evident in Table 2.4). The supply of bamboo and hardwood was in the ratio of 75:25 and was claimed to be adequate till 1986, only small quantities of wood being purchased from private parties.

Since the nationalization of bamboo forests the mill procures additional quantities from Bengal and Assam (Table 2.5 and Figure 2.1). It also uses unconventional material like rags and grass besides hard pulpwood which it tries to cultivate (see Figure 2.2). OFDC continues to supply both bamboo and hardwood (see Table 2.6) cultivated in its plantation. This will be discussed later.

JK claims that its position was comfortable as long as it was permitted to harvest bamboo from its own leased area. Its argument is that enlightened self-interest kept the bamboo forests in good health by following correct silvicultural practices of felling, but that bamboo has been doomed since nationalization of forests, it claims, was not a defensive measure to save the forest from pillage through over-feeling but a political dicision to generate revenue for a poor state which has no industries but is rich in minerals and still has some forests. It also charged OFDC with faulty techniques of felling and felt that the inexperienced, newly-recruited IFS cadre was ignorant and incapable

TABLE 2.3: BAMBOO HARVEST BY JK PAPER MILL FROM LEASE AREA, RAYAGADA (1973-4 TO 1987-8)

Year	Total production of bamboos		Royalty collected (in Rs.)
	Daba, B. arundinacea (in metres)	*Salia, D. strictus* (in metres)	
1973-4	Nil	2,51,56,120.00	3,50,000.00
1974-5	Nil	3,03,43,800.00	3,50,000.00
1975-6	Nil	2,62,23,200.00	3,50,000.00
1976-7	Nil	2,68,31,560.00	3,50,000.00
1977-8	Nil	39,67,040.00	10,34,828.54
1978-9	Nil	5,34,71,180.00	13,94,667.86
1979-80	Nil	4,40,36,338.00	11,48,582.57
1980-1	31,492.00	6,12,09,474.00	21,67,862.98
1981-2	67,855.00	7,76,29,070.00	27,62,427.76
1982-3	2,91,925.00	4,44,74,690.00	15,91,442.24
1983-4	Nil	4,85,740.58	17,14,664.26
1984-5	58,401.60	5,60,14,862.40	25,32,426.31
1985-6	75,706.80	6,66,40,885.50	21,60,397.63
1986-7	1,77,440.40	29,201,188.80	13,30,555.40
1987-8	26,551.20	2,97,81,962.70	13,42,658.00

Source: Forest Department, Rayagada.
Notes: 1 SU = 2,400 mt. up to 12 December 1992.
Salia Bamboo – *D. strictusd* – 1 S.U. = 1 notional metric tonne.

TABLE 2.4: PRODUCTION OF BAMBOO IN RAYAGADA FOREST DIVISION (1988-9 TO 1991-2)

1988-9	17,715.01 S.U.
1989-90	14,445.79 S.U.
1990-1	12,888.56 S.U.
1991-2	9,111.15 S.U.

Production of Bamboo in Rayagada Forest Division

Year			
1991-2	Industrial *Salia* Bamboo	=	8,879.83 S.U.
	Industrial *Daba* Bamboo	=	70.81 S.U.
	Commercial *Salia* Bamboo	=	157.66 S.U.
	Commerical *Daba* Bamboo	=	2.85 S.U.
			9,111.15 S.U.

(*contd.*)

TABLE 2.5 (*contd.*)

1990-1	Industrial *Salia* Bamboo		
	Industrial Full Bamboo	12,605.35 S.U.	12,616.83 S.U.
	Industrial Half Bamboo	11.48 S.U.	
	Industsrial *Daba* Bamboo		
	Industrial Full Bamboo	78.41 S.U.	78.48 S.U.
	Industrial Half Bamboo	.07 S.U.	
	Commercial Bamboo (Sales)		
	long bamboo	8.68 S.U.	193.25 S.U.
	Commercial Bamboo	184.57 S.U.	
		12,888.56 S.U.	

Source: DFO, Rayagada.

TABLE 2.5: JK'S SOURCES OF RAW MATERIAL IN MT (1991-2)

		1991-2	1 April 1991 to 31 March 1992
Bamboo		72,543	OFDC
		18,600	North Orissa
		26,800	Assam
		2,700	Bengal
Hardwood			
	Andhra Pradesh		
	Casurina	13,500	Andhra Pradesh
	Eucalyptus	4,100	Dept. OFDC
	OFDC		
	Eucalyptus (Karnataka)	4,900	
	Haryana (Eucalyptus)	3,900	
	Tamil Nadu (Eucalyptus)	840	
	Andhra Pradesh (Eucalyptus)	8,000	
	Madhya Pradesh (Eucalyptus)	1,300	
	OFDC (Eucalyptus)	330	

Source: J.K. Paper Mills, Forest Department.

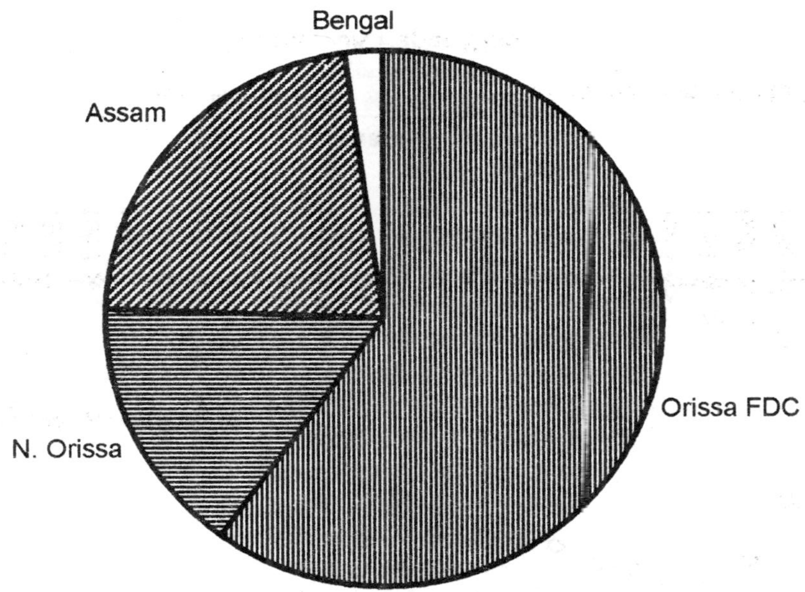

Source: JK Paper Mills

Figure 2.1: Sources of Raw Material (in tonnes) for the year 1991-2

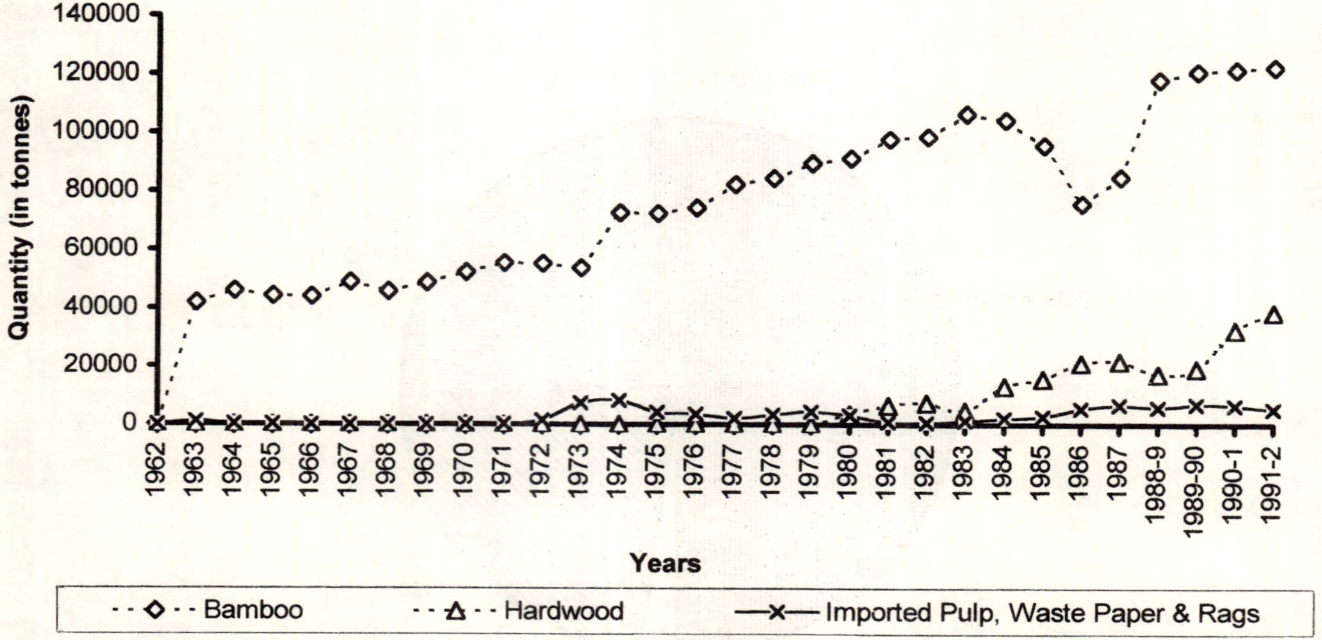

Source: JK Paper Mills

Figure 2.2: Raw Material Consumption by JK Paper Mills 1962 to 1991-2

TABLE 2.6: SUPPLY OF BAMBOO AND HARDWOODS BY OFDC TO JK (IN S.U.) FROM 1988-9 TO 1992-3

	Supplies of Bamboo		
	1988-9	-	1,07,882 + 1,600 (S.U.)
	1989-90	-	1,21,000
	1990-1	-	11,10,000
	1991-2	-	96,000
	1992-3	-	96,000
1989-90	Industrial *Salai* Bamboo	=	13,984.42 S.U.
	Industrial *Daba* Bambo	=	339.43 S.U.
	Commercial Bamboo	=	121.94 S.U.
			14,445.79 S.U.
1988-9			17,715.01 S.U.
	Supply of Hardwoods		
1986-7	1133.69 tonnes	}	Eucalyptus & *Acacia auriculiformis*
1987-8	1124.20 tonnes	}	Eucalyptus & *Acacia auriculiformis*
1990-1	7535.00 tonnes	}	Eucalyptus & *Acacia auriculiformis*

Source: Orissa Forest Development Corporation.

TABLE 2.7: OFDC'S SUPPLIES OF INDUSTRIAL BAMBOO TO JK PAPER MILLS (CROP YEAR) IN S.U. (1988-9 TO 1992-3)

Name of Paper Mill	1988-9	1989-90	1990-1	1991-2	1992-3
J.K. Mills	1,07,882 1,600	1,12,000	1,10,000	96,000	96,000

Bamboo

1988-9	–	2,48,783 S.U.
1989-90	–	2,28,454 S.U.
1990-1	–	2,12,332 S.U.
1991-2	–	2,37,745 S.U.
1992-3	–	75,219 S.U. (up to Dec. 1992)

Source: Orissa Forest Development Corporation.
Note: 1 S.U. = 2,400 mt. S.U. = Sale Unit

of properly training its forest labour force to reduce depletion and destruction of bamboo forests; however, Table 2.7 indicates a fall even before OFDC took over felling, and also despite opening up of inaccessible areas.

Observation of felling Practices in JK Paper Mill's

Earstwhile Leased Area by OFDC

An observational tour in January 1993 of a recently felled area in Koraput circle 80 km away from Jaykaypore—the mill site on the road to Behrampur—revealed a variety of faults. On either side of the road were a series of denuded hills, no bamboos appearing till one was well past 75 km. from the mill. At one point, half the clump, about 15 bamboos were clear-felled at the ground level and the other half left totally intact regardless of the age of the culms. Everywhere along that route only clumps on the roadside had been harvested, high cuts about three and a half to four feet high were common and even current-year culms treated in this manner. The mill official complained about the still standing three and a half to four feet high bamboo which he said was precisely what the paper industry needed; however, it meant a loss to the exchequer, being of little use to anyone. An occasional culm that had been felled was sometimes left lying near the clump that had been worked and cleared. A clean slant cut at the suggested low height was not the rule. Jagged edges were not infrequent, a straight ground level cut was more the rule, despite the fact that the bamboo species were the only easily manageable *D. strictus*. The felled bamboos had evidently been lifted; they were not stacked on the roadside as is the general practice. The debris around the clumps had not been cleared completely, dried leaves and unwanted bits of bamboo were lying in tangled heaps around the clumps, creating the danger of possible forest fires.This would be aggravated by the humidity going down and by the summer heat expected in full blast, when the forest became tinder dry a couple of months later.

The mill official complained that OFDC labour harvests only bamboos growing on the roadside, leaving untouched those on the higher reaches of the hills on both sides of the road, because of somewhat difficult terrain. This may or may not be quite correct, as it is generally the practice in Orissa to first fell the roadside bamboo and then go into the interior. The interior in this case, was rocky laterite hills on either side of the road. Here the soil was lost through erosion and the hills were now covered by sparse dry deciduous scrub forests which offered easy entry. Much of the defective felling by OFDC, according to the mill management, is due to its inexperience and also because bamboos are worked by only one set of workers. A JK official claimed that its practice was to engage two sets of labourers. One set first felled bamboo in the interior and then cleared the debris, while the other hauled the material down-hill to the road side and stacked it. He also claimed that their labour was closely supervised. If this were correct, the bamboos should have continued to flourish. OFDC took on felling only in 1989, and observes a four-year cycle. If this is the real picture, why are the bamboo forests not as healthy as they should be? Why are there no clumps around the mill site? Also JK's own use of raw material does not tally with its claim of devoted care and concern for bamboo forests. As early as 1972 it had to augment its bamboo with Kenaf and Sabai grass that grows plentifully in Orissa; sometimes it was even driven to using rags.

OFDC's Explanation for Wrong Felling Practices

OFDC refuted the charges of irresponsible felling. It argued that the present labour force in this area was trained by JK and that 5,000 of these labourers who work in JK's erstwhile lease area had inherited their old employer's technique *in toto*. OFDC finds it hard to believe that the labour could have suddenly changed its felling practices and become so callous and unconcerned, praticularly when they were exploited earlier by low wages of just Rs.10 per day which OFDC had now hiked up to Rs. 25, besides giving other benefits. It conceded that training is necessary, involving enormous resources which the profit-oriented, industry would not be willing to share in the form of enhanced price of bamboo, and it would have the advantage of a sustained yield of raw material. A third impartial opinion comes from a much respected, retired and very senior conservator of forests. He found that only one industry (the old Titagarh before it passed into the hands of the new management) had scrupulously followed bamboo felling rules. He rated JK as 'not too bad', but OPM, he said, had 'ruthlessly destroyed the forests'.

The area around the JK factory is totally bare although the site was carefully chosen for its richness in bamboo—the nearest area for bamboo now is 80 km away. True, JK's production began to rise, but there was a parallel expansion of its lease area and if bamboo had been treated with as much care as Titagarh had done, the areas under JK's control would have continued to yield and fulfil the demand.

Augmenting of Forest Resources by OFDC

There was an acute shortage of wood, particularly fuel wood after the Second World War. The Five-Year Plans gave it an important place as it did to the rehabilitation of degraded forests. Plantation programmes began with both state and central resources and were originally meant for soil conservation. As the years rolled by, a variety of plantation programmes were introduced. From the point of production of industrial wood the three most important programmes in Orissa were (a) Quick growing species, (b) Economic plantations, and (c) Rehabilitation of degraded forests. The most important species that were grown were teak, cashew, eucalyptus, casuarina, *Acacia auriculiformis*, *Cassia siamea* and *Semul*. The first two species are not pulpable for paper. The rest particularly eucalyptus, casuarina and *A. auriculiformis* are meant for pulp. In Orissa pulpwood and firewood plantations are synonymous. But eucalyptus has no local demand, collieries refusing to accept it for mine props. It is now used only for pulp; also, 3,600 lakh mt of firewood is expected to be consumed by local consumers and by paper mills (Padi 1991: 47, 49, 51). Almost any hardwood is pulpable (see appendix, Ch. 4 for BILTs long list of hardwood trees that it had demanded from its local forests).

In all other states except Orissa, FDCs were set up specifically at the suggestion of National Commission of Agriculture in the mid-1970s to augment industrial and especially pulpwood. In OFDC pre-dates NCA by a decade. Its interest after nationalization of forests in 1988 was specifically to harvest bamboo and pulpable species. Till June 1989 it had harvested 2.67 lakh sale units of bamboo (OFDC, *20th Annual Report*, 1982-3).

Nationalization of Orissa forests was mooted in 1982 due to a major 'structural weakness', viz., surplus of unskilled labour. OFDC proposes to train this unskilled labour. Its other tasks were to undertake commercial plantations of different species, grasses, bamboos, etc. But bamboo had not matured till 1993. Interestingly it also planned to start manufacture of pulp and paper with *sabai* grass, *kenaf*, and bagasse (OFDC, *19th Annual Report*, 1981-2: 2, 3, 4, adopted on 18 August 1988).

TABLE 2.8: LIST OF INDUSTRIAL PLANTATIONS RAISED BY BHUBANESWAR (A) DIVISION OF THE ORISSA FOREST DEVELOPMENT CORPORATION 1982-3 TO 1992-3

Year of formation	Location of the plantation	Plantation raised (in ha)	Reduced area based based on survival % in ha	Species planted
1982-3	Champagarh	110.00	9.18 ha	Acacia
1982-3	Nalsingh	51.00	8.66 ha	Acacia
1984-5	Ratanpur	60.00	4.85 ha	Eucalyptus
1984-5	Aiginia	10.00	0.84 ha	Eucalyptus
1984-5	Kandambajhar	53.00	6.92 ha	Acacia
1984-5	Dhani (South)	110.00	63.28 ha	Acacia/Cassia
1984-5	Bhusandapur	190.00	101.01 ha	Acacia/ Casuarina/ Eucalyptus/ Cassia
1985-6	Ratanpur	38.00	2.95 ha	Eucalyptus
1985-6	Jariput	40.00	7.92 ha	Eucalyptus
1985-6	Dhani (South)	50.00	24.00 ha	Acacia
1985-6	Bhusandapur	50.00	18.74 ha	Acacia
1985-6	Nargoda	20.00	12.00 ha	Acacia/ Eucalyptus
1985-6	Nalsingh	96.00	17.00 ha	Acacia/ Eucalyptus
1985-6	Adal	20.00	1.62 ha	Eucalyptus
1986-7	Bhusandapur	32.00	9.97 ha	Casuarina
1986-7	Nargoda	20.00	15.00 ha	Eucalyptus
1986-7	Jariput	60.00	36.80 ha	Eucalyptus
1987-8	Jariput	35.00	20.47 ha	Eucalyptus
1989-90	Bhassara	17.00	16.92 ha	Mixed with 18 species
1992-3	Dhani RF	100.00	100.00 ha	Eucalyptus
1992-3	Patia RF	85.00	85.00 ha	Eucalyptus
1992-3	Mal RF	115.00	115.00 ha	Eucalyptus
1992-3	Kodalpalli BRF	40.00	40.00 ha	Eucalyptus
1992-3	Kotagarh BRF	60.00	60.00 ha	Eucalyptus
1992-3	Kutunisila BRF	20.00	20.00 ha	Eucalyptus
1992-3	Bankamunda RF	80.00	80.00 ha	Eucalyptus
1992-3	Khandagiri RL	27.00	27.00 ha	Mixed Plant
1992-3	Dhani (N) RF (Teak)	1.20	1.20 ha	Mixed Plant
		1590.20	906.60 ha	

Source: Divisional Manager Plantation (A) Division, Bhubaneswar.

SECTION 3 (A)

Orient Paper Industries Brajrajnagar (OB)

The Orient Paper Mills belong to the Birlas. They are the largest holders of forest based industries in India. In fact, Birlas have three paper units and three rayon units spread over six states, and veneer and saw mills in the north-east. The one in Orissa—Mangalam, was closed down after the nationalization of forests. Only Tamil Nadu actively opposed the entry of Marwari capital in its paper belt. Division of the property and split in the control on paper and rayon units are later developments. One branch monopolized paper and the other rayon. But the two industries are clubbed together, as the source of their raw material is same. How did the Birlas of Pilani in Rajasthan, enter into manufacture of paper which had been the exclusive preserve of the British rulers? It certainly was a struggle. A glimpse into early history will provide a clue. History also gives a clue to the strategy and the tactics used by an Indian merchant family to transform itself into a pre-eminent industrial business house.

Entry of the Birlas

It is important to remember that colonial domination destroyed Indian agriculture, while permitting the growth of lower forms of industry, like rice-milling, oil-pressing, spinning and ginning. In fact the Indian moneyed classes, mainly the trading classes, were encouraged to invest in small industry (Shirokov 1973: 14). The Birlas were one such family.

If we go a step backwards in time and look briefly into the history of the growth of the Birla family, we find that Sheonarayan Birla migrated from Pilani to Bombay in 1860 and made his money speculating in opium (Jaju 1986: 19). The British stranglehold over China was through opium. Marwari Vaishyas monopolized the opium trade to China. Opium grew in Rajasthan and Marwar but was exported from Bombay (Jaju 1986: 25). Sheonarayan's son Baldeodas found it difficult to get a foothold in the Bombay share market which was then controlled by the Gujaratis and Bhatias (Jaju 1986: 16). He therefore moved over to Calcutta in the late nineteenth century (Hazari 1967: 148); to work as a share broker to the British traders, while still continuing to speculate in opium (Jaju 1986: 25). It was the ambitious grandson Ghanshyamdas (GD) who was finally able to take off.

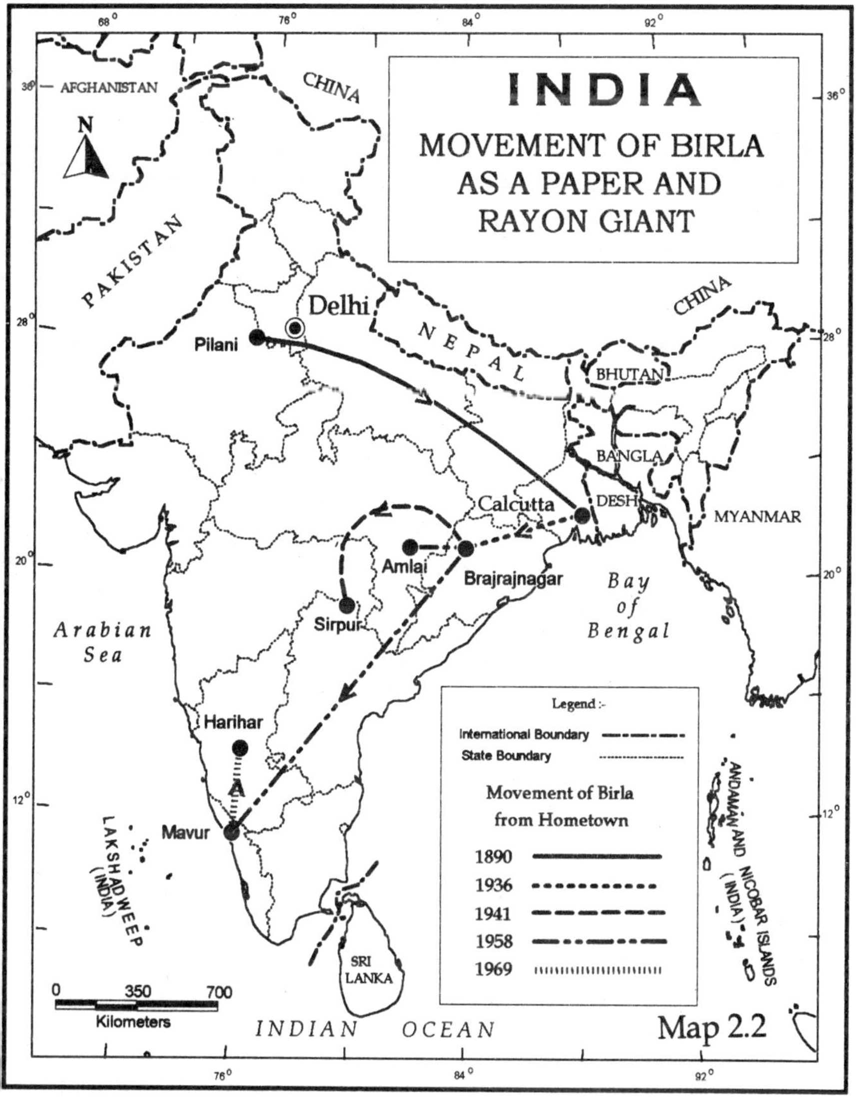

INDIA
MOVEMENT OF BIRLA AS A PAPER AND RAYON GIANT
AFGHANISTAN
CHINA
PAKISTAN
NEPAL
BHUTAN
BANGLA DESH
MYANMAR
SRI LANKA
Delhi
Pilani
Calcutta
Amlai
Brajrajnagar
Sirpur
Harihar
Mavur
Arabian Sea
Bay of Bengal
INDIAN OCEAN
LAKSHADWEEP (INDIA)
ANDAMAN AND NICOBAR ISLANDS (INDIA)
Legend :-
International Boundary
State Boundary
Movement of Birla from Hometown
1890
1936
1941
1958
1969
0 350 700
Kilometers
Map 2.2

The concern of this study is not to sketch the growth of the house of GD Birla within 40 years, but to highlight certain relevant points. The most important of these is to show how Birla, the industrial policy maker and the partner of the Tata-Birla plan of 1945, often broke the basic regualtions of the policies he himself prescribed. His actions appear both socially and ecologically destructive. The study also highlights how, with the waning of British power in the world arena, when the situation was ripe for the upsurge of national liberation, Birla and the newly—emerging capitalist class aligned themselves with the leadership of the nationalist Congress and benefited from the movement. Jaju, G.D. Birla's biographer, states that Baldeodas came into close association with Gandhi as early as 1915, when they discussed such issues as 'Religion and the Rise of a Nation', and tells how Baldeodas gave donations for Harijans (Jaju: 38). In 1915 Gandhi had just stepped into the Indian political arena.

Post-War Industrial Structure

The British had thwarted the Indians from freely investing in industrial enterprises. It is important to remember that after World War I, the British government in India was compelled to revise its industrial policy out of sheer necessity, in order to hold India as a part of its empire. By the Act of 1919, industries were made a provincial subject, which R.C. Majumdar considered an instalment of constitutional reform (Majumdar et al. 1973: 912). Although the responsibility of the representatives of the people was limited, yet this act permitted the emerging bourgeoisie to grow enormously rich particularly during the inter-War years, due to the selective protectionism given to light industries. Such protection of course, was aimed at its imperialist rivals, but incidentally benefited the up-and-coming Indian industrial entrepreneurs and soon the emerging capitalist class became a force to be reckoned with during the world economic crisis (Shirokov 1973: 24).

Paper Protection Act—A Double-edged Weapon

Nevertheless the British were still the rulers of India. They built the paper industry, initially with the help of the *Kagzi* families, strengthening it later with the use of bamboo. Bamboo was discovered as a cheap and inexhaustible raw material at the Forest Research Institute (FRI) after years of intense research. The paper industry closely guarded its own monopoly in which it brooked no interference either from other imperialists or enterprising Indians, till almost the mid-1930s. As soon as the British administration discovered the wonder raw material, bamboo, it began to feel threatened by foreign competition it passed a series of protective acts. One such was the Paper Industry (Protection) Act of 1925. Five years later, Act XX of 1927 was amended and the *ad valorem* rate of 15 per cent was raised to 20 per cent from 1931. In November 1931, the Indian Finance Supplementary and Extended Act imposed a surcharge of 15 per cent on the existing revenue, and protective duties were levied on different classes of paper for a further period of seven years. Also, a specific duty was imposed on wood-pulp. The Bamboo Paper Industry Protection Act of 1932 continued to give protection to all classes of paper (Ahuja 1980: 36) and a duty of Rs. 45 was imposed on imported pulp (Poddar 1979: 6). Paper protection acts became a double-edged weapon making an entry into this area difficult. But once the Indian bourgeoisie stood up to them, and entered the area of paper production, as Birla did in 1936, the protective acts became a boon against fierce foreign competition. Birla shrewdly chose the newly-created bamboo-rich Orissa state, a province where the Congress was powerful, unlike its base at Bengal. British intelligence reports suspect that Birla supported the Contress right from 1931 (Chopra 1991).

Choice of the Site

Orient Paper Mill was incorporated on 30 July 1936, with an installed capacity of 20 tonnes per day at Brajrajnagar where the river Ib flows. This was done in the fond hope that it would satiate the thirst of the mills. The site was shrewdly selected for other reasons too. Brajrajnagar is on the south-eastern railway track on the Howrah–Bombay route, the national highway was probably situated to run close to Brajrajnagar to be particularly useful for the transport of raw material.[21] Coal mines of the south-eastern coal fields are located at a radius of 10 km from the plant, to meet fuel and steam for power generation. It is perhaps not surprising that the Hirakud Hydel Power station was commissioned in the very First Plan period although there was a great deal of doubt regarding the suitability of the location. The energy requirement for paper production is enormous and a hydel power station nearby is convenient. There are sufficient natural lime deposits nearby, lime being a critical raw material for paper production. And finally, the surrounding forest areas were rich in bamboo, the main raw material for paper and paper board. Production started in 1939 at the rate of 20 tonne per day with a Vioth's Fourdiner machine (MF). No historical records on the early growth of the mills are available. Their history was reconstructed by a very senior official of a mill. Interestingly, the handout supplied on request does not quite tally with the oral account.

War-Time Statutory Controls and After

Orient made enormous profits during the War. When prices started rising due to scarcity, the paper industry was brought under the control order for the first time in 1942. Except for four Indian-held paper units, production of paper was still under the control of British industrialists. The government arranged with the industry in 1942 for prices to be negotiated and fixed for its purchases (Dagli 1988: 995). A series of orders, nine to be precise, relating to price control were issued during 1944-6. Although these pertained mainly to wartime economic measures and government purchases, by 1945 the difference between the government purchase prices and the prices of civilian markets widened so considerably that the problem had to be referred to the 'Commodities Price Board' in 1947. The Board merely recommended a marginal increase in prices. The problem was again referred to the Tariff Board in 1949 and a new set of prices was introduced in 1949 for civil purposes, viz., for the non-government consumers. The statutory price control order of 1945 was withdrawn in 1950 on an assurance by the industry that it would not raise the prices without consulting the government. But it did revise prices five times. As a result, the government once again, on the recommendation of the Tariff Commission, imposed control over paper prices in January 1960 for both government and civil purchases on 24 varieties of paper and paper products. The prices remained stable till 1968 when the control over prices was removed.

Status of Paper Industry in Independent India

Independent India's industrial policy was first announced in 1948. This envisaged a mixed economy with an overall government responsibility for planned development of industries and their regulation in the national interest. Birla was one of the architects of the notion of a mixed economy for India (see Bombay Plan 1945). The government, recognizing the importance of paper as vital for sectors such as formal education, banking, communication, etc., listed the paper industry as one of 17 basic industries of importance in the Industrial Policy Resolution of 1948 (and later classed it under Schedule 'B' in the new Industrial Policy Resolution of 1956). It is also important to remember that paper was covered under the Essential Commodities Act of 1955 so that the producers of paper could fulfil their social obligations. The main social obligation was to produce writing and printing paper. In the course of discussion one of the main questions this study raises is: Does Birla, the policy maker, fulfil his social obligations? The other question is: Does the pattern of production of paper by Birla augur well for the forest eco-system from which his industry draws its basic raw material?

The government occasionally issued control orders when the Paper and Pulp Industry (PPI) flouted certain basic rules and regulations. PPI's reaction was to reduce its production. On the contrary, lifting of paper control orders always coincides with increase in production. What was the pattern of production by Orient's seven paper machines till 1977? It is known that in 1960 the Price Control Act was introduced and paper was included for the first time under the heavy investment sector and brought under the purview of licensing.

The Early Phase of Development

Paper production increased steadily. The mills have no records of production figures for the early years. The rate at which the paper machines were installed is an indication of the rate of growth.

In April 1939, the Tariff Board reviewed the situation, and finding it satisfactory, recommended continuation of protection for three more years for wood-pulp, all writing and printing paper, other chrome marble, flint—stereo and super calendar newsprint containing 70 per cent or more fibre content and *badami* mechanical. The Second (Amendment) Act of 1939 imposed a duty on wood-pulp at the rate of Rs. 30 per tonne or 25 per cent *ad valorem*. The duty on paper was fixed at 9 *pies* per 1 lb coupled with an alternative rate of 25 per cent *ad valorem*. The industry made progress even in those varieties which were not granted protection. In some cases indigenous varieties were available at about half the cost of similar imported paper, so protection was discontinued from 1947 (Ahuja 1980: 37).

By and large, the industrial potential of the country was put to wasteful use in armament production and carrying out wartime orders at the cost of drastic reduction of production for civilian needs from 1939 to 1945 (Nehru 1946: 347). Nevertheless, certain industries like paper stabilized and expanded, first as a result of cessation of imports, and secondly because the profitable industry attracted sizeable investment. Production thus increased.

According to an oral account given by a senior official, production began in 1939 on a single Voith-made machine which was used to produce all varieties of paper. The raw material was bamboo and production was around 30 tonnes per day. When the demand for paper began to rise in 1944-5, two second-hand machines with production capacities of 10 and 5 tonnes per day were installed. The first machine was modified and its production capacity raised by another 10 tonnees per day. In the early 1950s, a fourth machine was set up, the expansion programme including the production of writing and printing paper. The total production reached 100 tonnes per day. One of the machines was modifies to produce craft liner paper and all other machines were upgraded. The next expansion programme began in 1957-8 when two Yankee machines with 100 tonne capacity each, were installed and the seventh, a 20-5 tonne capacity machine was installed for production of duplex board. The total installed capacity was 200 tonnes per day while the actual production was about 80-5 per cent of the installed capacity. DGTD agreed to this arrangement. The entire production worked out to 76,000 tonnes per annum, giving Orient the status of the largest paper mill in India.

After 1958 there was a lull in Orient's growth in Brajrajnagar. Its attention was diverted to setting up a second unit in Amlai in MP with a single large paper machine. This does not imply that the Brajrajnagar unit stagnated. Renovation and 'modernization' began in two stages when Indian-fabricated balancing equipment and boilers were added and some other equipment renovated. In the second phase between 1974-85 new equipment like double disc refiners and machine nos. 5 and 6 were modified as a part of its modernization programme.

In 1965-6 production increased substantially, but the company bitterly complained that though the cost of production was rising rapidly, the price of paper could not be raised and the higher cost had necessarily to be absorbed by the company. Yet, in the same year, it happily announced a rise in its export of paper which amounted to

Rs.16 lakh. Exports fell drastically to Rs.10 lakh the following year (*Stock Exchange Directory*). In 1966 recessionary trends were evident and under US pressure the rupee was also devalued. That year paper and pulp were exempted from licensing provision to enable entrepreneurs to expeditiously set up additional capacities. But the response was poor as the investment was high and the gestation period long. In 1968 Orient's production fell, and similarly that of the other paper mills (*Stock Exchange Directory*, 1978). Price control over paper was also removed in 1968. Orient gives an interesting explanation for the fall in its production. The reason given was shortage of water and an inadequate supply of power from the Hirakud, besides trouble with its own captive power plant which was being renovated. True, Hirakud was on a controversial site and there were reports that the site was prone to flooding, but one cannot ignore the clout Birla used to successfully pressurize the government to build the dam on an unsuitable site.

Contradictions between Rapid Modernization and Declining Production

It might seem illogical that when prices were increasing and production decreasing, any company should announce plan to increase production capacity by 11,000 t.p.a. or place orders for new machinery, as Orient did in 1966-7. This was just a short time before the government lifted its price control order. Viewed retrospectively, we now know that the forest-based paper industry was quietly shifting its 'product line' from ordinary paper to high value-added speciality paper. It is true, in the Second Plan period that the government had offered to grant licences to import MG machines and had also put a ban on the import of expensive speciality papers. But PPI did not take up the challenge. By 1968-9 most of the balancing equipment was received and installed the very next year. By 1970-1 the company decided to renovate and 'modernize' its old equipment in the pulp section and the power plant at a cost of Rs. 4 crore. The very next year , a boiler and other auxiliary equipment were ordered, and work in the pulp plant began. In 1972-3 construction work began and by 1973-4 'erection of the new boiler' was completed and a 'high consistency high pulp storage tower' installed.

Note should be taken of certain issues regarding paper production and price control acts which were introduced in the mid-1970s. They

TABLE 2.9: ORIENT PAPER MILLS: BRAJRAJNAGAR: DETAILS OF PAPER AND BOARD MACHINES

Installed capacity	Year of installation	Make	Average per month 1988-9	Type
16,000	1939	Voith	1,377	M.F. Fourdrinier
6,500	1942	Sandy Hill	238	M.F. Cylinder Mould
3,500	1946	Airthrey	133	M.F. Fourdrinier
20,000	1950 (Commissioned in 1951)	KMW	1,235	M.F. Fourdrinier
8,000	1957	Bruderhaus	679	Yankee Fourdrinier (Drum Machine glazed one side)
8,000	1958	Bruderhaus	604	Yankee Fourdrinier
14,000	1959	Bruderhaus	1,015	MG Cylinder Mould

Source: Orient Paper Mills, 1992.

are relevant to the issue raised in the introduction, viz., paper as a necessary commodity for education, versus production of paper as a destroyer of the eco-system.

Distribution of Paper for Educational Needs

This issue was raised by the Dagli Committee in 1979 which pointed out that by the Paper (Production Control) Act 1974, the distribution of white printing paper made available at a concessional price of Rs. 2,750 was controlled only informally (Dagli 1979: 299). The estimated need was 2,00,000 tonnes of which roughly 80,000 tonnes was to be supplied to government and the remainder was to meet the requirements of the student community. A Central Coordination Committee was constituted in the Ministry of Industry with representatives of the Education Ministry, paper manufacturers, publishers of text-books, manufacturers of exercise books and paper dealers. The allocation was to be made with reference to the number of students in each state, along with the estimated requirement of paper for production of students in each state, along with the estimated requirement of paper for production of text-books, but in actual practice allocations were *ad hoc* and somewhat arbitrary.

At the Centre, only state-wise allocations were made. State-level committees were constituted, usually headed by Education Secretaries, to work out the quarterly allocation of paper to text-book producers, manufacturers of exercise books, etc. This system worked reasonably well in respect of text-books as they were printed and published under the aegis of government, and also for paper for examination answer books directly required by educational institutions. At the same time, black marketing of paper released for production of exercise books was reported. The Dagli Committee noted that the 'Essential Commodities Act has eliminated the leakage of white paper for unauthorized users'. Another defect of the system, it noted, was 'general lack of control over actual despatch by bills in accordance with the allocations of paper made by the state-level committees, which has anyway been arbitrary. The allocations of white printing paper have not only varied sharply between states, but also from year to year for different purposes and for different states'.

Considerably more than half the allocation of white paper went for production of exercise books as well as for production of text-books by private publishers. The actual use of concessional paper was not monitored, while black marketing of white printing paper released for the production of exercise books was noted (Dagli: 299-300).

In 1975-6 a letter of intent was received for expansion of the paper mill from 76,000 to 91,000 t.p.a. (*Stock Exchange Directory*). The same year Orient invested K£16,39,988 to set up a joint sector pulp and paper mill in Kenya. Compare this picture of apparently breathless speed of modernization and investment, with the depressing picture of its falling production and the numerous tales of woe that Orient reported routinely in its *Annual Reports and Accounts*.

Orient reported that production began 'to stagnate' in 1970-1 although the demand for paper was high, and that even its exports declined to Rs. 67 lakh, though no reasons were assigned. In 1971-2 'production declined further' due to what the company reports as go-slow tactics of the workers and an inadequate supply of wagons, though the working results were considered as 'satisfactory'. The exports for that year declined 'due to competitive market abroad'. The company explained that production continued to decline even further in 1972-3 for a strange combination of reasons, viz., shortage of water 'due to the river Ib changing its course' and the 'workers' tactics to go slow'. Yet the refrain continued—'the working results were satisfactory'. In 1973-4 the reason given for declining production was 'heavy rains'. In 1974-5 the production was reported to

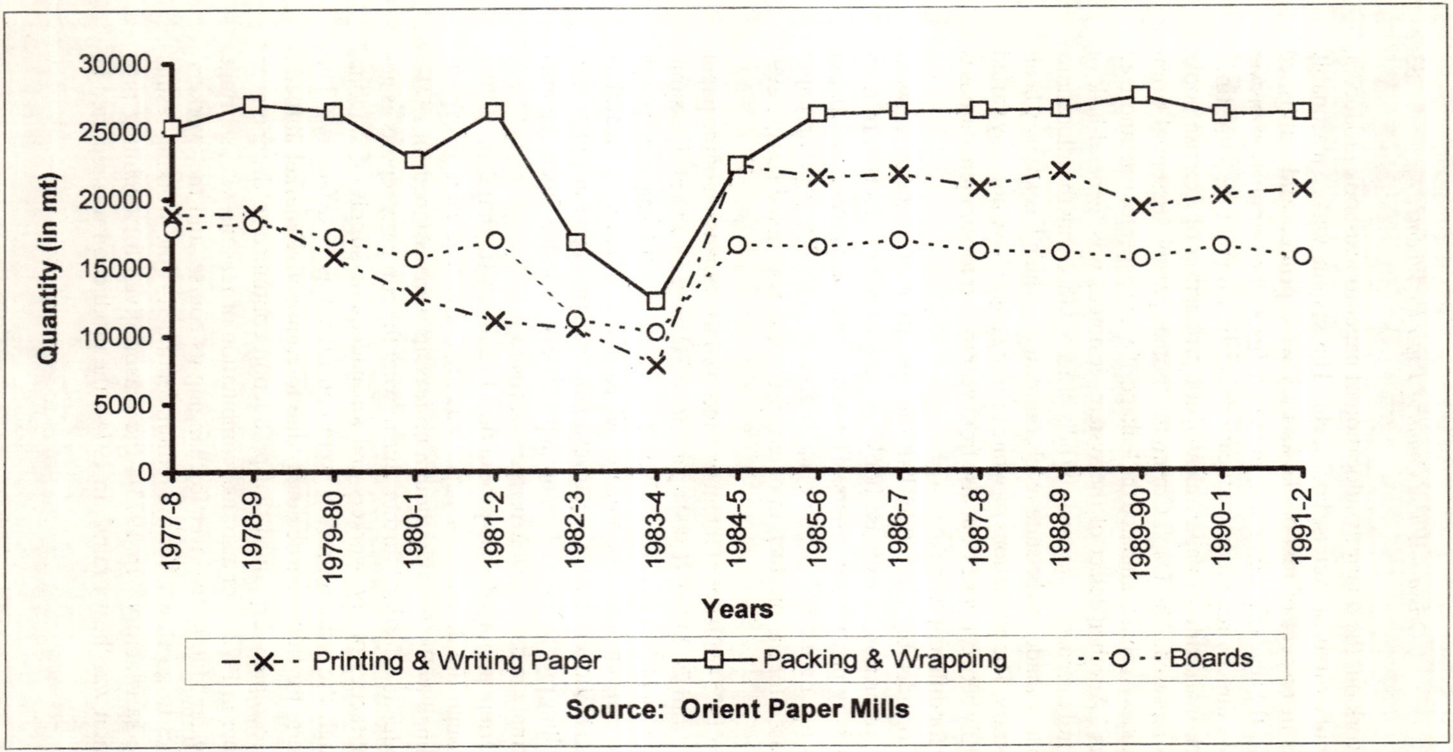

Figure 2.3: Year-wise and Quality-wise Finished Production of OPM-Brajrajnagar 1977-8 to 1991-2

have stagnated at 1,31,800 tonne as against 1,31,428 metric tonne the previous year. In 1975-6 production paper was reported to have declined mainly due to acute shortage of water during the summer months resulting in the closure of both units in Sambalpur and Amlai, as also due to 'go slow by workers' (*Stock Exchange Directory*, 1978).

Decrease in the production of levy paper in the early 1970s leading to an acute paper famine in 1974 coincides with the general pattern of behaviour of all the large forest-based paper units. Incidentally, paper production was controlled by seven business houses at this point of time. Orient did not explain why the river Ib or whichever river which also fed the other paper mills, and whose monsoon patterns and industrial relations differ, also had similar problems during the same time. In other words the real reason for the decline of cultural paper has to be sought elsewhere. Figure 2.3 shows Orient's year-wise and quality-wise finished production from 1977-8 onwards. Writing paper includes expensive value added bond paper (see Figure 2.3).

Reasons for the Decline in Production

There are two plausible reasons deducible from the prevailing situation: (a) The modernization process and the type of machinery installed in the case of Orient and the types of paper produced indicated that there was a slow but definite shift in the pattern of production away from socially useful cultural and printing paper to highly value-added industrial paper, contribution to acute shortage of cultural and printing paper in 1974. For that matter Orient even exported paper during these years until this was banned by the government. (b) The price structure of paper which had remained relatively stable rose by 11 per cent in the next two decades with the lifting of the price control order in 1968.

Even after an informal agreement between the government and the industry was worked out, the price of paper continued to rise. The wholesale price index of paper showed an increase of 35 per cent between 1968 and 1973 and in 1974 it rose to about 60 per cent. The informal structure of regulation broke up. The rapid rise in the price was in order to mop up the 'on money' that prevailed in the market (Rao 1989: 151). The 'on money' ranged from Rs.1,500 to Rs. 2,500 per tonne on all varieties of paper (Dagli: 297). With the combination of rising prices and acute shortage of paper in the market, publishers and printers in the educational sector began to complain. The Directorate General of Supplies and Disposal (DGTD) reported the reluctance of the paper mills to honour their commitment. 'There was a suggestion that Defence of India Rules should be invoked' (Rao 1989: 154). But the government took a soft stand and at a joint meeting in May 1974, the industry was prevailed upon to revert to the 1968-9 production pattern particularly regarding commonly used writing and printing paper, and to supply 2,00,000 tonnes of white printing paper at a concessional rate of Rs. 2,750 per tonne to meet the needs of the vulnerable sections. Finally, in order to compel the industry to abide by its commitment, the government issued the Paper (Control of Production) Order in 1974, exercising the power conferred on it by the Essential Commodities Act of 1955. Further, to ensure that the concessional paper would reach the educational sector and government departments, mills were issued orders to use a particular shade of a dye in order to facilitate detection in case of misuse. Within a year, Orient and several other manufacturers complained of accumulation of stocks. Orient also complained that it had to allow extensive credit facilities to customers to lift stock, and this, it claimed, adversely affected its profits.

Government's Relation with the Paper Industry

Government's ambivalent attitudes or rather its real position vis-a-vis the large paper industry becomes clear when industries which began to produce profitable 'speciality' paper were granted industrial licences for import of new MG machines. Orient introduced new MG machines. MF machines which produced 'cultural paper'were quite often replaced by some MG machines which produced speciality and industrial papers.

The government, instead of penalizing such units, exempted, wholly, or in part, those companies which had partially or fully lost their capacity to produce cultural or levy paper.

In other words, the errant companies were let off from their social obligations. There are other instances which show the government's strong bias towards the large wood-based industry. In fact, at this point of time when the country faced both acute shortage of paper and rising prices, it was the new small entrepreneurs who responded to the government's call and fulfilled it. The government encouraged their growth through a series of concessions and grants, provided non-conventional raw material like bagasse or straw were used. Orient, on its part, complained of sluggish trends in the market in 1975-6 and accumulation of stocks. It also complained that the paper that it had informally agreed to supply at concessional rates against allotment made by the government had not been lifted, so that the company was compelled to allow extensive credit facilities to its customers, which adversely affected its profit. Thirdly, it complained that the cost of production was rising above the selling price of Rs. 2,750 per tonne set by the government by the 1974 Act (*Stock Exchange Directory*). On the one hand, the industry represented its case to the government, while on the other it started hiking up the prices of writing and printing paper. The price of a common variety like the cream wove/laid paper was hiked from Rs. 4,000 to Rs. 4,700 per tonne during 1977. The government reviewed the situation and modified the Paper (Control) Order 1974 with respect to production and prices with two new control orders.

New Control Orders

Initially the government introduced the Paper (Regulation of Production) Order 1978 on 8 March 1978, in relation to large mills. The order provided that white printing paper should be manufactured to the extent of 30 per cent of the total production of paper and on 30 June 1979, through the Paper (Control) Order 1979, it imposed formal statutory price control on white printing and cream wove/laid papers. It prescribed Rs. 3,000 and Rs. 3,785 per tonne respectively when the latter was being marked at ex-factory price of Rs. 5,300 per tonne. Orient challenged the legality and the validity of the Paper (Control) Order 1979 in the High Court at Calcutta. The court passed an interim order, staying the operation initially for a limited period giving liberty to the company to apply for extension of the interim order, while at the same time directing it to supply white printing paper at Rs. 3,000 per tonne to educational institutions. The company agreed (*Annual Report Accounts*, 1979).

But the paper mills began to default in the production of white printing paper on various grounds (Rao: 159). Orient's reasons included the usual ones: rising price of raw material, loss of production due to power cuts, water shortage, labour unrest, etc., till all of it began to sound like a liturgy, and this chant continued in every one of its annual reports. Yet the industry was not penalized for this behaviour. However an investigation committee headed by R. Rajagopalan, Chief Adviser, Costs, Ministry of Finance, was appointed to investigate into the affairs of Birla's paper company under Section 15 of the Industrial (Development and Regulation) Act of 1951. Mismanagement was detrimental to then paper industry (*Economic Times*, 30 March 1977). Meanwhile, Birla had fallen foul of the Janata government. The company immediately challenged the investigation and a stay order was issued by the Calcutta High Court (*Economic Times*, 19 April 1979). There is, however, no further information on this matter.

Very soon in March 1980, the representatives of Pulp and Paper Industry met the Minister of Industry, and it was agreed that the industry would honour its commitment to supply white printing paper to the education department and the PPI would withdraw the writ petitions filed by it. In return, the government lifted the controls on all other types of paper except printing paper. The price of printing paper was negotiated and reviewed and fixed at Rs. 3,500 per tonne from 29 November 1980. The supply of concessional paper to educational departments thereafter improved for a while but was disrupted again in 1982-3 by stay orders passed by the court (Rao: 162).

The price of open market paper was rapidly rising to reach dizzy heights, while the price of levy paper was pegged at Rs. 3,500. The sharp difference between the two led to leakage of levy paper into the open market. The delay in fixing of prices confounded with the variety of product mixes made monitoring difficult for the government. The forest-based paper mills again made a representation to the government that a uniform levy of 30 per cent of white printing paper was too high considering their low MF capacity of production of cultural paper. It is important to remember that all the large wood-based mills had made a deliberate decision to install MG machines and renovate their old MF machines, in order to profitably produce value-added industrial paper. What is surprising is that government gave them licences to do so.

The Production Control Order was amended on 11 April 1983, to provide different rates.[22] But the industry was never satisfied. Even in 1987 when the government allowed the selling price of paper to rise from Rs. 6,400 (1986 price) to Rs. 7,200 in 1987, Orient still complained that the price was inadequate to cover the cost of increasing prices of inputs (Orient: *AR*, 1986).

Finally, in June 1987 control orders were lifted and the forest-based paper industry was freed from its social commitments. It could now develop as it pleased—but at whose cost. Simply stated, this was at the cost of small paper mills which were made sacrificial goats and at the cost of the eco-system. It is important to remember that it was the government which encouraged small entrepreneurs to produce paper with non-conventional material like bagasse and baited them certain concessions in the mid-1970s when the country was facing a paper famine. They responded, took up the challenge and within less than a decade their total production outstripped that of the total production of the recalcitrant large forest-based mills. In fact, in 1986, there was a glut in the paper industry and the large mills were in the red, while some of the largest and the oldest such as Titagarh, Bengal, Rohtas, Punalur, Ashok, Straw Products (Bhopal) and Rayalseema even began to fold up. In 1987 the government not only lifted the paper control order which 'reinforced the large forest-based units, but it also imposed 25 per cent excise duty on imported waste paper which was a very important component of production for the small mills. Finally, by extending Modvat schemes on the paper industry the government effectively pulled out the carpet from under the feet of the small paper units who were upset to make room for the resurrection of the big. Orient and other big units rose from the ashes of these unfortunate mills. Could all this be mere coincidence? The government bent back even further, and in April 1988, at a time when newsprint was stock-piling at NEPA due to wrong pricing policies, it permitted Orient and other large wood-based industries to sell surplus stocks of cream wove paper to newspapers. This had never been done before. And the newsprint industry, particularly NEPA, suffered. Its newsprint was naturally poor in quality because of the government's persistence in using *salai*, while the newspapers naturally preferred cream wove paper to NEPA's newsprint. The stocks which burdened the large wood-based industry lightened and made way for it to rise again. Selling prices of cream wove were Rs.11,400 (1988); Rs.17,950 (1 April 1989); Rs.19,500 (1 December 1989), and Rs. 21,460 (1 October 1990) (ICICI, n.d.: 21).

The government also took care to absolve the industry of its social obligation and took upon itself or rather on the public sector units, mainly the Hindustan Paper Corporation, to supply concessional paper to the educational sector. The government had thoughtfully set up the Hindustan Paper Corporation a decade earlier. The gestation period of all the three units was over and they had started producing paper. This is evident from the fact that Orient expressed great satisfaction that the government had repealed the paper control orders, and it would now be possible for it to 'manufacture the quality of paper required by the market'. Nevertheless, it reported that the paper market was sluggish in 1988 (Orient: 4, 5). Interestingly, its old refrain—shortage of water, power, labour, etc., etc., also ceased. Its industrial relations were reported to be normal (*AR*, 1987: 4) and production went up though the factory had to be shut down for four weeks due to poor quality of coal. Finally, exports jumped to Rs. 77.04 lakh as against Rs. 1.89 lakh in the previous year (*AR*, 1990: 6) while the sales output grew by 30 per cent (*Business Standard*, 20 November 1990). Right from 1988 it had announced its plans for modernization which were reported to have been finalized in 1990 to 'improve the quality of the products, economise on costs of production, reduce consumption of chemicals energy, improve raw material yield, control pollution' all at an 'estimated cost of Rs.100 crore' which it expected to complete in 1994 (Orient, *AR*, 1991). But in practice the situation was quite different. In January 1993, the mill was involved in perpetual crisis, the boiler threatening to burst any moment. After all, Orient was 56 years old, other post-independence units were gearing up to capture the market with a variety of products. At the moment Orient cannot afford to be left behind—the Brajrajnagar unit is limping along. The Amlai unit probably helps to maintain some equilibrium.

The Raw Material Question

Orient chose Brajrajnagar in Orissa to set up its paper mill mainly because the bamboo forests in the Sambalpur district seemed inexhaustible. Even as late as 1973, the survey conducted by IDBI, ICICI, and others prospecting the state for industrial development felt sure that it could take on more paper mills.

The number of paper mills Orissa supports has fallen from seven to just two—Orient and JK. In 1991 Thapar obtained control over two fallen units. Yet Orient complains that the supply of bamboo from its leased area was inadequate in 1977-8 and that it had to secure bamboo from outside the state. It is not clear why it should be so, when the installed capacity of the plant has not increased. Bamboo, after all, is a renewable resource and theoretically inexhaustible. It tends to proliferate rather than shrink, a fact which alarmed the British because their major interest was in teak plantations and *sal* forests. Total control should have been by the local population, failing which, proper silvicultural practices should have kept the bamboo forests intact and flourishing. The question is, has Orient followed silvicultural rules?

To begin with, Orient was allotted bamboo-rich forests in the Sambalpur division. With increasing production more areas were added—Radaco, Deogarh, Kariar, Bolangir divisions. Nine forests in Gumsur north division, one forest each in Sundargarh and Phulbani division and two forests in Baligurha division and more recently in Mayurbhanj. Earlier, it also used to draw bamboo from Vindhya Pradesh in MP, which area was allotted for Orient's Amlai plant. Orient could afford to be indifferent. This argument is based on the fact that it got bamboo at throw-away prices. Secondly, Birlas have the reputation of being environmentally indifferent (Senior Retired Forest Official, personal communication).

Although the Orissa Government had suggested Rs. 15 per tonne as the royalty rate while the Forest Department wanted it to be raised to Rs. 25 per tonne, Orient was eventually allotted a large expanse of Orissa's forest at a royalty rate of a mere Rs. 4 till 1989. This reversed in 1988 when forests were nationalized and leases ceased. This was reported by a knowledgeable forester. He did not hold all the paper industries equally responsible for the destruction of forests. He was emphatic that the British-owned Titaghur applied silvicultural rules to the letter. So much so that the Forest Department suggested that this area be a 'control area' for our observation of felling practices. A retired senior forester rated Orient as a destroyer of forests. JK stood a little higher than Orient in his esteem but nowhere near Titaghur. (For details regarding bamboo forest and assessment of felling practices, see Section two, entitled OFDC's Reaction to Orient's Felling Practices.)

Except for some heavily dust-laden eucalyptus plantations, Brajrajnagar is totally devoid of any natural vegetation on the way to Jharsiguda or on the 70 km. stretch to Sambalpur. Occasionally healthy clumps of *D. strictus* and *Bambusa newtonnes* or *Dendrocalamus hamiltonnei* are noticed. Around them are always a cluster of huts indicating that these are well-maintained homestead or village community bamboo.

The company consistently challenged the payment of its dues such as the sales tax on bamboo[23] and the 10 per cent purchase tax on bamboo royalty from the year 1977. Royalty rates rose as a result of rough and ready methods of estimating tonnenage on the basis of running feet. Orient's contention was that such levies add to the cost of production. The High Court of Orissa admitted its writ petition. Orient not only claimed a 'refund of minimum royalty amounting to Rs. 25 lakhs in respect of the bamboo forests leased out' to it 'under arbitration of the State Government of Orissa', but it also complained of the imposition of Rs. 55 lakh as compensation by the DFO Bamra, which is still pending with the state government of Orissa (Orient *AR* 1977: 6). But there has to be a limit to such behaviour. On 17 March 1978, both members of the Opposition and the Treasury benches (Janata government) in the Orissa Assembly demanded cancellation of the contract given to Birla for collection of bamboo. This demand was made when the Minister of Forests said that compensation amounting to over Rs. 2 crore, had been claimed from the mills for violation of the contract it had entered into for collection of bamboo in the years 1974-5 and 1975-6 (*Financial Express*, 17 March 1978).

With the rise of royalty rates many paper companies played a variety of dirty tricks to get access to cheap raw material. Orient Paper Mills closed down its Brajrajnagar plant in Orissa, ostensibly because of the workers' agitation for more bonus. It was widely rumoured that the management had done so deliberately to send out cheaper bamboo from Orissa and pressurize the MP government which had suddenly jacked up its rates to Rs. 102 per tonne in 1981 (*The State of India's Environment*, 1984-5: 5), as Orient has another unit at Amlai in Madhya Pradesh. SOS telegrams to the Forest Department to send bamboo immediately for fear of closure of the plant seems to be routine behaviour for Orient even today, a fact that the study team noticed during a field visit in April 1992. Orient, however, complained that shortage of bamboo and hardwood was due to illicit felling and therefore it had to purchase wood from distant places at higher prices (Orient, *AR*, 1985: 71). Tables 2.10 and 2.11 show the year-wise consumption of raw materials from various sources by Orient.

A senior official of Orient conceded that for the quality of fibre, the L:D ratio of good quality mature eucalyptus or casuarina is preferable to bamboo. But the quality of both these hardwoods depends on the soil and climatic conditions under which they are cultivated. Haryana, he said, produces superior eucalyptus, but the mill finds the transport cost too prohibitive and has to settle for poorer quality material from Orissa and Bengal, and some quantity of eucalyptus from MP because it is cheaper. He complained that social forestry in the state was a failure and that local farmers were not interested in cultivating tree crops.

TABLE 2.10: ORIENT PAPER MILLS: BRAJRAJNAGAR—YEARWISE BAMBOO RECEIPT, CONSUMPTION AND BALANCE 1984-5 TO 1991-2

Year	Lease (mt)	OFDC (mt)	OFC (mt)	Baster (mt)	Sirincha (mt)	Private Party (mt)	Total (mt)	Consumption (mt)	Balance (mt)
1984-5	83,591.811	—	296.006	2,859.406	—	29,710 150	1,16,458.150	1,10,278.865	8,523.423
1985-6	70,429.149	—	87.185	2,886.947	6,468.222	18,556 322	98,427.825	1,06,479.719	71.529
1986-7	88,534.572	—	188.299	—		32,905 498	12,1628.360	1,13,490.900	8,608.988
1987-8	84,175.217	1,203.909	—	—	—	30,873 724	1,16,252.850	1,23,425.848	1,436.000
1988-9	59,238.055	15,409.415	2,027.847	—	—	51,407 708	1,28,083.025	1,21,249.358	8,269.667
1989-90	91,81.848	74,659.415	500.399	—	—	35,017 695	1,19,359.783	1,22,463.016	5,166.434
1990-1	—	83,301.619	36.435	—	—	39,378 380	1,22,716.434	1,27,163.823	719.405
1991-2	—	87,173.861	1,710.144	—	—	47,289 753	1,36,173.758	1,30,446.527	6,446.276

Source: Orient Paper Mills, Brajrajnagar, 21 January 1993.

TABLE 2.11: ORIENT PAPER MILLS: BRAJRAJNAGAR—YEARWISE HARDWOOD RECEIPT AND CONSUMPTION

Year	OFDC	PURCHASE			Total (mt)	Consumption (mt)	Balance (mt)
		Hard Wood (mt)	Eucalyptus (mt)	Casuarina (mt)			
1984-5	11,990.422	16,977.990	—	—	28,968.412	31,115.166	83.000
1985-6	7,763.752	20,012.965	—	—	27,776.707	27,422.707	437.000
1986-7	4,097.783	19,176.870	31,34.258	30.089	26,876.000	22,321.000	4,555.000
1987-8	5,932.644	99,29.340	2,02.251	—	16,064.235	19,429.240	1,190.000
1988-9	7,411.632	13,650.774	5,59.535	3,148.581	24,770.522	22,800.531	3,160.000
1989-90	59.218	10,368.199	2,30.895	6,114.452	16,772.764	19,932.764	205.550
1990-1	—	6,548.390	7,39.692	8,725.120	16,013.202	15,579.560	639.192
1991-2	—	1,477.128	15,41.269	12,025.197	15,043.594	15,662.410	20.376

Source: Orient Paper Mills, Brajrajnagar, 20 January 1993.

A paper mill overcomes shortfalls in raw material by 'buying wood in the open market' but where this open market is located, and how it functions, has never been made clear by the industry. Since 1986. Orient has also been trying to overcome inadequate supply of wood by negotiating with the government for setting up joint sector plantations of pulpable trees in suitable forest areas (Orient, *AR*, 1987: 5 and 1990: 6).

The government has examined this issue and is reluctant to permit the private sector to get control over forest or other land (see National Commission of Agriculture, 1976). In October 1988, bamboo trade was nationalized and lease holdings cancelled. Bamboo is now collected/worked from the various forest circles by the Orissa Forest Development Corporation (OFDC). The government, however, has asked OFDC to make an adequate supply of bamboo and wood available to Orient (Orient, 1988: 5).

The dream of a joint sector has made no headway (Orient, *AR*, 1990: 6). NCA had recommended Social and Farm Forestry, in collaboration with NABARD as the financing agent as an alternative. In 1990 NABARD agreed in principle to give loans to farmers on both bipartite and tripartite bases, the only condition being that Orient will give a buy back guarantee of the produce. A tripartite agreement has also been made between the farmers, the company and the bank, ensuring cheap supply of wood with no danger of some other firms outbidding (Orient, *AR*, 1990: 6). Compare this system of 1977 with that under operation with NEPA in which farmers had the right to sell wood output to whomsoever they choose. It would appear that the govern-ment operates with a sustained bias against its own public sector industry.

There is, of course, no guarantee that Farm Forestry (FF) and Social Forestry (SF) would succeed, while the natural forest raw material is squandered and the forests ravaged. Orient is waking up to this issue rather late, not due to environmental reasons, but because all cheap raw material resources are drying up. It has therefore taken steps to first change the raw material mix and then find alternate pulpable materials like *sisal*, *kenaf* and old gunny bags. *Sisal*, says a senior executive, is fast growing and can be cultivated on wastelands. It can yield excellent value-added paper. According to a FAO specialist *sisal* yields strong paper (Kurkland, personal communication, Oct. 1995) but the mill's major problem would be investment in new machinery to suit the fibre. This executive concluded that there would be no demand for *sisal*-based paper. Jute on the other hand is inexpensive, but yields only inferior paper and therefore only small units can use it. Though straw is abundantly available, being a cumbersome raw material it calls for new machinery. The Birlas are not willing either to change their raw material or invest in raw material-saving technology because for the necessary heavy investment. Yet investment for diversification was part of the game in the seventies. Despite an acute raw material crisis, Orient as well as most other mills continue to use the outdated sulphite process despite the fact that this process yields only 35-40 per cent of the cellulosic raw material. A shift to a somewhat later sulphite process would have raised the yield to 75 to 80 per cent. The modern oxygen delignification process would be unthinkable for such cost-conscious manufacturers. It would mean a change in digesters and installation of totally different recovery units. So too if chemi-thermo mechanical pulping technology is used, saving on the raw material by this eco-friendly technology is not acceptable although it would mean better products. In America solvent pulping was introduced only because of pressure on the industry to reduce pollution. Birla had interests in the jute industry as early as in the thirties. There is intensive research going on in its Paper Research Institute at Amlai. The Birlas have not given up hope of joint-sector plantations. In 1991 negotiations were on (Orient, *AR*, 1991: 7) and by end of 1993 the Orissa government had conceded this, against the opinion of the Forest Department and the NCA.

Water is another critical component that the paper industry needs. The river Ib is the main source OPM's demand is going up but it claims that the shortfall of water is due to the introduction of minor irrigation schemes in the state. Orient has therefore learnt to cut down its need for water from 300 to 220 ml and even 210 ml per day mainly through recycling, better discipline and house management. Orient, however, has managed to get permission to build a mud and clay (*kuccha*) bund to store water for 60 to 70 days. Winter rains are critical; if they fail, the factory has to be closed, as was the case in 1976 and 1989. Fresh water is drawn from upstream and spent water let out downstream. But now OPM's technologist claims that 100 per cent of the water can be recycled (see section on effluents for another viewpoint). The requirement is 300 ml per tonne of paper. With recycling it will come down to 250 ml per tonne. With proper equipment 80 to 90 per cent of waste water can be saved. The river is now graded 'C' with its

BOD level at 3 in fresh water, which should rise to more than 5 (OPA's executive, personal communication, January 1993).

Effluent Treatment and End Use of Effluents

Neela Chaudhary of Jadhavpur University has developed a filtration technique which is being tried bout by Orient to salvage spent water. Its main principle is to use top soil as a lignin filter which should have zero discharge. Orient has its own 85 acres of land sloping downwards; the soil is rich in calcium and carbon which help in decoloration of the sludge when it is mixed with chlorinated water. The land is insufficient for this sort of land-filtration treatment. Therefore Orient has asked the government for another 85 acres to act as land filters. But unfortunately for Orient, the area proposed by OPM has been earmarked as forest land and therefore not available to Orient.

In and around the effluent water filtration area are huge mountainous mounds of black spent-out coal which is sold to contractors. It is also used by the local people as fuel. But it is mainly used for civil road construction work. Fly ash is made into brickets with lime sludge mixed into it. Nothing goes waste. The waste water with sludge is first given alum and chlorine treatment and then run into a huge basin on the top of the hillock where it stays for 15 days. When the sluice gates are opened the water overflows into trenches where the solids accumulate and the liquid flows down the grassy slopes. The gradient of the slop is 3 degree. Here the water falls into a second trench to trap the remaining solids and the liquids run further down. The trench-cum-slope treatment is repeated several times till the water reaches ground level apparently colourless and odourless. At this point the water is used by farmers to grow a second crop while the water trapped in pools is used for bathing by the villagers. The technician claims that the water is almost 100 per cent pure, and that the mill intends to build a dam to trap the water to be recycled for manufacture of paper.

An environmental scientist at Sambalpur University, who is familiar with Neela Chaudhary's method of water purification, insists that her method removes only organic, not inorganic matter, and that the Mahanadi into which the Ib flows has been polluted by mercury released from Orient which still uses mercury cathodes for electrolysis. His water analysis report, however, has not yet been made available for confirmation.

SECTION 3 (B)

J.K. Paper Mills—Jaykapur

The JK industrial empire is named after Jugilal Singhania (1857-1922), and his son Lala Kamalpat Singhania (1884-1937). They are credited as its founders by R.K. Hazari (1967: 152). The founder of its paper units, first Straw Products (1938) and then JK Papers (1959) is Sir Padampat Singhania (1905-79) (Straw Products, *AR*, 1989-90: 1) There is no indication when and why Padampat was knighted. Such seemingly unimportant information is useful for a fuller understanding of the growth of the paper industry in India, especially since the first Indian-owned units were set up when the industry was still an almost exclusive preserve of British industrialists in India.

Origin and Growth of the House of Singhanias

The Singhanias hail from a village called Singhana in the Shekhavati region of Rajasthan. The family belongs to the Marwari community and to the sub-sect of Aggarwals. Aggarwals are Vaishnavas, and by tradition, merchants and moneylenders. Biographer P. Chintsal Rao (1986: 5) writes that this community was a product of a fruitful relationship between warriors and merchants; the paintings on the walls in Shekhavati are a living testimony to this amalgam. Sir Padampat's shrewdness is said to match the clout of Harishankar Singhania, the present Chairperson and Managing Director of JK Paper. The latter is believed to pace the corridors of power (Nag 1990). He does show the hybrid vigour of the awesome combination that P.C. Rao writes about. This is not to suggest that the vigour is genetically acquired, but that it was socially possible because of certain clan practices which were taken advantage of and cleverly made use by the Singhania patriarch for the family's upward mobility.

The patriarch of the Singhania clan was Binodini Ram (BR). The entry of the British was said to be 'an unsettling factor, which fanned the ambitions of BR, giving him a free hand to fly high by night operation' (Rao: 6) whatever that may mean. All that we can discuss from the romanticized biographical version is that BR, who was already a rich man, made up his mind to seek a still brighter future, moved out from his desert homeland and migrated eastwards towards 'green pastures fed by rivers'. Shorn of romance, the place was no other than drab Kanpur, the British-held oasis of development in Uttar Pradesh.

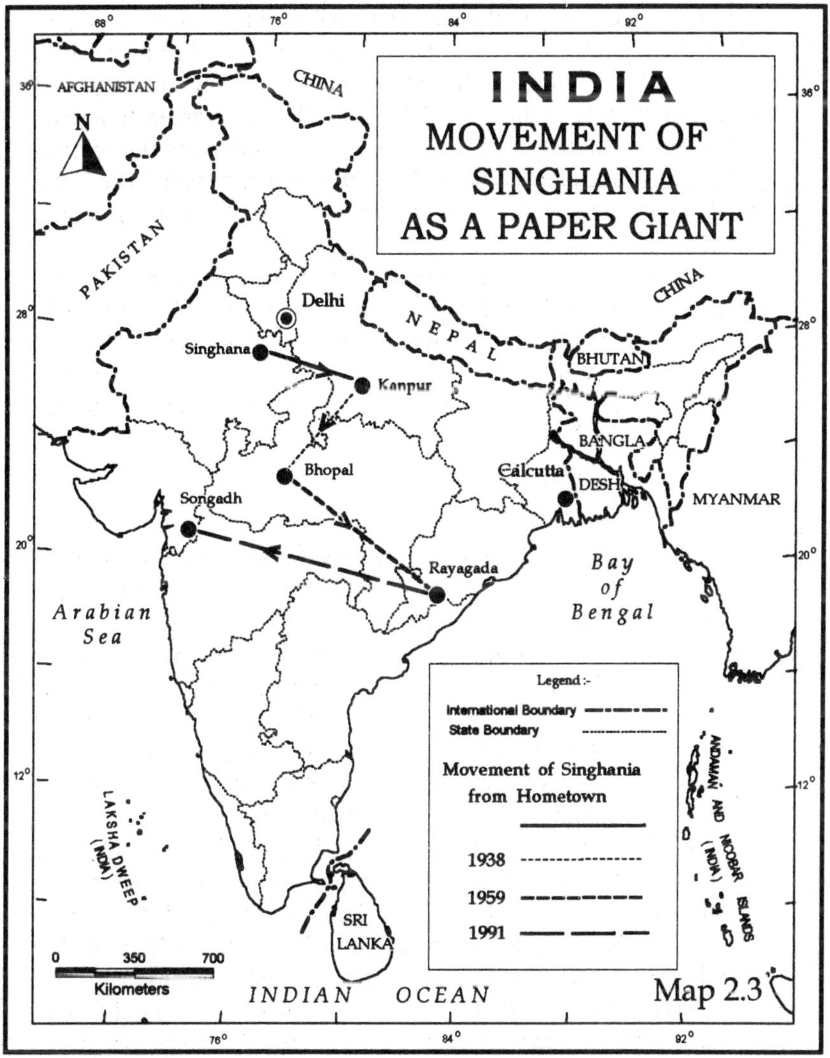
INDIA
MOVEMENT OF SINGHANIA AS A PAPER GIANT
AFGHANISTAN
CHINA
PAKISTAN
NEPAL
BHUTAN
BANGLA DESH
MYANMAR
SRI LANKA
Delhi
Singhana
Kanpur
Bhopal
Calcutta
Songadh
Rayagada
Arabian Sea
Bay of Bengal
INDIAN OCEAN
LAKSHA DWEEP (INDIA)
ANDAMAN AND NICOBAR ISLANDS (INDIA)
Legend :-
International Boundary
State Boundary
Movement of Singhania from Hometown
1938
1959
1991
0 350 700
Kilometers
Map 2.3

BR had made an astute move. In Kanpur, the Vaishya community had developed various institutions to help both trade and the trading community. It had established a communal credit network to enable smooth trading activities. It was the only door open for the Singhania's to grow. Under the nineteenth-century colonial system, industry was the tight monopoly of the British in the north and north-eastern part of the country. Industrial growth was largely confined to the setting up of jute mills and a few paper and pulp factories in Bengal, close to its seat of power. The British also established a couple of textile mills in Kanpur.

For the Singhanias the movement forward was not from trade to industry but from trade and moneylending to merchant banking. BR started banking business first in Kanpur and over the years he and his son set up a network of offices in Kanpur, Banaras and Mirzapur (P.C. Rao: 6). In order to understand the process it is necessary to go back in time.

JK's Second Step

The Singhanias like the Birlas sharply seized whatever opportunities that came their way during the twentieth century (refer to the earlier section and to Chapter 1 for details).

When the political and economic stranglehold, was loosened, Singhania immediately set up a number of units, and Hazari's findings (p. 152) indicate that he set up cotton gins, oil presses and flour mills beside a sugar factory and distillery. Every one of these were set up in 1905 the year when a separate department of industry and commerce was created (Mazumdar 1973: 147) in spite of strong opposition from Whitehall (Palme Dutt 1979: 147). With the First World War there was a complete reversal of government policy. Industrialization was officially set out as the aim in the economic field (see Lord Hardinge's Dispatch to the Indian Secretary, 26 November 1915 and in 1918, the Montague–Chelmsford Report on Indian Constitutional reform was released). As Palme Dutt had pointed out, the proclaimed first reason for the change of policy was military and strategic. The war had laid bare the weakness of the old style of the Indian empire. Second, foreign competition was beginning to break down the British monopoly in Indian markets. This had combined with the weakening of Britain's industrial position. Protective tariffs had to be raised to prevent India becoming a dumping ground of manufactured goods from various nations. The third was an inner political reason. In order to maintain control during the disturbed period and immediately after the war, the British found it essential to secure the cooperation of the Indian business class and win its support (Dutt: 150-1). Accordingly, the Fiscal Commission was appointed on 30 June 1921, and a Tariff Board set up in July 1923 to work out discriminatory protection for British industry. Paper was one of its important and closely protected industries. It is within this scenario that JK set up with its first cotton mill in 1919 and managed to establish its first limited company in 1924.

At this time the hopes of Indian capitalism were raised high. This was the period of the Swaraj Party, or the party representing the nascent progressive capitalism which defeated the non-cooperation policies of the Gandhian leadership at the National Congress in 1923, dominated it between from 1923 and 1926 with policies, the first of these being entering the councils to conduct the fight from within and eventually of 'honourable' cooperation. But their hopes were belied (Dutt: 151-2). It was during this period that father Jugilal and son Kamalapat won fame as the sole selling agents of the British-held Elgar and Cawnpore Cotton Mills.

In 1924 fiscal autonomy was given to India. That year JK entered into jute and engineering, also setting up an insurance company (Hazari: 152) but not paper as yet. Chintsal Rao (p. 11) writes that the Singhanias received tough opposition from the British industrialists, the alien government, and the policies it encouraged. British goods through preferential tariffs added to Kamalapat's frustration. Gandhi's call of boycott appealed to him, and the biographer states that Kamalapat associated himself with Pandit Malaviya and Motilal Nehru. There is no other indication of any close relationship between the Singhanias and the National Congress. Before the Singhanias entry into the paper industry is discussed, it is necessary to briefly highlight the protective legislation raised by the British for its paper industry, which at a more opportune time was utilized by the Singhanias.

Singhanias' Entry into Paper

Paper industry in India became the monopoly of the British first by having utilized and then displaced the *Kagzi* families. Discovery of bamboo as raw material for paper opened out the possibility for rapid growth. Its first bamboo-based unit was started in 1924. But it faced stiff competition from outside. A series of protective acts were passed starting with the Paper Industry (Protection) Act of 1925. Five years later, on 31 March, Act XX of 1927 was amended by which the *ad valorem* rate of 15 per cent on existing revenue, and other protective duties were levied on different classes of paper for a further period of seven years and a specific duty was imposed on wood pulp. What is important to note is that the Bamboo Paper Industry Protection Act of 1932 continued protection for all classes of paper, a duty of Rs. 45 per tonne being imposed on imported paper. Padampat was interested in paper, but the time was still not favourable despite fiscal autonneomy being granted to India as early as 1924. In the face of the stiff opposition from the British industrialists and its government in India that the House of Singhanias had faced in their earlier ventures, both the time and place had to be carefully selected for setting up its paper industry.

Sir Padampat waited till 1938. Britain had barely got over the long period of depression. However, when it seemed that it was on its way to recovery, dark war clouds had begun to loom. In India, by the Act of 1935, 11 out of the 14 provinces had elected the Congress to power. It was no doubt a meaningless power, but would nevertheless be a little more protective to its own people. Both industry and forests were provincial subjects. One would expect that an enterprising Indian entrepreneur would select a good forested area in one of the 11 Congress-held provinces. But, nowhere in or close to the well-established British-dominated paper belt. In 1938, he chose the 'native' state of Bhopal for setting up his unit Straw Products, which would use not bamboo but straw. The 1932 Paper Industry Protection Act, unlike the earlier one, was applicable to all classes of paper whether made out of bamboo or not.

The question is, why did he choose Bhopal? Again, why straw and not the well tested bamboo, the wonder raw material that the British had put to rigorous tests and hard research, and then held as its own monopoly raw material to manufacture more and more better paper for the world market. There are no clear answers to these questions. Did Sir Padampat want to avoid stepping on the toes of the British? There would have been some clue if information of the status of the bamboo forests in the British-held catchment areas of the paper industry were available. It is known that Titaghur Paper drew its material from bamboo-rich Madhya Pradesh and Orissa, and that Bhopal state was in MP.

The choice of Bhopal for a totally new venture is intriguing. We know that the British, whether it be Lord Minto or Lord Hardinge, had adopted a conciliatory attitude towards what they called the 'native states'. This was deliberately done to get greater cooperation from the rulers, who Lord Hardinge described as the 'Helpers and Colleagues' in the great task of imperial rule (Majumdar 1973: 923). The nawab of Bhopal was particularly close to the British rulers and had been invited to attend the Round Table Conference. Was Bhopal the half-way house for Sir Padampat who feared the British and mistrusted the Congress? The answer is not known. Even British intelligence suspected that the Singhanias link with Congress was only much later. Its letter dated 6 November 1942, from CIO Lucknow, states, 'The Singhania group of Cawnpore was reported to be the principal financial backer in Lucknow, but none of the information received was definite' (Chopra 1991: 60). According to its earlier, August 1942 Intelligence Report, 'two Congress workers received a sum of Rs. 25,000 from Sir Padampat Singhania for Congress movement and to set its rival Victoria mill on fire' (Chopra 1991: 59). But what mattered was that the protective legislation that was given to the British-held industry would also be applicable to the Indian-owned industry.

Singhania's Straw Board was incorporated in 1935 and commissioned in 1939 to produce 5,000 tonnes of paper board per annum. The raw material was straw and not bamboo. Within a few years its production increased with the outbreak of the Second World War. The British were now fully absorbed in the production of war goods. This opened out some possibilities at least of growing rich for the Indian industrialists, even if it did not develop their industrial base. The Singhanias snatched whatever opportunities were available. In 1940 though JK came into prominence, it would set up only a number of investment and trading companies and a bank. It was Singhania's business instinct to the fore. Later it acquired textiles, cotton mills, woollen mills and innumerable companies in Calcutta. The British stranglehold over the industries of Bengal had weakened. The concern here is neither with Straw Boards which was leased out by JK in the 1980s, nor with the growth of the JK group *per se*, but with the paper unit that Padampat set up in Raygada in Orissa in 1959.

Post-1947: Singhanias in the New Setting

Singhanias who had made their way through, survived, and grown in a highly restrictive setting in British India by making full use of every single opportunity that came their way, had now nothing to fear. With the transfer of power and the new Government's concerns and its policies geared for rapid industrialization the Singhanias forged ahead in the industrial field. JK acquired Hoyle paints and jute mills and entered in a modest way into mining in the 1950s (Hazari: 152). It also prepared itself for an expansion in paper production.

Newly-independent India had dreamed of a total development where not only industries would flourish but where the land would go back to the tiller, where every child would go to school, where every adult would learn to read, and the demand for paper would soar. The dream remained a pipe-dream and a dream recorded in statute books. But the growing demand for paper was a reality and that was what mattered to the Singhania family.

The new state rapidly built up the infrastructure for industries to grow, set up financial institutions to provide the resources, and announced an Industrial Policy Resolution in 1948 to maintain industries in a healthy state.

Padampat watched and waited till 1959. He had a good twenty years experience with the manufacture of paper boards in the mills in Bhopal, and was even able to export the product in 1951 (Ahuja: 21). The status of the paper industry was good and the prospect of the continuous rising demand for paper was attractive. The Government of India had listed the paper industry as one of the seventeen basic industries of importance in its Industrial Policy Resolution (1948). Even earlier, all the wartime regulations on paper production had been lifted. The Statutory Price Control Order (1945) was lifted in 1950 with an understanding that the industry would not hike prices. And, unlike as in the British period, the agreement was coolly ignored and prices were revised arbitrarily by the industry five times within a brief span of eight years between 1950 and 1958. The situation seemed perfect for setting up a profitable bamboo-based paper industry and Padampat did so in 1959. The only hitch and a small one at that, was the government's demand for a certain quantum of paper as levy paper. At this point of time, it seemed to Padampat Singhania and to the PPI that it was a reasonable deal for the enormous support the government gave for the rapid and more importantly independent growth of the paper industry.

JK Paper Mills—Jaykapur

In 1959 Padampat Singhania embarked on a new integrated paper and pulp mill project in Raygada in the backward district of Koraput in Orissa. As with all other paper mills the site chosen was close to a small perennial river, Nagavalli, though two other sites with more abundant water had been suggested. The present site Jaykapur was deliberately picked as it was surrounded by inexhaustible bamboo forests. The present management proudly claims that in 1975-6 it learnt to cope with limited water by recycling it. Besides, the Visakapatanam-Raipur link railway line was only half a mile away from the site. It only needed a railway siding, which was completed in 1961. A power grid nearby was an added attraction. So, too, were the ravines around in which to dump waste. Orders for a plant and machinery were placed with Messrs J.M. Voith, G.M.G.H. of West Germany, who acted as engineers for the entire project. Erection of the entire pulp and 18,000 tpa paper mill, including the recovery section comprising 5,000 tonnes of plant and machinery was completed within 10 months. The entire township including residential quarters was ready by 1961. The Rs. 5 crore plant was commissioned on 18 October 1962.

The first machine to be installed was an MF one meant to produce 18,000 tonnes of writing and printing paper, also referred to as cultural paper, from which levy paper is drawn. This was in general the pattern of paper production in 1962.

Expansion of Production Capacity: Fall in Production

In 1965 the company received a letter of intent for expanding the capacity of the mill by 25,000 tpa. This implied an incredible jump from 18,000 to 43,000 tpa. But what Padampat had not anticipated was the control over paper prices that the government imposed at the recommendation of the Tariff Commission in 1960, at a time when his new plant was being installed. The Government of India realized that the industry's behaviour had to be regulated. The industry did not take kindly to this, and its reaction was to create a paper famine. JK Paper mills reported that its production fell slightly by 1,874 tpa in 1966, the reason given however was a strike in the mill for eight days. Its sales and profits were not affected. JK cleverly diverted its paper to the Middle East and African markets. In 1967, once again, the company claimed its shortfall in production was due to a go-slow by the workers followed by another nine days of 'illegal strike' (*Stock Exchange Directory*, 1983).

It was not JK alone but the industry as a whole that was rebelling. The concerted action of the industry seems to have paid off. On 8 May 1968, the Government of India lifted the price control on paper.

Immediately the prices of paper were hiked, and again for the second time, in 1969. But the industry was still under an obligation to supply cultural paper for the benefit of the vulnerable sections. JK attempted to overcome this small barrier. The way out was to change its product line.

Modernization for Development or for Profit

Did the industry take the plunge or was the industry perceptive enough to see a new growing trend which would make the change in product line profitable? True, there was a clear shift in the National Education Policy from its old emphasis on primary education which would have ushered in education for all, to a new narrower vision focusing on higher education. Schooling and higher education had become a compulsive need for the middle class as a means to social mobility. The ambitions of the middle class were not hampered by the price of cultural paper. This is not to deny that the absolute number of children from the vulnerable sections entering the primary levels was large. Some got past into secondary and even into higher education, keeping the demand for writing and printing paper, large, if not as massive as one would have expected, and at a reasonable price. Writing and printing paper thus continued to be a necessity.

True, in a skewed economy, it is the tertiary sector that grows, but it was only in the 1970s and not the late 1960s that offices of banks, insurance companies and sophisticated corporations proliferated. These, generally, are the willing buyers of sophisticated writing and printing paper. Urban sophistication was only slowly creeping in, and the need for expensive, packing, wrapping and myriad 'industrial' papers was still very minimal in the late 1960s for several paper manufacturing units to take a plunge. But calculated risks and insightful moves were a part of the strategy of the sophisticated Indian industrialists.

JK industry, like the industry as a whole, began to insidiously change its product line. In 1970 JK installed its first MG machine for the purpose. And the government granted it a license to do so. It produced machine-glazed wrapping and craft paper. The trend gathered momentum and JK proposed to increase its production of paper by 3,000 tpa and boards by 2,000 tpa by installing balancing equipment at a cost of Rs. 1.40 crore. Meanwhile, in 1971 an imported plant and machinery was erected according to schedule. In 1972, an indigenous machine was set up, and paper machine no. 3, with a capacity to produce 75,000 tpa of value-added speciality and not ordinary cultural paper, was also commissioned. Incidentally, such paper has a rag content. In 1973 the project for establishing 4,500 tpa of value-added coated paper boards was started and commissioned in 1974. The company received a letter of intent for increasing MF or cultural paper by a mere 4,000 tpa while the capacity for the production of MG or value-added industrial and speciality papers was more than threefold, i.e. 13,500 tpa (*Stock Exchange Directory*, 1978).

MF Machine Develops Chronic Problems

The 'sickness' of JK paper industry began in 1970. In 1970-1 production which rose marginally by a mere 7.5 per cent mainly by MG machines. The first machine or the MF machine was shut down for several reasons, varying from maintenance, integration of the new pulp line with the existing one and a major breakdown in the 'machine drive motor'. Naturally its production further fell in 1972, while that of the MG machine improved over the previous year. In 1973 production was said to have been affected adversely by 40 per cent due to two main reasons: a 40 per cent power cut imposed by the Orissa Electricity Board and because machine no.1, the MF machine, was shut down to carry out modification. What sort of modification was required is not clear. What is strange and does not tally with the above explanation is that production in 1973 stood at 25,588 tonne while that of 1972 was 22,234 tonne. The MG machines had stabilized to produce commercial tissues, airmail, manifold and bond paper of good quality. All these products were highly profitable, some being exported (see Table 2.12 column 2). Meanwhile, the educational sector starved. JK is an excellent instance of how one of the units of the paper industry ganged up with the other units to pressure the government and create a second paper famine.

The Power of the Industry vis-a-vis the Weakness of the Government

The first informal agreement and the paper industry's promise to provide reasonable quantities of paper for the vulnerable section were not honoured, second time only of levy paper. One would have expected a strong government to have slammed down an order under the Defence

TABLE 2.12: VARIETIES AND COMPARISON OF PRICES OF PAPER MANUFACTURED AT JK PAPER MILLS

Varieties	Substance	Price (excl. exice duty and cess) (Rs./tonne) As on 15 Oct. 1975		As on 18 Jan. 1993
1	2	3		4
Cream Wove	40	—		18,373.57
	45/47	—		19,869.65
	54/56	5,250		17,128.33
	60/above	4,850		16,680.40
SS Maplitho	54/56	6,300	56 to 69 GSM	19,515.79
	60/65	6,300		19,336.62
	66/80	6,100	70 and above	18,839.42
	81/above	6,100		18,615.45
SS Map Printing	54/56	—		19,605.37
	60/65	—		19,426.21
	66/80	—		18,929.00
JK Exclusive	75 GSM	75	—	22,449.72
SS Pulp Board	185/above	6,100	185 to 253	18,012.09
Delux Pulp Board	300/500	—		18,907.95
Colour Printing	(44)	—	5,750 for 50 GSM	20,590.82
(Standard Shades of Pink, Blue, Green & Yellow)	42/47			
JK Offset Printing	54/56	—		18,705.04
JK Ledger	60	—		19,242.55
	70	—		18,834.94
	80/above	—		18,387.01
NG White Poster	29/31	5,850		19,918.92
	32/35	—		19,471.00
	36/40	5,400		18,436.28
	41/55	5,200		18,028.67
	60/above	5,050	-60 GSM	17,222.39
		4,950	-70 GSM	
MG Colour Poster (Standard shades of Pink, Blue, Green, Yellow & Orange)	29/31	6,400	-30 GSM	20,033.15
MG Colour Poster (Deep Orange & Rexona Green)	38	6,350		22,033.15

(Contd.)

TABLE 2.12. (*Contd.*)

1	2	3		4
MG TDL Poster	38/40	6,700		21,406.05
	50	6,600		21,088.02
MG White Poster (AR Grade)	50	—		18,301.90
White Airmail	26	—		26,216.80
	29/30	7,500	-30 GSM	24,913.33
Coloured Airmail (Standard shades of Pink, Blue, Green & Yellow)	29/30	7,900	-30 GSM	26,037.63
JK Bond	58/80	7,000	-58 to 66 GSM	20,774.47
JK Chancellor Bond	70/80	—		31,766.63
JK SS Magnum Print	54/56	—		21,580.74
	60/65	—		20,590.82
JK SS Magnum Board	190 to 250	—		20,963.94
JK Copier (Cut size)	70/above	—		19,859.80
JK Copier (Full size)	70/above	—		20,658.01
JK Colour Copier				
(Cut size)	70/above	—		20,576.48
(Full size)	70/above	—		20,658.01
Wood Free Printing	50	—		21,540.42
JK Literature Printing	35	—		28,048.83
	38/40	—		25,535.95
MF Tissue (Acid Free)	21	11,000		4,1478.10
	44	—		28,366.85
JK Cote Chromo Paper (HG)	75/above	—		27,426.65
JK Cote Chromo Paper (X)	75/above	8,000		26,620.38
JK Cote Art Board (HG)	175/300	—		28,638.75
JK Cote Art Board (Normal)	175/300	8,000		27,742.89
JK Cote Enamel Board	158/300	—		28,102.23
JK Cote Ivory Board	175/300	—		30,340.87
JK Magnum Chrome Paper	75 & above	—		29,5.6336
JK Magnum Art Paper	90/above	—		38,714.45

Source: J.K. Paper Mills.

of India Rules (DIR) to bring the recalcitrant industry to heel. Paper had been declared an essential commodity as early as in 1955. But the government chose to enter into an informal agreement. This time the industry conceded, but only for a brief period. It realized that the boat could not be rocked too violently. 1974 was a year of great political instability for the ruling party. Complaints were pouring in from publishers, printers and from the educational sector regarding the non-availability and exorbitant prices of common varieties of paper.

The Directorate General of Supplies and Disposals (DGTD) also reported that paper mills were reluctant to honour their commitment. The total production of white printing paper had fallen from 1.22 lakh tonnes in 1972 to 88 lakh tonnes in 1973. The Dagli Committee noted that there were reasons to presume tacit collision between the seven major producers to manipulate the markets to create a temporary shortage of cultural paper which in no way affected manufacturer's profits. The total output along with the speciality paper remained roughly the same between 1972 and 1973 (Dagli: 273). The Commission also noted that the circumstantial evidence proved that major manufacturers collected 'on money' ranging from Rs. 1,500 to Rs. 2,500 per tonne on all varieties of paper, the black money thus generated was estimated to be as much as Rs. 80 crore (Dagli 1979: 298).

In August 1974, the government, put on the defensive, finally issued a control order on paper. This was done under the Essential Commodities Act and not under DIR. The first order was the Paper (Control of Production) Order 1974, which stipulated that factories with installed capacities of more than 25 tpa would have to produce at least 30 per cent of their output in the form of white printing paper, unless exempted wholly or partially. The order also stipulated production of another 27 per cent by the way of other varieties of cultural paper. Instead of handling the situation firmly, the government wriggled out of its responsibility and granted exemption to those factories whose output was now mainly from MG machines designed to produce speciality and industrial paper. MF machines were rendered sick. It is not clear whether the government was taking a soft stand towards the industry or colluding with it. The latter seems more plausible, otherwise it is difficult to explain why it freely granted licences for not only one but several MG machines for the asking, so soon after the industry had created a paper famine in the mid-1960s. JK too had naturally moved towards production of the more profitable industrial paper. The price of levy paper was, after informal discussions between the government and the industry, fixed at Rs. 2,750 per tonne.

With the wide difference in price and profitability between the levy paper and the newly introduced speciality papers it was tempting to ignore the Control Order. But JK did not resort to confrontation due to the political instability and the threat of the Emergency. It cautiously tested the government's strength to withstand its pressure.

Effect of Paper Control Order

In 1974, JK's production of paper immediately rose to 32,872 tonnes 'due to improved performance of all machines'. MF's health recovered with the shock treatment. Production of MG did not fall either. In 1975 total production rose marginally to reach 33,927 tonnes, but the take-off of the speciality paper was poor (*Stock Exchange Directory*, 1983). The Control Order had done the trick. The supply of cultural paper eased considerably and the companies began complaining of accumulation of stock.

The government than 'reviewed' the position and modified the Production Control Order in August 1975, retaining control only on white printing and on the inexpensive cream wove/laid paper, and fixed the price of Rs. 2,750 per tonne. The agreement continued on an informal basis.

The paper industry was displeased. The game that it would play this time had to be subtle. JK's paper production was reduced marginally to 33,375 tonnes in 1976 and the sales of coated paper, a value-added paper, stood at 1889 tonnes as against 175 tonnes in 1975. Indira Gandhi lost the elections and the Emergency was lifted. The industry was now ready for a confrontation. JK recycled its cultural paper to manufacture 2,341 tonnes of coated and board paper. By 1977 it began to raise the price of writing and printing papers, the price of the common variety of cream wove/laid going up from Rs. 4,000 to Rs. 4,700 per tonne. The industry stated that the levy price Rs. 2,750 was far below the cost of production, resulting in heavy losses.

Whether the Congress or the Janata Party was in power, was in a sense, of no great consequence. The most powerful component of the ruling party was the bourgeoisie, paper industry being an oligopoly controlled by eight powerful business houses. The government toned down its demands. On 8 March 1978, the government modified the Paper (Control of Production) Order 1974 and replaced it with Paper (Regulation of Production) Order 1978. The order provided that white printing paper should be manufactured to the extent of 33 per cent of the total production of which not less than 20 per cent should be cream wove/laid paper. This act was amended on 16 October 1979, under which it excluded mills with less than 10,000 tpa capacity and those which used unconventional raw material. Newly commissioned mills were also out of the purview of the 1978 control order for a period of five years. These were the small mills that the government desperately promoted with a wide variety of incentives to produce ordinary cultural paper while the big industries were displeased and defiant in turn.

JK's Style of Functioning

JK was unhappy but always cautious. On the one hand, it colluded with the other paper units, while at the same time it maintained a low profile and an even pace of production, albeit a low one. In his speech, the Director (JK 1978), stated that the mill had reached a 107 per cent capacity utilization, a clear 3 per cent higher than the previous year of 36,881 tonnes of varied speciality papers. He justified it on grounds of inflated prices, cost push of the various inputs, chemicals, fuels as being on the squeeze. He was still sore that the hike on levy paper was inadequate and fell short of the actual cost of production. Like the

TABLE 2.13: RANGE OF PRODUCTS OF JK IN 1987

PAPERS	
Printing Paper	Writing Paper
Writing Paper	Cream wove
Coloured Printing	Coloured wove
Maplitho Printing (Surface Sized)	Bond
Offset Printing	Airmail
Art Paper	Typewriting
Chrome Paper	
Manifold	
Super Printing Paper	
Opaque Printing Paper	
Commercial Acid Free Tissue	*Speciality*
Carbonising Base Tissue	Dyeline Base Paper
While Poster	
Coloured Poster	
BOARDS	
Pulp Board (Surface Sized)	
Art Board	
Enamel Board	
Ivory Board	
Black Centred Art Board	

Source: JK Paper Mills (1992).

other paper industrial units, JK too challenged the validity of the Paper Control Order 1979. The stay was granted. Nevertheless, JK and the industry initiated a dialogue with the government in 1980. It was agreed that the industry would honour its commitment to supply white printing paper to the education sector and withdraw their writ petitions. While the government on its part would consider revision of prices. In November 1980, the price was raised to Rs. 3,500 per tonne, controls over paper were lifted and the price of white printing paper began to rise and then soar. Padampat Singhania chose a dual method of development for JK Papers diversification and internal expansion—which paid off a decade later. The JK house had only to bide its time. Its reduced profit in paper due to deliberate underproduction was made up through diversification in various industries other than paper.

JK's Development through Diversification

Already in the mid-1960s and 1970s all the large and even small industrial houses had begun to diversify and enter into collaboration with multinationals. JK became a co-promoter of an automobile tyre and tube project in technical collaboration with General Tyres International Inc., of USA, 1973. Its Plant was set up in a backward district in Rajasthan. JK Batteries Division was formed in 1977. In 1979 it set up a cement project in Sirohi district in Rajasthan. The company holds equity shares in its subsidiary Bhopal Oil and Flour Mills (now Bhopal Udyog). In order to augment its working capital the company issued debentures in 1979, 1981 and 1983, while Dena Bank was amalgamated with JK in 1975.

Renovation, Modernization and Expansion

What is important to remember is that though production did not soar, the company was protesting covertly till 1977 and overtly in 1978-9, but the government did not react negatively. It did not stop issuing licences for buying and installing MG machines which produce only varieties of speciality paper. It is interesting to note that in 1977 JK received a letter of intent for increasing the productive capacity of its MF machine by a mere 4,000 tpa, while it was as high as 1,35,000 tpa for its MG machine. In 1978 again, the company received another letter of intent to expand the capacity of the paper mills by 15,000 tpa. This was converted into an industrial licence in 1979. In the first

TABLE 2.14: PAPER PRODUCTION AND RAW MATERIAL CONSUMPTION BY JK PAPER MILLS

Year	Paper production		Raw material used		
	Wts, Pts, MG & MF Paper (T)	Coated Paper & Board (T)	Bamboo (T)	Hardwood* (T)	Import of Pulp, Waste Paper Rags
1962	1,312	—	n.a.	n.a.	n.a.
1963	15,737	—	41,907	—	1,118
1964	18,031	—	45,838	—	236
1965	18,929	—	44,126	—	213
1966	18,474	—	43,884	—	62
1967	18,993	—	48,678	—	45
1968	18,577	—	45,613	—	2
1969	19,598	—	48,667	—	—
1970	20,851	—	52,195	—	7
1971	22,424	—	55,230	—	290
1972	22,235	—	55,188	—	1,388
1973	25,588	—	53,537	—	7,459
1974	32,842	30	72,743	—	8,299
1975	33,556	371	72,368	—	4,253
1976	33,375	2,046	74,363	—	3,553
1977	35,652	2,341	82,426	—	2,207
1978	36,881	3,106	84,426	—	4,362
1979	37,799	3,069	89,507	—	4,351
1980	38,342	2,969	91,289	3,540	3,202
1981	40,169	3,867	97,854	6,498	870
1982	40,097	3,695	98,643	7,385	548
1983	41,619	4,049	1,06,379	4,671	1,236
1984	45,014	4,046	1,04,375	13,149	2,210
1985	43,605	4,082	95,735	15,567	2,843
1986	44,610	4,019	75,731	21,189	5,526
1987	48,909	4,666	84,817	21,755	6,833
1988-9 (Jan. 1988 to Mar. 1989)	61,001 15 months	6,141	1,18,000	17,371	5,853
1989-90	57,680	4,875	12,0,709	19,149	6,875
1990-1	60,631	5,619	1,21,395	32,581	6,541
1991-2	61,473	6,980	1,22,199	38,642	5,312

Source: JK Paper Mills.

Note: *Hardwood consists of casuarina, eucalyptus, *subabul* wood, etc.

phase of the expansion project pulp production was increased by 12,000 tpa and paper by 5 tpa. It involved a cost of Rs. 5.5 crore and was completed in 1982. That year the paper productive capacity was increased from 20,000 tpa to 66,850 tpa. Although the actual production was maintained at a low rate, around 23,000 tpa from 1977 to 1981, it then ceased while MG and other speciality paper increased.

During 1982 productive capacity was further increased from 20,000 to reach 66,000 tpa was maintained only at a certain level between 1979-81. It certainly reached a record level in 1984—known as a hard year for the forest-based paper industry and a year when it was facing a glut. The small sector, on the other hand, was doing well and would soon catch up to outstrip the forest-based big industry. A year earlier, in 1983, the industry and JK too was smarting under the restriction. It again initiated a dialogue with the government and pleaded that a uniform rate of 30 per cent levy paper irrespective of MF capacity was unduly harsh on the mills. What the industry fails to record is that all the while that it was introducing only new MG machines and increasing their productive capacity on a massive scale, the productivity of MF was marginal or not at all.

The 'dialogue' paid off. On 11 April 1983, the Production Control Order 1979 was amended to provide for differential rates. It broadly stated that the percentage of production of white printing paper would be reduced to 25 per cent in the case of mills with MF capacity below 50 per cent. This most large mills manipulated and arrived at the percentage comfortably over the decade since it began introducing MG machines. The quantum of levy paper was further reduced to just 20 per cent in 1984. Y.A. Rao who is very familiar with the paper industry commented that after an initial improvement in the supply to the educational sector, there was a progressive deterioration in spite of periodic hikes in prices, the price increasing from Rs. 2,750 per tonne in 1979 to Rs. 7,200 per tonne in 1983. Besides, the stay order issued by the High Court disrupted the supply of printing paper.

JK's Foresight Pays Off

In the mid-1980s when the large forest-based PPI industry was in deep crisis, the Singhanias as well as the other paper tycoons complained that they were passing through 'difficult times'. Yet Singhania boldly proposed a Rs. 23 crore scheme for renovation of his plant. After all, as

Business India rightly pointed out, Singhania had much clout, and as an ardent spokesperson of the industry, he must have used his diplomacy and clout, to promote his business (Kilwala, 17 March 1990). Within a year, financial arrangements had been made and plans were afoot to complete the project by 1987. As one senior executive mentioned, when the demand was low on both material and labour, and prices down, it was the right time for renovation. Renovation began in 1984 and moved headlong through 1985, 1986 and 1987. By 1986-7 most of the process equipment and the spreader stroker-type coal-fired boilers were commissioned and so too were the recovery boilers, when lo and behold, the government lifted the Paper Control Order and paved the way for the big paper units to succeed beyond expectation. The nagging problem, the old social obligation to supply cultural paper, had been taken care of by the government.

The small-scale paper industry which the government nurtured, had done its task well and with great enthusiasm, having produced more than half the total production of paper. Its services were now becoming redundant or almost redundant, since the government's other 'prong' the Hindustan Paper Corporation with its four units has matured. The long gestation period of these writs was over and they had started production of both cultural paper and newsprint, thus taking the weight off the large forest-based industry's 'conscience'. The small industry which had proliferated and had used bagasse had not taken the weight off the old established sector. Production-wise and price-wise, the small industry posed a potential threat as a competitor, and the small paper sector based on bagasse and straw, even if it did not destroy the dwindling forest stock, could not be permitted to grow freely and easily and had to be stopped in its tracks. Oligopoly will tolerate no competition. FAO advised that small industry is polluting and it should go (B. Kurkland: personal communication, October 1995, Rome). The government obliged. With the lifting of the Paper Control Order for the benefit of the large forest-based paper industry, the concessions given to the small paper industry were withdrawn almost simultaneously. The Modvat scheme dealt the final death blow to the small industry paving the way for the rise of large mills. And they zoomed.

Singhanias said modernization during the dark period of stagnation alone saved its industry from going into the red, and sent it spiralling upwards. Singhania gloated 'there is room for both small and big

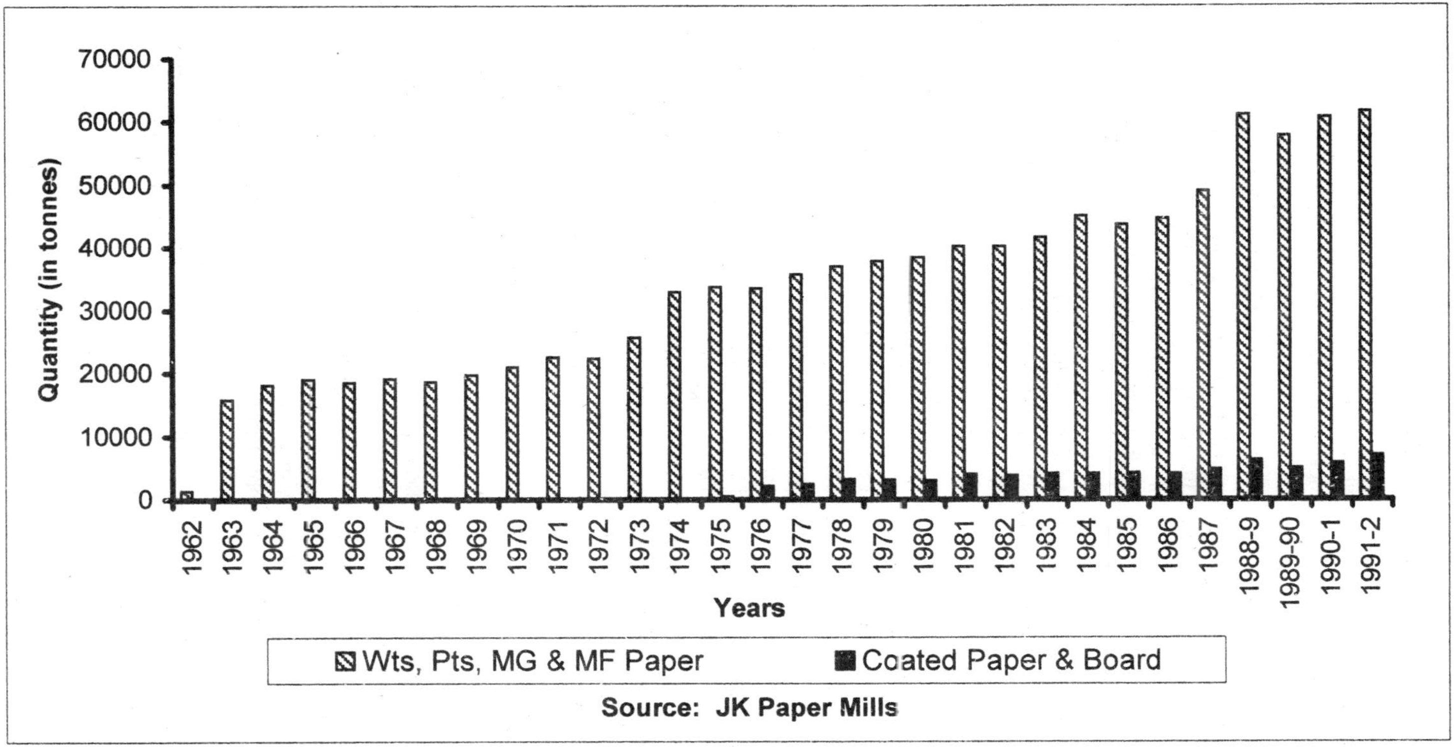

Figure 2.4: Paper Production by JK Paper Mills 1962 to 1991-2

industries, there is no glut whatsoever, we are in the process of expansion and acquisition of large sick mills' (Roy, *Business Standard*, 24 Sept. 1990). Many large mills, like Dalmia's, Rohtas and Punalur, the formidable Titaghur and Bengal Paper Mills had tumbled and fallen by the wayside and several small paper mills too collapsed (ICICI, 1992). But in the take-over spree that Roy speaks about, it was Thapar and not Singhania who got hold of Titaghur and Sewa, which Singhania was coveting. But it is doubtful if OFDC would be willing to supply the raw material to Thapar.

It was only in 1992 that JK got hold of the sick Central Pulp Mills Limited. It would be the sole claimant to the bamboos flourishing in the Dangs in south Gujarat. it has later come out with a Rs. 700 crore public issue part of which will be to rehabilitate this mill (Vishwanathan, 24 Sept. 1993). True, Singhania's paper mill was only the third largest, Birla's Orient and Thapar's BILT ranked first and second in their installed capacity at that point of time. As JK possessed clout in Orissa, the OFDC officials in January 1993 resented that the two mills that Thapar had bought should be supplied the quota of bamboo. The OFDC's argument was that bamboos were now in short supply and such an action would lead to closing down of all the four paper mills. In 1989-90 the major renovations carried out at a cost of Rs. 45 crore were completed. A new paper machnine, a fourth one for the manufacture of airmail, bond, and speciality paper was commissioned and production jumped to 61,000 tonnes in 1989-90 as compared to 48,909 tonnes in 1988-9. But the Directors' report considered that sluggish conditions still prevailed for the most part (JK, 1989-90). The industry was indeed beginning to recover; in 1990-1 production reached 67,680 tonnes 'despite erratic short supply of bamboo'. And the company began to export paper; diversification into speciality papers was specially aimed at the export market. In 1992-3 this company and others too, confessed that the paper industry was facing a glut. It regarded this glut as a part of the business cycle and made light of it as the profits were high. The West was facing a recession and the demand for paper was low in the western capitalist countries.

The powerful paper lobby, unlike the public joint sector newsprint, has been able to prevail on the government to prevent international paper companies from dumping their paper in India. The Government of India obligingly slapped a 65 per cent import duty, making it prohibitive. The threat has not ceased. The new economic policy has opened the doors for powerful American and European paper mills to set up their units in this country. The government has already given the green signal.

Their entry into India is 'no threat', says the brave chief executive of JK, though concedes that Western technology is superior, and so too is its pulpable raw material (personal communication, New Delhi, March 1994). He however failed to say that JK's profits were diverted into other profitable ventures cement, tyres, polymers, etc., rather than ploughed back to upgrade pulping technology that would have saved the raw material and improved the quality of its paper. Only post-1980 public and joint sector units introduced them, as soon as the technology was available in order to save the raw material. His immediate concern now is to upgrade the quality of pulp by improved hydrogen peroxide and oxygen bleaching and to intensify production of superior value-added paper. Conservation of energy and recycling are the twin methods of saving scarce water resources and energy. Another senior executive added that the real threat was inadequate raw material. Their multipronged method is to induce the government to yield degraded forest lands to the company to cultivate pulpwood plantation with the help and cooperation of the Forest Department, the industry providing effluents to grow the raw material (see SPB use of effluent in TN). JK is also attempting to make a deal with both Vietnam and New Zealand to raise joint-sector plantations.

The company plans to manufacture over one lakh tonnes of paper from its plants soon. As T.S. Vishwanathan (*Economics Times*, 23 Sept. 1993) says 'it is either consolidate or perish for PPI'.

The need for cultural paper for the growing number of people to be educated in this country persists. But that is a matter with which the government, and not Singhania, will have to cope (see Ch. 4 for details). In his annual report for 1990-1 Singhania proudly announced that JK Paper Mills had been ranked as the top 50 safest industrial units by the British Safety Council and awarded a Safety Performance Citation and Gold Grade in the Sword of Honour competition in 1990. It has to its credit National Safety Awards for 10 years, 1977-86, awarded by the National Productivity Council. The same organization awarded it the Best Productivity Performance, and a year later in 1987 it had been given a certificate of merit for capacity utilization by the Indian Paper Association (1989-90). Hari Shankar Singhania is proud of his achievement. The moot point is, is the government happy to give it a helping hand?

The first impression one gets of the industry is that it is managed by a set of professionals recruited on the basis for their training and ability rather than on their community, as so often happens in a Marwari-owned unit. However, it resorts to old techniques of recruiting and controlling over 2,000 workers, both skilled and unskilled, through contractors. The number was even higher when it harvested its own bamboo. The company boasts that its industrial relations are most cordial and were marred only by a 12-day strike followed by a brief lock-out for a few hours by management in 1966. But Susanta Acharya (1983: 13) writes that she was unable to get the desired response from many workers because of fear of the management. She also says 25 per cent of the workers are in perpetual debt for amounts ranging from Rs. 2,000 to 5,000. The main reason for borrowing money is to buy land (p. 38). In other words this industry has not contributed to the development of labour. Workers continue to live either in the taluka town of Raygada 10 km away or in the surrounding villages, but not in the paper township which provides homes for about 80 per cent of the permanent workers.

Health care and schooling are statutory necessities in an industrial township. A small dispensary managed by a single doctor takes care of immediate small problems. Major problems are resolved cleverly by having a state-run ESIS hospital immediately outside the township which is open to all employees who earn less than Rs. 3,000 per month, others being sent to the hospital in Raygada. The most common ailment is stomach infection, possibly due to both water and air pollution, despite the mill's built-in system of pollution abatement and alternate source of water provided. In 1964-5, seventy tube-wells were sunk in 72 villages and during 1978-83 six tube-wells sunk at a considerable cost to the company (JK handout 1993: 9, 10). The employees' children have the option of attending either the English medium school or a trilingual Oriya, Telugu, and Hindi medium school run by JK. Raygada has a college which provides courses in arts, science and commerce. Incidentally, Raygada is only 20 km away from the Andhra border.

JK's Augmentation of Raw Material

Much of R and D work at the prestigious PAPRI (Pulp and Paper Research Institute) set up by JK in 1971 has found that hemp gives better quality fibre than wood, and is even superior to imported softwood pulp as it needs only to be bleached. It is also pollutant free. This is the theoretical viewpoint of its chemist, but the industry's problem is cost. Although a 50 acre plot was set up at the Mesta Research Institute at Srikakulam in 1972 and the plot yield three crops, the determining fact was that the farmers hiked their price as soon as they realized that hemp and mesta were needed by a prosperous paper industry. Hemp which belongs to the same family as mesta and jute would have also been worthwhile but Orissa farmers refused to grow either of them. The price of hemp from UP is as high as Rs. 28,000 per tonne, whereas it is available at a mere Rs. 600-800 for the local Star Paper Mill in western UP. JK has two jute mills in Kanpur, Bambaria and Calcutta, but as in the case of hemp the cost of transport would be heavy. Nevertheless, it is regarded as a standby when other raw materials dry up. Orissa forests yield *sabai* grass; because its durability is very poor it was given up. Very heavy investment was made in a rag pulping plant but was eventually abandoned.

PAPRI is also trying out plantations with *E. tereticornis* clones. PAPRI finds *E. tereticornis* yields as much as 43-7 per cent cellulose. Casuarina, it found, had low lignin and its alpha cellulose content was 48 per cent, but JK's tragedy is that the species is not suited to Koraput's site. It also worked out the comparative yield of various species that can be cultivated as pulpwood trees. It is found that the per acre yield of *subabul* is 14.5 tonnes in five years and per acre yield of eucalyptus is 7.8 tonnes in five years which rises to 12.5 tonnes in seven years. JK has acquired usufructury rights to eucalyptus coppice plantations at Muchkund. Unlike as in other parts of India, any attempt to make a private citizen grow pulpwood trees either under social forestry or farm forestry is not workable, as the individual needs a clearance from a District Forest Officer (DFO) to fell the trees and this becomes a deterrent. JK has several 15 acres demonstration plots taken on lease from local farmers to grow eucalyptus (clones). In spite of all these interesting experiments by PAPRI, the JK official thinks that eventually it would have to take to bagasse. Though there is no sugar cane in Orissa, some sugar mills are coming up at Sitarampur and Bahuli a mere 10 km away, and 103 km away at Sirkakulam and at Titaghur 200 km away. This question is discussed in the next section.

Effluent Treatment and Solid Wastes

Effluents are subjected to conventional treatment and the treated water is let out in the downstream of Nagavalli only between 7 p.m. and 3 a.m. 'as a precautionary measure'. This is done to prevent any 'unnecessary protests' by the people.

Solid effluents are put to use. The plant has a 4-stage fibre-recovery unit which controls fibre recovered from the black liquor at the effluent treatment plant. The treated solid effluent is bought by manufacturers of inexpensive cardboards, such as those used for packing shoes. The sludge is also dried. After a lapse of 4 months either fungus or earthworms are allegedly used for converting it into compost. Its sodium content is reduced by profuse use of water. Its nitrogen value is said to be 9.2 per cent, phosphorus 0.04 per cent and potash 0.02 per cent. The team's observation of the 'compost' was that it was so hard and dry that it is doubtful whether any earthworm could survive in it. But it was reported that in PAPRI nursery, the yield of onion doubled, and that of brinjal by 1.5 times, but there was no feedback from the farmers who were provided with the compost. Bananas failed to grow because of the oil content in the sludge.

The organic content from the sludge is burnt as fuel. This does not tally with facts. Bamboo has a heavy component of lignin, the only known non-biodegradable organic matter. Only intestinal juices of some animals like elephants and cattle can degrade lignin; according to a very senior official at the factory lignin is used only as a landfill. The eucalyptus component has created further problems. A team of FAO and Indian scientists discovered that this hardwood is non-combustible. Although this information was available to the Ministry of Industry by 1980, it chose not to share it with the industry. With the threat of entry of powerful multinational paper giants, JK decided to pull itself up, not quite by its boot straps, but in some useful ways. It now claims that 90 per cent of the chemicals are recovered from black liquor and re-used, supposedly the highest in Asia.

Summary and Conclusion

The rugged terrain on the west had preserved Orissa from frequent invasions. The undisturbed western chain of mountains harboured rich forests where the gentle local folk lived in harmony as a part of the eco-system. This harmony was totally destroyed by the British colonizers who punished the resisting Adivasis by pushing them out of their homelands. Simultaneously they plundered the forest for its wealth and rather late realized the folly of their faulty forestry practices. The Indian foresters inherited the British tradition and the forest department hitched on to the powerful revenue department feels powerless vis-a-vis the state bureaucracy. Despite colonial plunder, western Orissa still remained the backwater, somewhat out of the focus of the colonizer's major interest. It proved a haven, a secluded 'spot' for the emerging Indian enterpreneurs to set up their dream industry. The post-colonial Government of India extended every help to the industry to flourish. But the response from the industry was negative, its perspective was not a rounded development. The forests were devastated thanks to the forest mafia and the timber industry and the understorey fell victim to PPI's reckless felling. Bamboo was available at a throw-away price.

Tiger Thapar has now joined in to take control of the two dying mills. The powerful trio once again ganged up to demand joint forestry projects to produce pulpwood. The Conservation Act of 1980 and the Forest Policy of 1988 are deterrents. But nothing seems to have deterred the then Chief Minister J.B. Patnaik's enthusiasm from carrying this project through. The anguish of the foresters of Orissa and the embarrassment of the foresters in the Ministry of Forests and Environment who were gearing up to fight the government which repeatedly succumbs to the powerful PPI lobby, is a matter of grave concern. This was very visible in March 1994.

Recommendations

Since Ordra-desa once belonged to the Adivasis, their rights to their homeland need to be restored. But it is important to remember that almost two centuries of deprivation have resulted in a long break in their history and to ask them to merely return to their exhausted forest lands, to live as primitive forest dwellers, is also an unfair deal. True development is possible only when opportunities open out to move on, and it is the National State's responsibility to create such opportunities. Orissa continues to be one of the most backward states in India. Neither education nor paper for the first step in development, has reached the Adivasis.

Foresters have made useful recommendations and these need to be honoured. The first one is to have the forests mapped by competent authorities in order to guard against further encroachments.

Bamboo is an excellent ground cover to protect the tropical forest eco-system; it has sprung up wherever the forests have retrogressed. Bamboo is also a useful raw material for manufacture of paper. Attempts must be made to regenerate it in bamboo-devastated areas which are its natural home ground instead of joint-sector pulpwood plantations which should never be permitted. This is also the opinion expressed by the foresters. Bamboo unlike the exotic species is eco-friendly. The felling cycle of the bamboo should be increased to prevent too large an outflow of the biomass.

OFDC which now harvests the bamboo should re-educate and train its labour force to use the right techniques. Since bamboo is the highest revenue earner, profits should be ploughed back to the forest department and used for training the bamboo harvesters as well for regenerating the forests.

Use of hardwoods should be banned in the production of paper, as its sludge is non-combustible and black liquor pollutes the river, the main sources of drinking water for the people and the cattle in the downstream area.

Encouraging farmers to cultivate hemp and *kenaf* which JK mill has found to be superior even to soft woods for manufacture of paper, is an alternate in development for a socially useful industry. It is practically pollution free. Farmers should not be expected to peg down its price.

The raw material production should be matched to modernization of pulping technology. The state must make it an imperative. Modern technology can yield as much as 95 per cent of the cellulosic raw material in place of the 35-40 per cent yield obtained by the currently used sulphate technology of pulping.

The government can use the Defence of India rules to compel the industry which is a core industry to produce levy paper. Orissa has five paper mills today but it continues to be the most backward state with lowest educational facility. Finally, as a significantly useful industry, it needs to be protected by stiff tariffs to prevent any inroads that could be made by powerful multinationals.

NOTES

1. Recent archaeological excavations in the thickly forested mountainous district of Sundargarh, Bolangir, Kalahandi and Dhenkanal have yielded palaeolithic implements as well as neolithic agricultural tools. Particularly notable are the cave paintings and the still undeciphered writings (written from left to right) in Soundergarh (Senapati and Mohanti 1971: 41) and Gundahandi and Jagimathin Kalahandi district (Senapati and Mohanti 1980: 4) (see respective *Gazetteers of Orissa*, new series, for details). Tank irrigation was widely prevalent in the above-listed districts. Then of course, there are the Kosthas the master artisans who weave intricate motifs in cotton and *tussar* fabric which once tempted the European pirate merchants to conquer their land. The British and even the Indian writers of the *Gazetteers* want us to believe that they were ignorant tribes and some even destructive.

2. The Hindu prejudice readily found an Aryan root to it, and interpreted it to signify filth, see Brahmavadhita Sri Sukhanada Nath *Sabdartha Chintamani* (p. 442 quoted in Hunter, 1956: 50).
3. In the midst of the tribes in Orissa there lived communities belonging to another stock representing a different stage of civilization. According to the Sanskrit text, one of the five sons of a holy sage founded a military kingdom beyond Aryan territory and gave his name Kalinga to it (Pargitan AIHT: 158, quoted in Hunter: 53). Perhaps people from the north-west had penetrated Orissa sometime in the sixteenth century. Interestingly, even today the coastal people are contemptuous of the people in the hills, and refer to the once-densely forested region as uncivilized and the people as backward Garhjats. At present 46.55 per cent of western Orissa comprises SC and ST. The main tribes are Gond, Binjhal, Khond, Savara, Munda, Bhuyan, Kisan and Oran. 86 per cent of the population is rural. 33 per cent are poor peasants with less than two ha of land. Another 33 per cent are landless. They are marginalized as a result of land settlement introduced by the British (Rao 1990: 26, 27).
4. This was an era of industry controlled by guns. The coastline was dotted with factories, fortresses to be more exact, by the Dutch, Portuguese, the French and the English merchants to weave exquisite fabrics by local weavers. The factories were set well within the reach of cannons from their ships (Hunter 1956: 184).

   In Ganjam, the British had already become landholders as well as big merchants and manufacturers. Why not then in Orissa? Ganjam was a part of the Madras Presidency and was transferred to Orissa after reorganization of the States. It was no easy task to rule Ganjam. Its residents pleaded for more and more soldiers to the Council of Chamber at Madras to collect taxes. The peasants resisted and were dubbed as 'incorrigible offenders' (proceedings, 2 January 1790, G.R.). The local battalion or the revenue core too broke out in open revolt in 1803 and had to be disbanded (Hunter 1956: 188-9).
5. There were several annual, triennial and quinquennial settlements, hastily introduced and inefficiently worked by highly corrupt and rapacious officials (Sahu 1956: 395-6).
6. The Marhatta Resident in Cuttack gave his full support to the rising (Sahu: 397). The aboriginal tribes continued to give trouble to the British administration throughout the first half of the nineteenth century.
7. According to Ewer, two-thirds of Oriya zamindars were completely wiped out and replaced by Bengali zamindars within the first 14 years of British rule. The landowners were also subjected to gross extortion and oppression at the hands of the *sarbarkars* and other underlings, to whom the government entrusted the collection of revenue, and were also subjected to the tyrannies of a corrupt and venal police (*Bihar and Orissa District Gazetteer*, Cuttack, p. 39).
8. The earth goddess, the Dharni Deota, has a central place in the Khond

Pantheon. She is always accompanied by Bhatburi Deota, the god of hunting, who is propitiated by sacrifice. The Khonds are hunters-cum-agriculturists. Strips of flesh of the sacrificial animal are distributed to all the villagers who bury them in the fields as a divine agent of fertility (Senapati and Sahu 1966: 106).

9. The Khonds are known for their kindness, and they are so hospitable that they will give shelter even to an enemy who may come as their guest. They also take great care to see that the non-Khond sacrificial boy—the Maria—is treated with great kindness and concern, and on the day of the sacrifice he is totally anaesthetised with alcohol so that he may feel no pain when he is sacrificed. Later, when a buffalo replaced the Maria boy, it was treated with equal kindness and concern. But buying a buffalo meant running into debt (Boudh Khondanmal: 1) which in turn meant a trap into bondage (Senapati and Salman 1986: 107). (Capt. W.W. Dunlop's report no. 20, dated 8 April 1947, and Brig. Gen. Dyce's report, dated 20 March 1847, quoted in N.K. Sahu: 397).
10. The gonds families are ancient. Their number seems to indicate that before the Oriyan immigration they held possession of the region, subduing the Munda tribes who were probably there before them (*Imperial Gazetteer of India*, 1908, vol. XXII: 9). *Orissa District Gazetteer Sambalpur* (1971: 1) confirms that the district was under Gond and Bijhal kings before 'historical' dynasties like Gangar, Kadambas and Chauhans appeared, probably sometime in the sixteenth century.
11. Surendra Sai's younger brother Chaibala and many brave fighters were killed in the field, Ujjal was seized at Bolangir and hanged, Surendra and his son Mitrabhanu were taken prisoners.
12. The Settlement Reports indicate a preponderance of Brahmins and Kultas. The Kultas are probably an admixtrue of Aryan and non-Aryan blood. They are referred to as the chief cultivating caste of Sambalpur in 1931 (*Sambalpur District Gazetteer*: 72-4) who through hard work and use of tank irrigation were able to grow good crops even on the poorest soils. King reports that they profited through a rise in the price of rice. By 1931 the aboriginal population in Sambalpur was reduced to 41 per cent (King 1931: 72).
13. In Sambalpur, according to the *Nethersole Settlement Report* of 1885-9, out of the 949 villages in Khalsa, 588 were occupied by Brahmins and Kultas, 89 villages by tribal Goantias and the remaining 281 by non-tribal Goantias. According to the *Hamid Settlement Report* (1925), of 918 villages Brahmin and Kulta Goantiship had increased to 681 villages and tribal control declined to only 15 (Tripathy 1990: 29).
14. A high school started in 1841 was converted to a secondary college, and degree classes opened 10 years later. The Orissa medical school was established in 1875 and a general hospital a year later (Sahu 1956: 399).
15. Bargarh tracts are undulating. They are drained by the Danta and Jira

rivers. But King (1932) reports that most of the land is too high or too broken for tillage.

16. Orissa forests harboured a wealth and variety of animal life. The forests were dotted with natural springs and criss-crossed by numerous mountain streams. Clearing the forest led to drying up of the water resources, and opened out the area for poachers to trap and kill animals in the few water-places in the hot months (Senapati and Mohanti: 30).
17. This is true for other districts too. In Orient Paper Mill's (OPM) first catchment area, Sambalpur, local households were charged commutation fees to graze their cattle or collect forest produce for household needs. This was modified to a licensing system which meant that villagers had to pay for everything that they removed. This system was in existence in the 1930s in a modified form as a supplementary system, the main system having been the sale to contractors of all or selected forest produce in coupes or demarcated areas in forests laid down in the working plan. In 1910 the working plan was revised after the end of zamindari. Coppice felling was reintroduced and selection felling was mainly for export. Revenue rose from Rs. 53,111 to Rs. 96,109 in 1920. Owing to the opinion that more detailed control would raise the revenue, the division was divided in two in 1928. Revenue was expected to rise as the forest became more accessible by road (King 1931: 112-16). OPM now draws its raw material from Bolangir too.
18. Some forests were worked under 'coppice with standard system' but with little success. All block forest working plans in Bolangir were revised only in 1938 (Senapati and Mohanti 1973: 14-15).
19. Forests continue to be a source of revenue even today. The compulsion to fell forests is however different.
20. Forest labour was organized by the Communist Party of India (CPI) in 1982 but not today. According to trade union leader Panda, forest labourers are drawn from landless labourers who work from October to April. During the monsoon they eke out a livelihood doing some hard temporary jobs like road building, construction, etc. OPM labour force which was once 6,000 strong is now reduced by 2,500, who are working in the forests. OPM labour was organized by CPI. Workers who used to be paid Rs. 450 per month as late as 1982, but bargained and had it raised to Rs. 1,800 per month for permanent workers and Rs. 25 per day for temporary ones. The strength of the labour force had fallen to 2,400 and no vacancy is filled up.
21. The highway proved to be particularly useful when the raw material in the vicinity began to dry up and had to be met by open market operations and private parties, whatever that may mean (Modernization Scheme for Orient Paper Mills submitted to IDBI 1990: 4-B II).
22. By 1984 the quantum of percentage of production of white printing paper was reduced by 25 per cent or 20 per cent in case of mills with MF

capacity below 50 per cent of the total. The capacity of small paper mills excluded from the order were defined as 24,000 tonnes per annum (Y.A. Rao 1989: 158).

23. The company, for instance, has protested against what it calls illegal levies, viz, the sales tax assessment of Rs. 5.71 crore for the year 1973-4, 1974-5, 1975-6 and has appealed for a stay. A stay has been granted on the payment of Rs. 50 lakh (*Annual Report*, 1977: 6). The company was also contesting the enhanced water charges for the year 1961-2 to 1976-7 which worked out to Rs. 2 crore as against the expected Rs. 24 lakh (*Annual Report* for the year ending 31 March 1977: 6). The revisional authority reduced the rate to Rs. 1.33 crore. The company, still aggrieved, filed a writ petition in the High Court which was admitted subject to a payment of Rs. 20 lakh (*Annual Report*, 1980: 5). The company was also contesting trade discount by excise authorities (*Annual Report* for the year ending March 1978).

REFERENCES

Acharya Susanta, 1983, 'A study of Quality of Life in J.K. Paper Mills', unpublished M.A. Dissertation, Utkal University.

Agrawal, A. 1984, *The State of India's Environment*, 1984-5, vol. I, Centre for Science and Environment, New Delhi.

Ahuja, S.P. 1986, *Paper Industry in India,* The Institute of Economics & Market Research, New Delhi.

Anon. 1949, *Economic Times*, 19 April.

——— 1971, *Economic Times*, 30 March.

——— 1978, *Financial Express*, 17 March.

Bombay Stock Exchange Directory, Reports on Orient Paper Mills and JK Industries.

Carmichel, D. 1896, *Manual of the District of Vishakapatanam*, Madras.

Chopra, P.N. (ed.) 1991, *Quit India Movement, Role of Indian Big Business, British Secret Documents.* Interprint, New Delhi.

Dagli, V. 1979, *The Committee on Controls and Subsidies*, Ministry of Finance, New Delhi.

Dutt, Rajini Palme 1979, *Indian Today*, Manisha, Calcutta.

Francis, W. 1907, *Madras District Gazetteer*, Visakhapatanam, Madras.

Gadgil, D.R. 1954, *Industrial Evolution in India*, Oxford University Press, New Delhi.

Hazari, R.K. 1967, *The Structure of the Corporate Private Sector*, Asia Publishing House, Bombay.

Headland, T.N. and A.L. Reid 1989, *Hunter Gatherers and their Neighbours from Prehistory to Present Anthropology.*

Hunter, W.W. 1956, *A History of Orissa*, vol. I, Sushil Gupta Ltd., Calcutta.
ICICI 1992, *A Sectoral Study of Paper Industry*, Bombay.
Jaju, Ram Nivas 1986, *G.D. Birla: A Biography,* Vikas Publishing House, New Delhi.
JK Industries 1969-93, *Annual Reports.*
King, F.C. 1932, *Bihar and Orissa District Gazetteer*, Sambalpur, Patna.
Mazumdar, R.C., H.C. Roy Choudhary and K. Dutta 1973, *An Advanced History of India*, Macmillan, New Delhi.
Nag, Kilwal 1990, *Business India*, 17 September, 30 Corporate Report.
Nehru, Jawaharlal 1946, *Discovery of India*, Signet Press, Calcutta.
O'Malley, L.S.S. 1909, *Bihar and Orissa District Gazetteer,* Sambalpur, Patna.
Orient Paper Mills 1978-92, *Annual Reports.*
Padhi, G.S. 1991, *Forest Resources of Orissa*, Udyog Printers, Bubaneswar.
Poddar, V. 1979, *Paper Industry in India*, Oxford and IBH, New Delhi.
Rao, P. Chintsal 1986, *Lakshmipat Singhania,* Vikas Publishing House, New Delhi.
Rao, R.A. 1989, *The Paper Industry in India—Status and Prospects*, B.R. Publishing Corporation, New Delhi.
Rao, R.S. 1990, *Towards Writing History: A Historical People* (n.d.), ISSC-Jyoti Vihar, Sambalpur.
Sahu, N.C. 1986, *Economics of Forest Resources*, B.R. Publishing Corporation, New Delhi.
Sahu, N.K. 1986, 'Orissa from the Earliest Times to the Present', in W.W. Hunter, *History of Orissa*, vol. II.
Sainath, P. 1994, *Times of India*, 14 February.
Satapathy, Rajaram 1994, 'Orissa Environmentalist in Jail', *The Times of India*, 6 March.
Senapati, N. and Durga Charan Kumar (1975) *Orissa District Gazetteer*, Sundargar, Cuttack
Senapati, N. and N. Sahu 1966, *Orissa District Gazetteer, Koraput,* Cuttack.
——— 1967, *Orissa District Gazetteer, Mayurbunj,* Cuttack.
Senapati, N. and B.K. Mohanti 1971, *Orissa District Gazetteer, Sambalpur,* Cuttack.
——— 1972, *Orissa District Gazetteer, Dhenkanal*, Cuttack.
——— 1973, *Orissa District Gazetteer, Bolangir,* Cuttack.
——— 1980, *Orissa District Gazetteer, Kalahandi*, Cuttack.
Shirokov, G.K. 1973, *Industrial Development in India*, Progress Publishers, Moscow.
Tiwari, D.N. 1991, *Forests and Tribals, Indian Forester*, vol. 117, no. 9, Dehradun.
Tripathi, P.K., *Transition in Political Economy of Western Orissa* (n.d.), ISSC-Jyoti Vihar, Sambalpur.
Wobst, H.M. 1978, *The Archaeoethnology of Hunter Gatherers in the Tyranny of Ethnographic Record in Archaeology, American Anthropology,* no. 43.

CHAPTER 3

Madhya Pradesh

PART 1

The Death of the *Salai* Forests of NEPA

An Ode to NEPA

The history of Nepa,
is a gory tale
of Death
of the splendrous Satpuras.
Where
the Salai was hacked;
where
the bamboos were ripped;
where
Sabai turned yellow and Tapti black;
where the red earth, weeping,
ran down the hills;
and the Gonds and the Korkus fled;
Here the brave lone wolf
Akela, stood his ground,
While
the treacherous Sher Khan
broke
the rule of the jungle
to incite

*In this chapter on Madhya Pradesh the section on NEPA and that on Orient Paper Mills have been made self-contained units. This is because, NEPA in Eastern MP and Orient in Western MP differ very considerably in ecology.

But as part of a single chapter, sub-sections in Part 2 will run continuously with those of Part 1.

the wolf packs of the north
to betray
and kill Akela
but,
the lone wolf
still stood his ground.

M.S.

SECTION 1

Social and Natural History of East Nimar

East Nimar, Hoshangabad and Betul were the districts assigned to provide *salai* and to some extent bamboos for NEPA in the mid-twentieth century.

Colonial Legacy

A century earlier, this region was referred to as the 'highlands' within the 'Nerbudda territory' of the Central Provinces (CP) that the British constituted in 1861. It was constituted for several reasons, the main ones being, to maintain strict control after the mutiny and to drive two main railway lines in order to rush troops in times of crises[1] (see Governor-General Hardinge Minutes, 28 July 1846), to reach black soil tracts on which 'Manchester had prophetically fixed its eye as its future fields for cotton' (Forsythe 1871: 10); and to control and organize the vast forests which would be opened up for commercial exploitation (Stebbing 1926: 344).

Perhaps the most intrepid chronicler of the natural and social history of this region was Capt. J. Forsythe, the explorer, a much respected settlement officer, Deputy Commissioner of Nimar and for sometime deputy to Pearson, the first Forest Conservator of the 'Nerbudda Territory' and of Saugar in CP. Forsythe unfortunately died at the age of 33 in 1871. Being an imperialist outsider as well as a knowledgeable insider, he had the dual advantage of being able to chronicle some of the follies of the Forest Department (FD) and raise certain issues which are of significance to this study. The highlands, as the Scot nostalgically refers to CP, became worthy of the colonizers' principal share of attention with their extensive forests, rich mineral wealth, limitless tracts of unreclaimed waste, great capacity of storage of water, where some 'scantily clad half wild population', who had no fear of the jungles, lived. For five successive years, several officers were unremittingly employed in the exploration of the 36,000 sq. miles covering the Central Hills (Forsythe 1872: 16-18, 155).

Colonial Expansion and the Status of the Timber Forest in 1861

Long before the British established administrative control over the highlands and the forest resources of the region, says a Forest Research Institute (FRI) document, they were strained to the utmost to meet the needs of their shipping industry and the railway (FRI, vol. II, 1961: 9). When the oak forests in Britain were depleted by 1800 (R.G. Albion 1926: Chs. 3 and 9), and teak with its combination of strength, lightness and elasticity proved to be a superior substitute, the Royal Navy had led expeditions to explore and exploit teak in the hardwood forests. But with the decline of wooden ships in 1840, the need for teak and *sal* did not slacken. The British were engaged in large scale railway building to consolidate their empire. Col. Pearson, the Conservator of Forests, remarked, 'It was only necessary for a contractor, whether European or native, to obtain a parwana from the civil authority to cut the timber for him to set to work.' As every Gond carried an axe, it was not long before every forest was strewn with fallen logs (FRI 1961: 9). Railway contractors and speculators, foreseeing the rise in the value of timber, went into the jungle with bags of rupees and 'broadcast it among the wild people', instructing them 'to slay and not spare, to fell every trunk larger than a sapling. It was only too faithfully done, but there was delay in the railway work and cut timber was abandoned' (Forsythe 1872: 32). *Salai* not being a favoured species was left alone.

The Origin and Growth of the Indestructible Salai (Boswellia Serrata) *in the Nineteenth Century*

Forsythe laments that in large tracts where valuable teak[2] once grew and was exterminated, 'worthless' *Boswellia* grew, which no one would think of cutting down, its wood being too soft and spongy and considered of litle value for building. It was even rejected as firewood. Since its specific gravity was so low, a great bulk would have to be transported. He regrets that foresters made the mistake of not forcing the common people to use it, instead of giving way to their cry for hardwood. He found *Boswellia* had the indestructible vitality of a weed; in the vast expanse of the trap country covered with grass and studded with small trees on which ninety-nine in every hundred were *Salai* 'a mere encumbrance'.

However, Forsythe discovered some of the potentialities of the *Boswellia*. It made excellent charcoal. The gum resin that it could yield was the expensive frankincense or *lobana* used in worship. His contemporary, Dr. Birdwood, had just discovered a fragrant gum resin in another species of *Boswellia*. Forsythe anticipated that the soft, woody fibre of that tree was capable of yielding cheap paper or cloth. There is no evidence that Forsythe was the source of inspiration for the founders of NEPA Newsprint. The English paper factories whether in India or in England did not try this wood, although just about this time the use of softwood for making pulp and paper was gaining ground in Europe. Forsythe, however, erred when he proclaimed that the supply would be practically unlimited 'should any economic value be found to attach to any portion of the wood'. This assertion was based on his observation that large stakes merely inserted in the ground during the rains shot up into trees. It was further confirmed by the fact that the railways took advantage of this quality of *Boswellia* for the construction of live fences to stretch the fence wire.

In the scorching heat of the central Indian summer the foliage of the poor *Salai* withers and droops and finally drops. 'The naked yellow stems of the Salai', says Forsythe, 'stand like a vast army of skeletons on the parched yellow grass turned into a dreadful bramble, armed with 'needles of steel and barbs of fish hook'. Yet, with the first rains, a remarkable metamorphosis takes place. The *Salai* springs back to life and the *salai* forest becomes a thing of beauty. He observes that at the height of the monsoon, the grass is green, the trees throw out a thin foliage of small bright green pinnated leaves. The river beds turn to foaming torrents. One might mistake the valley for some primeval wilderness in the North (pp. 205-8). The vitality of *Salai* has been highlighted only because it seems to have been completely lost in the twentieth century; *Salai* is of primary importance to this study. Grasses, especially *sabai* grass was used by the British in manufacturing paper products in the nineteenth century.

The Grasses on the Highlands

It is not known what grass Forsythe refers to; other reports speak of luxuriant growth of *sabai* grass in *Salai* forests. *Sabai* was used as raw material in paper mills in north India right from 1832, and continues to be used even today. But NEPA never did so, at any point of time.

The Imperial Gazetteers (1908: 8) mention that *babel* or *bhabar*, locally used for rope making, was diverted to paper making; also that reed pens were made from *bharru*. *Khoskhas*, a fragrant grass was woven into mats and hung on windows to keep out the blazing summer heat. The Gazetteers mention *masyal* and *dhadhara* as two other scented grasses; *crugallius* which grows round the tanks is called the 'God's gift', as its seeds are gathered and eaten by the poorer people; *darabordub* is the best grazing grass and *kushal* is an excellent fodder.

A Controversy—The Beneficial Effects of Firing Grass

The local people fire grass at the end of the grazing season to generate a fresh crop. This is strongly condemned by both the earlier British, as well as contemporary Indian foresters, as a dangerous practice which can set off a forest fire. But Forsythe gives a refreshingly different view. The universal practice is to fire the ground in the jungles to generate a lush growth of grass. He declares that it is beneficial in a great variety of ways. For one, it allows for manure out of the ashes, assists a crop of green and tender grass shoots to appear for grazing the vast herds of cattle, which form a great part of the wealth[3] of the people in the neighbourhood of the jungle tract. It kills a multitude of snakes and noxious insects. It probably prevents much malaria that would arise from the vegetation if it were gradually allowed to decay. Also, the ashes form a valuable ingredient in the deposits of soil carried down by the drainage from the hills to the lower region, and over the ages form a cultivable crust on the uplands themselves (p. 212). Forsythe dismisses the notion that burning of grass is injurious to the growth of saplings of precious teak and other valuable timber. On the contrary he holds that it is an undoubted fact that teak seeds germinate and produce seedlings better when the grass has been fired, than when it has not. Teak seed has a hard shell which must be decomposed by moisture and heat before it can germinate. He specifically mentions that he has not observed any adverse effect on the young tree by the firing of grass (Forsythe 1872: 211, 212 and 225). The bark of the live teak tree is very tough and withstands the heat of the fire when the grass is fired. This point has been highlighted mainly because the forest dwellers and forest village folk are held responsible for the destruction of the forest through forest fires. What the foresters fail to understand is that forest dwellers are a part of the forest eco-system and would not indulge in any activity which is detrimental, or adds to their woes. Their practices are based on long, keen empirical observation of nature.

Bamboo: From the Life-line of the People to Green Gold

Bamboos were plentiful in CP. The finest of them was found in the Bori forests of Hoshangabad, and much of this area was confiscated from Thakur Bhabut Singh in 1859 for his 'disloyalty'. It was the first region to be brought under systematic conservation (Corbett and Russel 1908: 207). In Betul bamboo was found in the teak areas, especially on the slopes of ravines (Russel 1907: 177). In Nimar this was confined to the Tapi and Narmada river tracts. Bamboos in CP were of two species, *bans Dendrocalamus strictus* and *katang Bambusa arundinacea*. These were of immense importance to the hill people.[4] Their houses were made of bamboo matting plastered with mud. Bamboo baskets called *dhulis* were used for storing grains in the forest tracts. Bamboo was used for beds instead of the cordage of hemp. Bonsors and Mangs made baskets, sieves, winnowing fans and mats for sale both within and outside the district (Russel: 107). The Bigas needed bamboo for bows, arrows, and also for bamboo wicker work for their habitation on hill tops (Forsythe 1872: 360). The young shoots of the bamboo as well as the seeds are edible. A certain number of bamboos flower every year. It is reported that a severe famine in CP and parts of Bombay in 1899-1900 coincided with the gregarious seeding of bamboo in some districts of CP. It afforded a food supply, so people could get relief for two to three months (the Inspector General of Forests, Mr. Hill, quoted in Stebbing 1926: 97). The alien ruler also noted that nature was very bountiful in its supply of food for its wild human denizens.[5] Many species of trees and bushes produce a wholesome and palatable fruit in the season, which supply the earth supplements with nourishing roots and tubers (Forsythe: 99). Their food also included coarse grain. But Forsythe did not quite approve of their slash and burn or *dhya* system of cultivation[6] (see report on Bamboo Forest of Chanda, Ch. 4). This method meant an immediate loss of revenue from commercially more valuable timber that was currently in demand for its railway expansion programme by the British (Thorner 1950: 11-12).

Forsythe put forth the hypothesis that when the hill people abandoned the *dhya* clearing after two years of cultivation it did not turn into a wasteland: on the contrary, the clearing was speedily covered up again with jungles which were qualitatively different from the earlier ones. He also felt that *dhya* clearings gave rise to impenetrable thickets of bamboo where neither people nor beasts could enter. But the point is that bamboo, after all, is the life-blood of the people and there is no reason why there should be any alarm at its growth. But Stebbing took up Forsythe's idea and popularized it in his four-volume work, emphasizing that forest people and their wasteful *dhya* cultivation were responsible for much of the exhaustion of the potential rich teak bearing forests of CP[7] (Stebbing 1920: 398). The Indian foresters unthinkingly kept up this refrain, without caring to investigate tribal needs and the tribal question. Stebbing seems to forget that the British experiment to raise teak plantations had also resulted in secondary bamboo forests (see Chs. 7 and 9).

True, the forest people, the 'wild aborigines', were already marginalized by the ruling parties and by the plains people of India, but the vast bountiful forests even as late as the mid-nineteenth century, that is even before the alien British seized them and called them government land or reserve forests, offered enough space for the forest people to live in.

Forests were an integral part of the lives of the people of forest villages as well as villages on the periphery. Their subsistence economy rested on the forest resources in many ways. Firewood was a major source of energy while hardwood was needed for construction of their homes as well as for their ploughs and carts, leaves were essential to enrich fields as the soils of the hills lack nutrients. Even more important was the grass for grazing cattle. And grazing rights would bring enormous revenue for the government as Table 3.1 shows. But it is with the passing of the Forest Act of 1878 that space shrank, and the system of *dhya* cultivation became disfunctional. The bounty of nature was totally cut off, its inaccessibility exposing the forest people to hunger and starvation. Bamboo was required not only by the forest

TABLE 3.1: YEARLY SOURCE OF FOREST REVENUE

(*in Rs.*)

Year	Timber	Fuelwood	Grass	Bamboo	Other minor FP
1881-2	5,000	5,000	24,000	4,000	7,000
1904-5	21,000	6,000	25,000	15,000	10,000

Source: *Central Province Dist. Gazetteer*, Betul, 1908: 176.

people but by the plains people, that too in large quantities as we have already stated. Although it was dubbed a minor forest product, the revenue that bomboo yielded did not lag too far behind timber. The fees collected for grazing too, maintained a high level. Available figures for the early years in Betul district of CP for instance are as given in Table 3.1.

Stebbing, an enthusiastic and much respected forester, confessed rather sadly that the Forest Department had come to be regarded by the heads of civil administration and by its own chiefs as a purely commercial concern. The rules then enforced were most minute and vexatious, the system of collecting revenue unnecessarily inquisitorial (Shrivastav 1926). FD's chief *raison d'être* was production of revenue; this runs through all the Dispatches, Proceedings, Memoranda (Stebbing 1926: 345). For that matter, even today, at the state level, the Departments of Revenue and Forests are still one.

Bamboo had new commercial and industrial potential. Enterprising Englishmen had set up paper factories in India as early as in 1832 with the help of *Kagzi* (paper makers) families. The raw material that was used right from 1832 was forest-based *sabai* grass and rags. Bamboo belongs to the family of grasses and appeared as a potential source for making pulp and paper. CP was rich in both *sabai* grass and bamboo. Colonel Pearson, the first forest conservator of CP and a forest economist, carried out a large scale survey of the bamboo resources and put W. Raitt to systematically work on its pulpable qualities (Pearson 1913: 159-279). Bamboo was potential green gold for paper making. But this is to go too far forward in time, while we need to go back to 1860 when the great forests tracts were still lying unexplored, although over-exploited by the British, towards expanding its naval fleet and railway network. Much wood was also drawn for building urban settlements for its bureaucracy, cantonments for its military and ports for its growing trade.[8] British administrators began to feel alarmed, and felt the need to regulate the outflow or 'outturn' as it chose to call it, and conserve the still untouched and inaccessible area for its own future use and profit. It called for forest 'management' which will be dealt with in the next section.

The Beginnings of Forest Management

Regular management commenced in 1861 and in 1862 'Wasteland Rules' were introduced. The numerous princely states pledged to conform to the forest legislations and the rules that the chief commissioner of CP was pleased to prescribe. The first Conservator of Forests of CP was Colonel G.F. Pearson; others who followed were also army personnel—Colonel Donetore, Colonel Bailey, Colonel Peryton, Major-General Michael, Major-General Cotton, Colonel Beddome, none of them trained in scientific forestry but 'who loved the forests and did magnificent work' says their compatriot Stebbing (1922: 392). On the other hand, Forsythe talks about the savage and intractable reputation of the foresters (1872: 158). As there was no trained Englishman, Brandis, a trained German forester, was appointed Inspector-General of Forests. The first thing he did was to arrange a systematic demarcation of the forests of CP.

Purpose of Forest Management

In 1865 the first Forest Act was passed. Under it the forests were categorized into Reserved and Protected. The government held full ownership rights over reserved forests and their products were not to be used without specific approval. The Protected Forests (also government-owned) were a temporary category, not yet systematically surveyed. That year the forests of CP were divided into six divisions and by 1886-7, the work of demarcation was nearly completed. A system of marking trees for felling was introduced. But experiments in plantation of teak and other important species failed (Stebbing 1922: 86). In 1870 the government found the financial output from forests unsatisfactory. Immediately the forest officials began to work on the best of the Reseved Forests to generate revenue. The forest officials worked as assistants to the District Commissioner/Collector, the chief revenue official in British bureaucracy at the district level. Hoshangabad, richest in forest resources, was made the headquarters of the 'Nerbudda' division, which was divided into Shagpur (on the east), Hoshangabad, Seoni (Malwa), Harda Unit; much later (1906), Kilbhit was added (Corbett and Russel 1908: 1).

In 1878 a new Forest Act was passed. By 1879 all government forests in CP were declared as Reserved. By the Forest Act of 1878 a Reserve Forest was put out-of-bounds for common people. For the forest people it suddenly ceased to be their homes. Forests were their source of livelihood. Forest people who had been spared from the deadly famines that afflicted India at the close of the century became very vulnerable. A mention has earlier been made of the merciful gregarious flowering of bamboo which was used to feed the tribals of CP for two months during the 1900-1 famine. The closing of forests by classifying them as reserved was purely for commercial use and to generate revenue. Brandis notes that very few offences were committed in the reserve forests (Stebbing 1924: 84). In 1882-3 Inspector General Schlich removed the distinction between Class I and Class II forests; both were to be managed in the same way (FRI, 1961: 9).

Betul was made a separate division in 1878 (Russel 1907: 176) for more systematic and intensive exploitation. A separate District Forest Officer (DFO) was appointed in 1906. The working plans were drawn up for the first time in 1879 for each range in Nimar, and were found defective. New plans drawn up by Taylor, were in operation till 1913. These were re-worked by Bell as high forest (teak) coppice and grazing circles, the latter meant to reduce pressure on adjoining areas under regular working. The presumption that forests were worked to yield revenue, lasted till 1931-2 (Shrivastav 1969: 25). Meanwhile, it must be mentioned that improvement in felling or cleansing and thining in the younger growing stock was undertaken in 1901 (FRI 1961: 9). In the late 1920s, departmental working was introduced as the local contractors defaulted in their contracts (Osmington and Champion 1965: 253) and it was not in the interest of the government to employ the large Calcutta-based ones due to heavy overheads (Hewetson 1964: 253).

The Debate on Forest Management

The concern of this study is not to review in detail the history of forest management but to highlight certain issues. According to C.E. Hewetson, the Conservator of Forests, the older tradition in Central Provinces (CP) was conservative management and light felling. The general feeling was that nothing should be allowed to slow down the natural process of regeneration. Others like Harlow argued for short conversion periods and energetic steps to liquidate poor standing crops. In practice, forests of lower quality were placed in coppice working circles for small timber, and fuel and rotations were fixed for 30 to 40 years. Overlapping circles were formed for bamboos and worked under a four-year felling circle (Hewetson: 252-3).

The two world wars took a great toll of the forests. All the theories of regeneration and working were thrown to the winds, nothing was spared. The *zamindari*, *malguzari* and *jagirdari* forests were also not spared because of the inflationary trends in prices (FRI 1961: 9). The demand for timber in the Second World War was intense not only for use in India but for export to war fronts in the Middle East and Egypt. India had to bear the brunt as Malaya and Burma were lost to the British empire and were under Japanese control. The demand was for poles and bamboos, then for 'hutting' timber. *Salai* ceased to be despised and was sacrificed towards the war effort. Teak was in great demand by the government as import from Burma had ceased, and *sal* was for making railway sleepers. The other main demands were for jetty poles, wharf timber, telegraph poles, ammunition boxes, lorry bodies, airfield towers. Sawing camps were organized and V.K. Maitland was honoured for setting up 12 saw mills. The British foresters' professed love for trees was thus thrust aside when the Empire was in danger. And the revenues from forest products shot up from Rs. 50,50,000 in 1937-8 to Rs.1,88,73,000 by 1946. During the same period while the expenditure merely doubled from Rs. 37,07,000 to Rs.74,57,000 the surplus increased more than tenfold, from Rs.13,43,000 to Rs.1,14,10,000 (at 1939 prices). Finally, just before the transfer of power, the authorities felt it was time to take stock of damage and assess what still stood. Rais started his work and between 1945-9, stock mapped the entire forests of Madhya Pradesh (MP) (Shrivastav 1969: 25). But the wisdom of hindsight shows that the regenerative power of what once seemed inexhaustible, of MP jungles, was being exhausted.

Whither Nineteenth Century Regenerative Power?

Why did the situation change so drastically within a century? Why has the indestructible *salai* tree of the nineteenth century, despised and left alone as an ubiquitous weed by the British and permitted to flourish in the drier tracts of Central Provinces (CP), now vanished? Why does this one tough little tree refuse to respond to the coaxing of the anxious silviculturists of Nepa Nagar? Why did the majestic teak which seemed equally indestructible, even after its jungles were slashed and burnt, or merely slashed clean for its timber, rise from its embers, even throw up a slim tall and vigorous coppice and suddenly earn for itself the name of 'the most difficult species to reproduce' by the British foresters in the second half of the nineteenth century? Why did its seed refuse to respond to the firing of grass and the cool rains afterwards? In short, the teak was protesting. But we know the teak was assaulted by the East India Company at the turn of the eighteenth century or even earlier. The same question may be asked about the bamboo which is a mere giant rhizomatous grass which grew into impenetrable thickets after slash-and-burn, much to the annoyance of E.P. Stebbing. It has suddenly turned shy despite its annual dual system of reproduction, through its numerous shoots from the rhizomatous spread, and periodically through seed, which makes it seem almost immortal. Who has hurt it so badly that it has vanished completely from its native home in the former Rewa state, and is not too happy elsewhere? Bamboos and hardwoods like teak were known to be used for a millennium by the hill people as well as in the plains, without causing any destruction to its regenerative power.

The readily available natural history of the region gives some empirical information on the habits of regeneration, growth, survival and the sudden unexplained disappearance of some hardy species. Only a fuller study of the social history of the region, mainly in terms of the harvesting patterns of the natives of the region, would provide a plausible clue for the failure to regenerate with equal ease. The social history of other regions (see Ch. 6, Karnataka), indicates that self-imposed rules are evidently based on long-standing empirical observation and experience and passed on as tradition rather than as a theory of silviculture. Due to the white man's contempt of the 'wild' inhabitants of the hills, the plains people's inhibitions and caste prejudices against the Adivasis, the traditional and empirical practices of cutting and lopping in silvicultural harvesting are now lost, much to the detriment of the forests.

Forsythe did not have answers but on the basis of his keen observation offered some useful suggetions and raised a couple of worthwhile questions. He insisted that not explorers, but only trained foresters be employed. Whether these were trained in France or Germany did not matter. He also warned of the danger of too professional a view, or of the blind application of continental practices of working forests in blocks by rotation, which is quite worthless. He regretted that this practice was still blindly followed in Indian forestry and was the cause of much wasteful expenditure (Forsythe 1872: 434). He also lamented degradation and a slow death of the once lush forests and erosion of soil. Forsythe faulted the foresters, who in their cry for timber totally forgot or neglected the demand of the masses of people for small timber for their houses and for many other purposes, and closed in every acre of teak-producing country. He noted that other species, when used young, generally decayed in the Indian climate. So out of the simple wisdom of an imperialist, rather than perhaps of any true concern for the common people, he suggested that the Forest Department should reserve only the very best of teak for itself. It should release the rest of the quickest produce of the coppice wood (teak) for the requirement of the local people, instead of pruning and preserving to make large timber grow out of the scrubby coppice wood which had once supplied their wants (Forsythe 1872: 203).

SECTION 2

Forests: NEPA's Catchment Area

Natural Resources and Forest Management in the post-Colonial Period

Newsprint production requires vast amount of forest produce—mainly bamboo. Yet NEPA, the newsprint mill, was allotted *Salai* and only some bamboo by the government. The site for a PPI mill was also generally selected where minerals, lime, coal and clays were readily available. Madhya Pradesh has an extensive stock of natural resources. The present extended catchment area of NEPA falls in the southern part of Madhya Pradesh.

Location of the Forests

NEPA's first catchment area is in the old Central Provinces or Mahakoshal. Strangely, it was here that forest management was started by Pearson. The forests of MP are vast, and the Forest Department (FD) is also very large. With the abolition of the *zamindari* system, degraded forests came under FD, its control now extending to over 18,470 sq. km. (FRI, vol. II, 1961: 203). The major concern here is only with the three forest circles of Khandwa, Hoshangabad and Betul, the first catchment areas. Much later, bamboos were also allotted sporadically from three other forest circles—Seoni, Chindwara and Balaghat.

It is also important to know the physiography, soil structure, mineral content, rainfall, etc., where the natural forests had thrived and formed the natural catchment area of NEPA. Such information will give vital clues to why the vast forest plantations that are tended carefully to supply pulpwood for NEPA have failed to respond.

Drainage

The water resources of these areas are the two major river systems, viz., the Narmada and Tapi, which, with their broad network of tributaries, nurture the forests in these region (Shrivastav 1969: 182). NEPA's water requirement of around 8 million gallons per day is met from the Tapi river.

Soil Structure of the Six Forest Circles

The soil structure of these regions is as follows:

Khandwa: The north-eastern part of Khandwa is covered with black cotton soil on the Deccan trap while the southern side has a red and brown forest soil with luxuriant forest growth.

Hoshangabad: Hoshangabad has a red and brown sandy soil on sandstone and shell in the northern part; black cotton soil is found on the Deccan trap in the western part and in the southern part it has red-brown forest soil, indicating a physiographic link with the Khandwa region.

Betul: The soil structure of Betul is of the mixed black, red and yellow soil type of Archaean and Dharwarian.

Chindwara: The soil type in the north-western side is similar to the type found in the northern part of Hoshangabad. The central part has a mixed, black, yellow and red soil redeposited in valleys. The eastern and the southern part, has black cotton soil, whereas in the south-eastern part, red and brown forest soil is found.

Seoni: Seoni has black cotton soil on the north-western part, on the Deccan trap, mixed black, red and yellow soil redeposited in valleys on the central strip, while brown forest soil is found on the south and south-eastern sides.

Balaghat: Red and brown forest soil exists on the northern part of the district. However on the southern side mixed black, red and yellow soil of Archaean and Dharwarian exists.

The NEPA Soil Survey Division was created in Khandwa in 1965 in

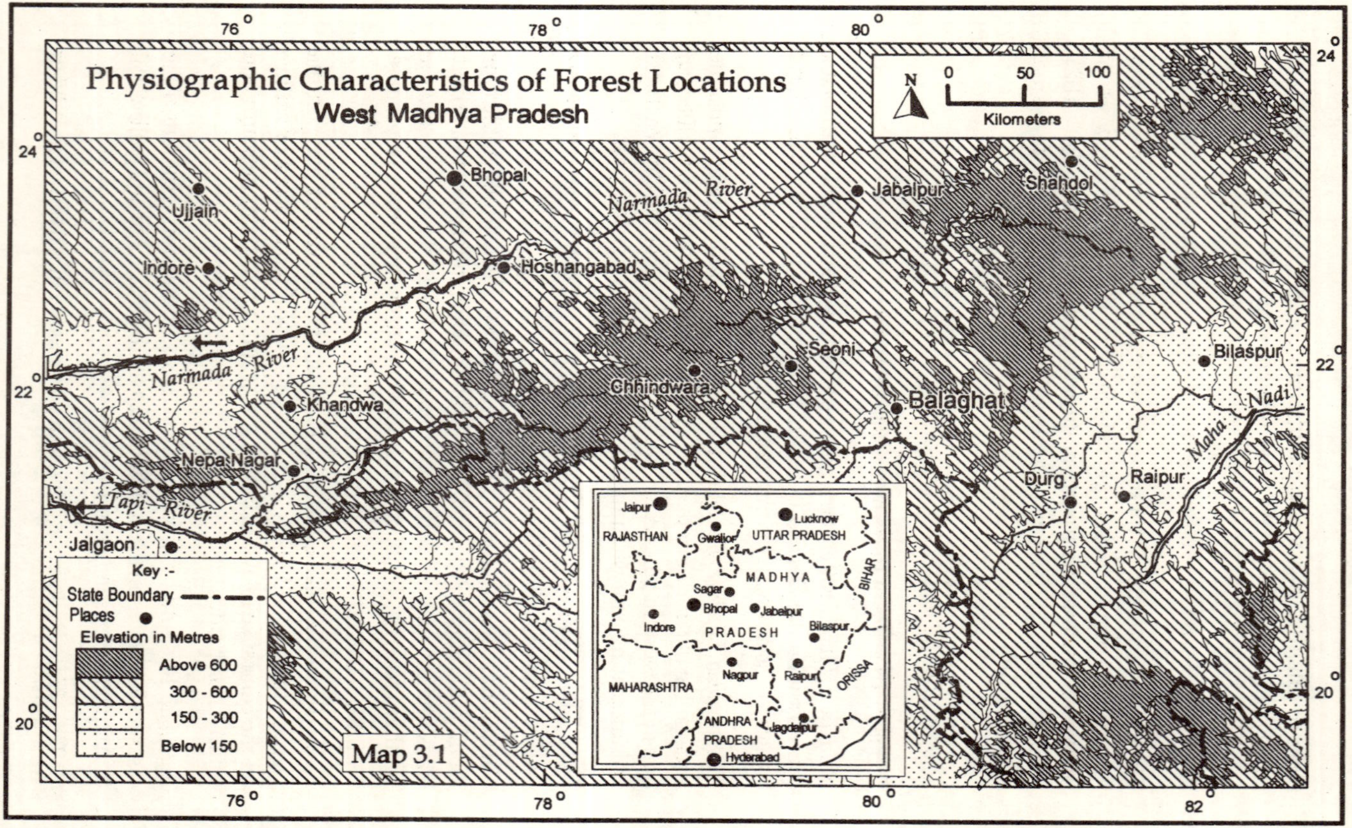
Physiographic Characteristics of Forest Locations
West Madhya Pradesh
N
0
50
100
Kilometers
Bhopal
Ujjain
Indore
Hoshangabad
Narmada River
Jabalpur
Shahdol
Seoni
Chhindwara
Balaghat
Bilaspur
Khandwa
Nepa Nagar
Tapi River
Jalgaon
Durg
Raipur
Maha Nadi
Jaipur
Lucknow
RAJASTHAN
Gwalior
UTTAR PRADESH
MADHYA
Sagar
Bhopal
Jabalpur
Indore
PRADESH
Bilaspur
BIHAR
Nagpur
Raipur
ORISSA
MAHARASHTRA
ANDHRA
PRADESH
Jagdalpur
Hyderabad
Key :-
State Boundary
Places
Elevation in Metres
Above 600
300 - 600
150 - 300
Below 150
76°
78°
80°
82°
24°
22°
20°
Map 3.1

order to survey the whole district for plantation of *Salai* (*MP Gazetteer Balaghat*, 1969: 337).

Mineral Resources

On the one hand, mineral content of the soil as trace elements influence the growth of particular species of flora, and on the other, certain minerals like limestone and coal are specifically required for the manufacture of newsprint. There are various kinds of minerals which are found in this region. In the Khandwa region are found limestone, dolomite and copper. China clay, lead, zinc, ochera, dolomite and coal exist in Betul district. Seoni has a good stock of bauxite, dolomite and mica. Balaghat is rich in manganese, limestone, bauxite, copper, kynite and mica, whereas Chindwara houses abundant reserves of coal, manganese, dolomite, asbestos, felspar, calcite and mica.

Relief

The present catchment area of NEPA can be divided broadly into two forms of natural regions. These are (a) the Satpura, Mahadev and Maikal ranges and (b) the Narmada-Sone trench.

Khandwa is in the East Nimar district. The major portion of the Khandwa region comes under the Satpura, Mahadev and Maikal ranges, while the northern part is in the Narmada-Sone trench. The height of Khandwa varies from 500-1,000 ft above sea level at some regions, to 1,000-2,000 ft in others. Hoshangabad is similar to North Khandwa. Betul and Chindwara on the west of Khandwa are contiguous, they have the same physiographic make-up of the Satpura natural regions, the natural region being the Satpura, Mahadev and Maikal ranges with heights varying from 1,000 ft to 4,500 ft due to the presence of the Mahadev hills. However, the extended catchment areas, Seoni and Balaghat, which form a continuous strip further westwards, are lower than the Betul and Chindwara region, the height varying from 1,000-3,000 ft above sea level. Physiographically they belong to the same natural region as that of Betul and Chindwara.

Status of Forests in the Catchment Area

Khandwa in East Nimar is the merging zone of the southern and northern form of dry deciduous forest. Government forests cover

41.5 per cent of the area of the district, since *Salai* and bamboo are the only two species we are concerned with, from the rich mixed forests; it need only be mentioned that teak (*Tectona grandis*) and *Anjan* are the only two other species of economic importance (Srivastav 1969: 17, 22).

The forest area in East Nimar district is broadly divided into:

(1) Northern belt with compact blocks on both sides of the Narmada river consisting of Punasa and Chandgarh ranges. Bamboo is found in this belt.
(2) The continuous central belt on northern branch of the Satpura forming the two Kalibhit forest ranges, Nepanagar, Piplod and Asigarh forest ranges, where bamboo is found.
(3) Scattered blocks between the northern belt and the continuous central belt consisting of Khandwa and Singaji forest ranges. This is the home of *salai*.
(4) The southern belt of Hutti hill range and the detached block of Samardeo forests on the Samardeo hills.

Salai forests are typical for the drier southern half of the district (Asirgarh, Burhanpur, Mardwa and parts of Singaji range and western portion of Kalibhit range), where they are found in an almost pure state with little undergrowth except grass. In other areas it is also widespread, particularly on the hills and in certain regions it occurs in association with the teak. The characteristic *Salai* type forests on the hill tops and dry slopes usually give place lower down to teak and mixed forests. It used to reproduce by seeding, and though it still coppices freely, yet now one finds that coppice reproduction is doubtful. It is also known to reproduce from root sucker (Shrivastav 1969: 21), but it is doubtful whether this kind of reproduction would solve the regeneration problem of *Salai* on which NEPA is largely dependent (Waheed Khan 1972: 438). Bamboos are absent in *Salai* forests. The area occupied by bamboo in the district is 62,120 acres. The indigenous bamboo occurs on steep well-drained slopes and in ravines. Bamboo is not generally found on the flat tops of hills. The best bamboos are found in Kalibhit and Chandgarh ranges to the north of the Narmada where they are most plentiful at some places, often so dense as to exclude all undergrowth (Shrivastav 1969: 22).

With the exhaustion of *Salai* in East Nimar, NEPA sought its raw material from neighbouring districts. The Betul reserve forest in the northern part of the district forms large blocks on hills and plateau.

Reserves along the southern border are smaller and those on the west, still smaller and isolated. The Protected Forests (PF) fringe on the Reserve Forests and are detached from each other except in the north. The PF in the north, where they form marginal chains, are large blocks in the valley of Tawa, the Machna and Morand. From the ecological point of view and for management purposes, the forests in this district are divided into five types: (a) moist deciduous teak (with or without bamboo), (b) dry deciduous with or without bamboo, (c) mixed, (d) *Salai*, and (e) bamboo.

In Betul forest circle *Salai* is scattered all over, small patches being found on ridges and flat tops. Extensive *Salai*, however, occurs only in the Dalska, Swalmendha and Tapi ranges. The factors influencing its abundant growth in these parts are low rainfall, southern aspect and shallow soil on rocky slopes or dry tops of hills (Shrivastav 1971: 23). The greater part of Patoli, Nilgarh and Swalmendha blocks of Dalska range and parts of Tapi range near Betul river are covered with *Salai* forests. The density of crop is generally low, the average being about 0.5. Bamboo as a rule is not found in *Salai* forests, the undergrowth of which is mainly grasses.

However, in certain localities as in Bakur and Lonai forests of the Amba range, bamboo grows so densely that the area can be classified as bamboo forests with the scattered over-wood of trees. Extensive bamboo areas lie in Saonligarh, Bhawargarh, Betul, Amla and Tapi ranges. In the Asir range bamboo is found scattered only in small units.

Sporadic seeding takes place every year in this region. *Dendrocalamus strictus* is known to have flowered gregariously in the year 1932-3 in the south-esastern part of Saonligarh range.

In the northern part of Saonligarh range gregarious flowering occurred in 1941-2, while in the western parts of Tapi, it occurred in 1950 (Shrivastav 1971: 24).

Bamboo also occurs as understorey in moist teak forests in Churna in Bori valley and Anhoni forests of Sohagpur range where it flowered gregariously in 1942 and later died (1979: 23). The flowering cycle of *D. strictus* is stated to be 21 years in MP (Krishnaswamy 1956: 308). Bamboo, mainly *D. strictus* occurs in Bori valley which has a basal rock formation belonging to the Damuda series of the lower Gondwana system with a capping of Deccan trap in the north. Sandstone, conglomerates and Deccan trap are the commonest rocks found in this region. The soil is rich alluvium and holds abundant moisture. Much of the bamboo for NEPA in the early years of its production came from Betul and the other neighbouring district—Hoshangabad. In

Hoshangabad division, the occurrence of another species of bamboo, *Bambusa polymorpha munro*, a native of Burma, was first recorded in the Bori valley in 1952 (Kadambi 1952: 506). The bamboo found in the Haida division is of superior quality. The size of the culm and the internal cavity of *D. strictus* generally depends upon the soil and climatic conditions (Sekhar and Bhartari 1960: 58). Culms of this species are dull greyish green when young, turning to yellow on growing old (Gamble, 1922). As compared to teak, bamboo *D. strictus* has in general higher basic strength (Sekhar and Bhartari 1961: 611).

According to one estimate of the Forest Department the area under bamboo in MP was nearly 18,300 sq. km in 1974 and the potential annual availability of bamboo around 8,00,000 mt (Seth 1974: 20). But for supply to NEPA the allotted bamboo growing area is a mere 6,860 sq. km. with a potential annual availability of around 30,000 tonnes in 1974. However, in 1990, the situation was such that the committed amount of 10,00,000 tonnes of bamboo to NEPA was not maintained by the Forest Department (Forest Officer, personal communication). Why did this happen? For the answer we need to look briefly at the structure of the forest management and locate structural problems.

Forest Department Structure

The entire forests of MP were stock-mapped in 1945-9 (Shrivastav 1969: 26). In 1962, K.B. Nair of the then two-year old statistical branch of the Forest Research Institute (FRI) used the live plot method in two pilot sample surveys made in MP in 1950 for establishing resources of *Salai* bamboo for the production of newsprint of NEPA mills in areas of 1,65,900 and 1,37,600 ha respectively (FRI 1969: 177). Till 1962 all the forests of East Nimar were under the charge of the Divisional Forest Officer (DFO). After the NEPA mills were set up, five ranges were grouped to form the south subdivision office at Nepanagar. In 1960 the Khandwa division was extended to include Chandgarh and Kalibhit ranges to increase efficient management. It was divided into North Khandwa and South Khandwa Nepanagar Forest Divisions, each under the charge of a DFO. In 1964-5 the forest divisions were reorganized. Besides the North and South Khandwa division, a Nepanagar Plantation Division under an independent DFO and NEPA Soil Survey division with headquarters in Khandwa were created to carry out a survey of the division for plantation of *Salai*.

Since 1964 these divisions have been placed under the control of

Conservator of Forests, Hoshangabad (Shrivastav 1969: 337). Meanwhile, research nurseries had been set up at Indore and Nepanagar especially for research on *Salai* and *Ailianthus excella* (FRI 1961: 177). A working plan division created in 1964 was placed under the administrative control of the Conservator of Forest (CF), working plan circle at Rewa, now based in Bhopal. In the first stage of the working plan (1949-62) Nepanagar plantations covered 61,425 acres in two forest circles, bamboo being worked in 4-year cycles. The forests of the district are a major source of revenue. Forest raw material is needed for newsprint industry, bidi-making and charcoal (Sinha 1979: 155). Hoshangabad is divided into three divisions and all the three supply bamboo to NEPA, which is its biggest consumer. Harda division is known for its excellent bamboo (Shrivastav 1970: 141). Betul, which had one division, was divided into two in 1961 (Shrivastav 1970: 312). In 1952 K.N. Sharma created a soft wood circle for increasing the *Salai* which was in demand by NEPA. The plan for 1952-66/7 also provided for the construction and improvement of earthen roads (Shrivastav 1961: 28).

Harvesting of Salai *for NEPA*

Salai constituted, till January 1990, the major raw material for newsprint production at NEPA. No sooner had the commercial production commenced in 1956, than 4,70,000 acres of *Salai* forests in South

FOREST DEPARTMENT STRUCTURE: NEPA

PCCF: Bhopal

CCF (Production): Bhopal

CF Khandwa Circle	CF Balaghat Circle		CF Betul Circle	CF Hoshangabad Circle		CF Chindwara Circle	CF Seoni Circle
DFO (Pro) Khandwa	DFO(N) (Pro)	DFO(S) (Pro)	DFO(W) (Pro)	DFO (Pro)	DFO (Pro) Harda	DFO (Pro)	DFO (Pro)
	DFO(N) (Pro)	DFO(S) (Pro)	DFO(W) (Pro)				

Note: PCCF–Principle Chief Conservator of Forests; CCF–Chief Conservator of Forests; CF–Conservator of Forests; DFO–Divisional Forest Officer; Pro–Production; S–South; N–North; W–West.

Khandwa Division were leased on a 90-year basis. The exploitation of *Salai* was at such an alarming rate that by 1973 one-third of the leased area had been worked and NEPA had to seek *Salai* from elsewhere. Reports indicate that the request by NEPA in 1993 for allotment of additional *Salai* bearing area received no response from the Government (Joshi 1973: 56).

The details of the harvestging of *Salai* were not made available to the team when it approached the Forest Department. The matter was treated as a thing of the past. But *Salai* is known to follow a 40 year felling cycle. The marked trees are felled and provided to NEPA as stacks. The most recent agreement, between the state government of MP and NEPA made in 1985 is valid for 12 years. It was agreed that 20,000 stacks of *Salai* would be made available annually to NEPA. However, since 2 January 1990, the felling of *Salai* has been banned. It would be interesting to see how NEPA supplements this loss, based on the directives issued by the Ministry of Environment and Forests, and if it does, one wonders: why was it not done earlier? Table 3.2 and Figures 3.1 and 3.2 indicate *Salai* exported for NEPA's newsprint.

Augmenting Raw Material

Large scale plantations of miscellaneous species have been undertaken, the marketable species being given preferential treatment except in Nepanagar where raw material was required for newsprint. The species tried were *Boswellia serrata*, *Ailianthus excelsa*, *Lannea coromandelica*, and *Dalbergia paniculata* (FRI 1961: 87).

Harvesting of Bamboo Resource

Prior to nationalization of forest produce in 1971-4, the forest resources were harvested by contractors or agents. They were allowed to enter forests and work in coupes assigned to them. The general opinion is that the contractors over-exploited the forest, but some retired forest officials are of the opinion that the quality of the working of forest areas was better when contractors used to carry out the felling. This was attributed to the self-interest of the contractors or mill-owners in maintaining a proper exploitable stock for the subsequent year. After nationalization in 1974, these contractors were prohibited from entering the forests and the working of forest areas was undertaken by the Forest Department (FD).

TABLE 3.2: SOURCE OF SUPPLY OF *SALAI* FROM 1975-82

1975-6	Burhanpur & E. Khandwa		
(Only in June, July)			69,611.68 mt
1976-7	Burhanpur	S. Khandwa	62,265.85 mt
1977-8		S. Khandwa	92,661.79 mt
1978-9	Burhanpur	S. Khandwa	
(July to Dec.)			79,353.30 mt
1979-80	Burhanpur	S. Khandwa	61,829.81 mt
1980-1	Burhanpur	S. Khandwa, Khargaon	95,089.00 mt
1981-2	Burhanpur	S. Khandwa, N. Khandwa Khargaon	47,250.50 mt

SOURCE OF SUPPLY OF *SALAI* FROM 1986-91

Nov. 1986-June 1987	Khandwa, Singaji Chindwara	31,000 stacks (1 m–2 m @ Rs. 151.50 per stack)
Dec. 1987–11-6-1988		20,353 stacks
From 27-1-1988 the royalty has been revised, from Rs. 151.50 to Rs. 165	20,000 stipulated	
27-9-1988–9-3-1989		15,550 stacks
19-9-1989–16-6-1990		22,722 stacks
28-2-1990 the royalty rate is revised from Rs. 165 to Rs. 201 per stack.		
27-6-1990–19-4-1991	Dewas	3939 stacks
Revised rate from 28-3-91 to Rs. 245		

Source: Forest Department.

In Madhya Pradesh, due to nationalization, bamboo is harvested completely by the FD for all purposes. From each harvest a committed quantity is supplied to NEPA and to mills from their respective allotted forest circles. The working of bamboo in MP now follows a three-year cycle, not a four-year cycle, as earlier.

It has been observed that in South Raipur division, clumps of various sizes, when worked under a particular combination of felling cycle and felling intensity, tend to attain a definite size in course of time, thereafter giving a sustained yield in terms of the number of bamboos.

If the interval of gregarious flowering is twenty-four years or more, the best combination was said to be a three-year felling cycle and four N felling intensity, where N stands for current season culms, i.e. retention of culms four times the number of current season culms (Tomar 1963: 9, 410). This was as against the current bamboo cutting prescriptions, applicable in the whole of MP, *Karlas* (the current

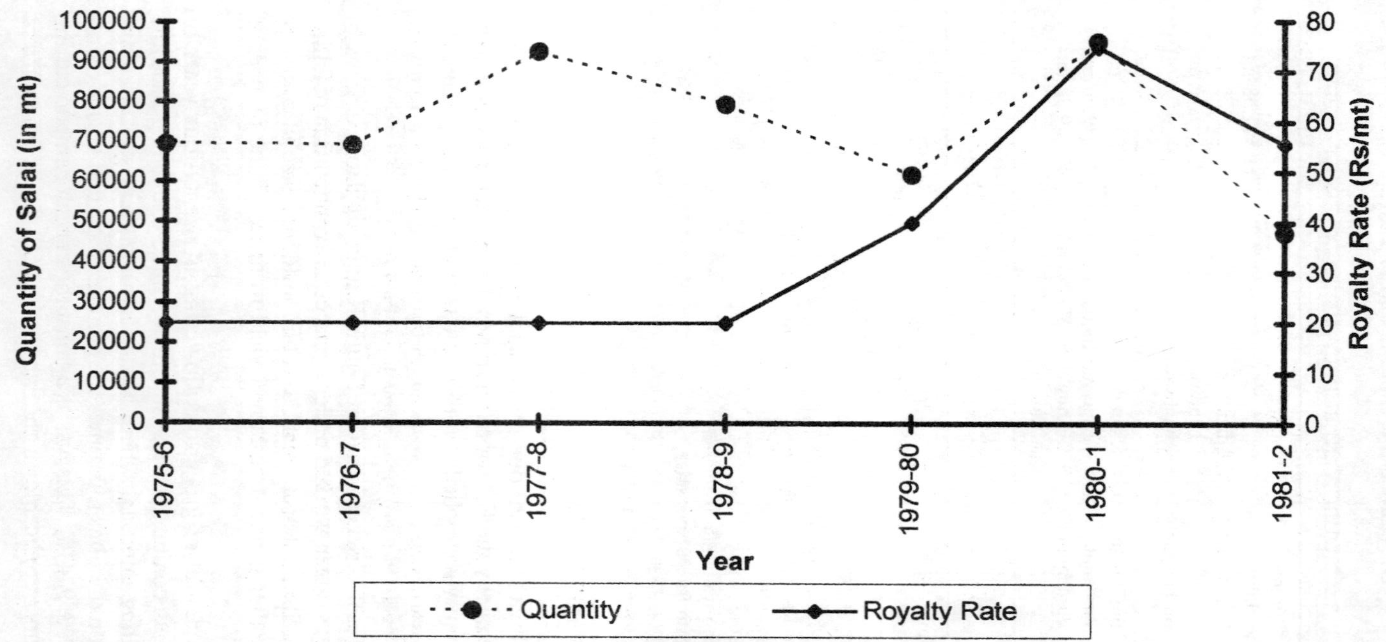

Source: MP Forest Department

Figure 3.1: Figure showing the Quantity of *Salai* Supplied to NEPA and the Royalty Collected (Rs/mt)

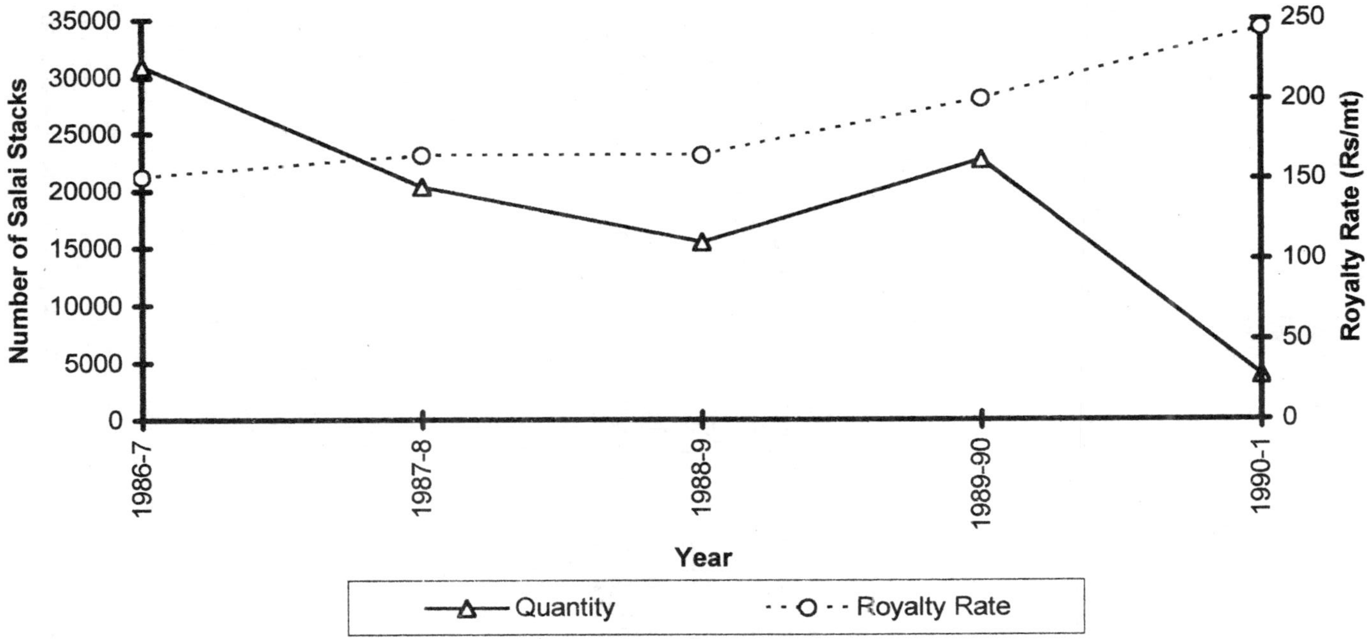

Source: MP Forest Department

Figure 3.2: Supply of *Salai* stacks to NEPA and the Corresponding Royalty Rate (Rs/stack)

season's culms) and no *Mahilas* (two seasons culms) must be cut, while the number of mature culms (more than two seasons old) to be retained is equal to twice the number of *Karlas* (2 x N), subject to a minimum of 10 live culms (SDO Khandwa, personal communication). Tomar's observation has been questioned by V.J. Chacko (Chacko 1963: 492).

The commercial and industrial felling coupes are the same as in MP. This decision led to progressive nationalizing of forest produce (1971-4).

Nationalization

In contrast to the trend of allowing contractors to work the forest areas, nationalization meant that no unauthorized persons would be allowed to collect forest resources. Nationalization procedures started in 1964 with *tendu*. In 1969 other minor forest products were nationalized. The nationalization of timber trade, begun in 1970, was carried out in four phases, ending in 1975. Bamboo and *Salai*, two important forest resources for NEPA and OP Mills, were nationalized in 1972-3 [Saxena (retd. IFS): personal communication: December 1991]. Since nationalization meant that the working of forests is to be done by FD, this led to an increase in the workload for the FD staff, he added, thus affecting the efficiency of foresters in carrying out their duties efficiently. The nature of work too changed from more of field-work to desk jobs. Though there has been expansion of staff and posts at higher levels, lower level staff are limited. A Conservator of Forests (CF) was of the opinion that location of new offices and new posts at the field staff level would bring down the workload, provide better promotion chances and keep the officials contented.

Agreement with the Pulp and Paper Industry

In 1985 the MP government entered into an agreement with both NEPA and Orient Paper Mills, Amlai (OPA), for the supply of raw materials from forest resources. The agreement is valid up to 1997, the government being committed to supply bamboo up to the amount of 1,00,000 sale units (S.U.), per year to each NEPA and OPM (1 S.U. = 2,400 running metres = 1 m). This is to be done in six instalments, the first of 20,000 S.U. and the remaining five instalments of 16,000 S.U. each. These would be done against the payment of the royalty (in similar instalments) by the respective units. The deadlines for the payment of instalments by PPIs is as follows:

1st Instalment	due on 16 November
2nd Instalment	due on 1 January
3rd Instalment	due on 1 February
4th Instalment	due on 1 March
5th Instalment	due on 1 April
6th Instalment	due on 15 May

In case of a delay in payment, the unit concerned is charged an interest on the amount payable for the duration of delay at the rate of 18 per cent or at the prevalent bank rate.

The agreement commenced on 1 July 1985, and is valid until 30 June 1997. A summary of the agreement with NEPA is given below:

1. Agreement	– 1 July 1985–30 June 1997 (12 years)
2. Product	– Industrial bamboo
3. Stipulated Qty.	– 1,00,000 S.U. (1 S.U. = 2,400 running metre)
4. Areas (Circles)	– Khandwa, Hoshangabad, Betul, Seoni, Chindwara and Balaghat

Royalty Rate

Each state fixes its royalty rate and the rate charged may differ from unit to unit. Unlike as in Maharashtra, the royalty rates are calculated differently in MP. After nationalization the royalty rates for NEPA and OP Mills are the same, i.e. bamboo is provided at the same rate which is revised every year. Since bamboos are worked by FD, the exploitation cost or charge (which includes labour charges, transportation and other incidentals of the operation) incurred by FD, is also included in the royalty rate charged. For example, in the year 1984-5, the royalty rate was Rs. 501 per mt of bamboo provided. This actually meant that, out of Rs. 501, Rs. 201 was the actual royalty and Rs. 300 was the exploitation cost incurred by the FD. Thus:

Royalty Rate	Exploitation Charge	= Net Royalty Rate
Rs. 201	+ Rs. 300	= Rs. 501 mt of Bamboo

Thus the royalty rate is only apparently higher in MP. A year-wise (1985-92) list of the royalty rates paid by NEPA under the agreement with the MP government is provided in Table 3.3.

TABLE 3.3: CHANGING ROYALTY RATE OF BAMBOO

Year	Net royalty rate (Rs.)	Date	Phase
1985-6	510.00	1.7.1985 – 30.6.1986	First
1986-7	551.00	1.7.1986 – 30.6.1987	
1987-8	595.00	1.7.1987 – 30.6.1988	
1988-9	643.00	1.7.1988 – 30.6.1989	
1989-90	725.00	1.7.1989 – 30.6.1990	Second
1990-1	783.00	1.7.1990 – 30.6.1991	
1991-2	846.00	1.7.1991 – 30.6.1992	

Labour Charges

Working the forests in MP requires a large amount of labour power. The workers employed for this purpose are basically local villagers—the Bhils, Gonds and also some migrant skilled labourers from adjoining Maharashtra. These workers work on piece-rate basis for bamboo felling, and are paid at per bundle rates which are fixed by the Conservator of Forests of the circle concerned. He calculates this piece-rate on the basis of the local job rate, which is fixed by the Collector of the district. Though the workers are entitled to daily wages, they are paid piece-rate because it is felt that the workers when paid a fixed amount per day, do not put in adequate work. Paying them their daily wages in the form of piece-rate might ensure proper working speed.

For example, in Khandwa, the daily job rate fixed by the Collector is Rs. 21 per day. However, based on this rate and the average number of bundles made by a worker in one day, the Conservator of Forests of Shahdol circle, has fixed the number per bundle as follows:

1 bundle = 20 bamboo pieces of either 1 m or 2 m
for, 1 m – Re. 0.90 per bundle and
for, 2 m – Rs. 1.80 per bundle

The work includes: felling, conversion into pieces, stump-dressing, bundling and transporting the bundles to a truckable spot (forest depot). In the years 1971-4, the wage rates were:

1 m – Re. 0.01 per bundle, and for
2 m – Re. 0.02 per bundle

Apart from the daily wages paid to the workers, the FD also utilizes the funds from the World Food Programme (WFP) a foreign-aided agency. Under this programme the workers are supplied with essential commodities (food) at concessional rates, i.e. for Rs. 6/unit of food. A unit of food consists of 2 kg of rice, 200 gm of dal and 200 gm of soya bean oil per unit for a week.

Therefore, if

7 days work is equivalent to 6 units,
At the rate of Rs. 21 per day the wage = Rs. 7 x 21 = Rs. 147/-

Since the labourers are given Rs. 6 per unit under WFP, their net payment would be:

Rs. 147 – (6 x 6) = Rs. 147-36 = Rs. 111/-

In addition they would be provided with:

6 x 2 kg = 12 kg of rice
6 x 200 gm = 1.2 kg of dal
6 x 200 gm = 1.2 kg of oil (soya bean).

The amount of Rs. 6 per unit, charged from these labourers is utilized for the welfare work of the area.

Supply of Bamboo to NEPA

Bamboo is supplied to NEPA from the six forest circles mentioned earlier in this report. The division-wise data for the supply was not made available. Hence, the total figures for each year have been quoted from 1985-6 to 1990-1. See Table 3.4 for the same. Figure 3.3 indicates the quantity and also the change in the royalty rates.

Augmenting Raw Material

The National Commission of Agriculture (1976), recommended two sources for augmenting industrial raw material: (a) Forest Development Corporation (FDC) and (b) Social Forestry (SF). At that point of time social forestry included farm forestry and agro-forestry.

MP's Forest Development Corporation is known as Madhya Pradesh Rajya Van Vikas Nigam Limited (MPRVVNL). We approached the Nigam but it was unwilling to provide precise data. The data presented here is gleaned from the seventh and eighth annual reports, i.e. for 1981-2 and 1983-4, reluctantly made available to us by the Nigam. Both the annual general meetings were held six years after

TABLE 3.4: BAMBOO SUPPLY TO NEPA

Supply Year	Actual (in S.U.)	Stipulated (in S.U.)
1985-6	96,617.230	100,000
1986-7	102,657.169	100,000
1987-8	68,000.000	100,000
1988-9	78,916.144	100,000
1989-90	77,753.336	100,000
1990-1	52,000.000	100,000

Source: Forest Department.

the due dates, these reports therefore contain updated information up to 1990.

On the basis of these reports it seems that land to MP FDC was transferred from the Forest Department (Government of MP letter dated 14 November 1979) and part of the cost of plantation of wood was met by the MP government (MP FDC *Ninth Annual Report and Account* 1983-4). It was just about that time, i.e. 1979-80 that the participation of the central government had ceased (MPRVVNL *Seventh Annual Report and Account* 1981-2: 1), and the capital expenditure for plantation of teak and bamboo for Phase I & II, amounting to Rs. 600 lakh, was met by NABARD (MPRVVNL Report: 1) Elsewhere, the Managing Director and Manager-Costing of Nigam reported very positively about the organization's achievements in terms of planting, productivity and profitability (Seth and Jaju 1979: 613). They claimed that they had evolved a method of monitoring performance which would be applicable for all FDCs (Jaju and Seth 1979: 613). But the updated information of Nigam does not quite warrant such optimism. The total area planted up to 1990 was 2,015 ha (MPRVVNL 1983-4: 4). The work began in 1982-3. The cost, Rs. 328.61 lakh, was shared on a 50:50 basis by the Nigam and NEPA mills. The project had to be closed down because a monitoring committee reported that due to paucity of water resources the plantation would fail and further investment would mean a loss (MPRVVNL 1983-4: 3). No further plantations were carried out because climatic and biotic factors had raised doubts about the ultimate success of the project (MPRVVNL 1983-4: 4). NIGAM was only to maintain the plantation for which NEPA would share 50 per cent of the costs. The degraded plantation area was to be transferred to the Forest Department in 1992.

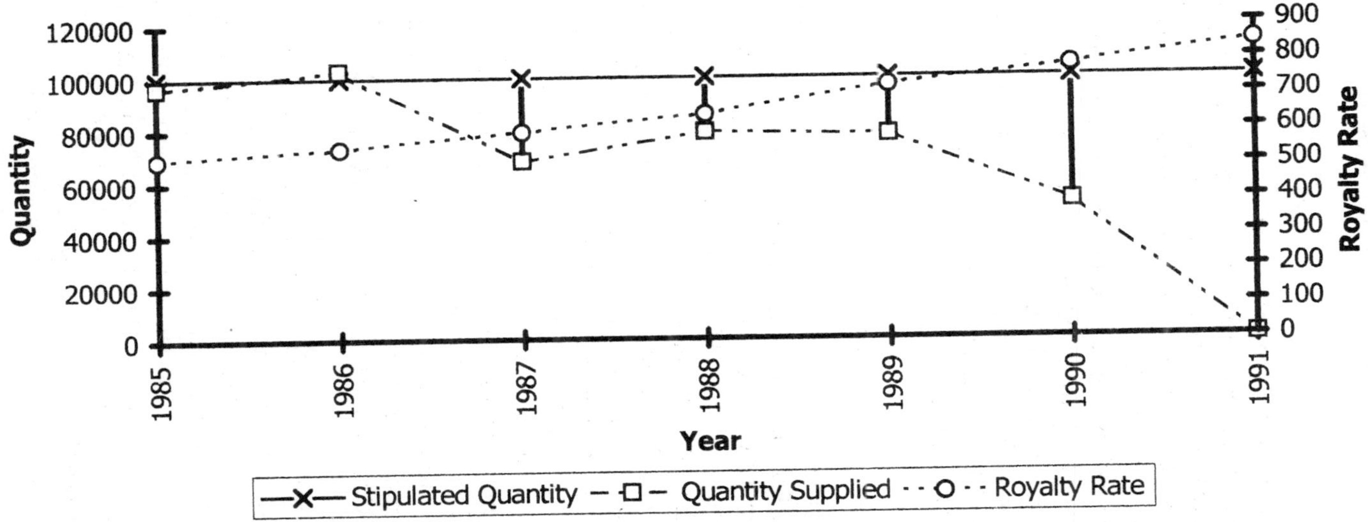

Source: Forest Department

Figure 3.3: Figure showing Exploitation Pattern of Bamboo by NEPA (1985-91)

MPRVVNL (MFDC)

It is important to remember that the National Commission of Agriculture (NCA) had categorically suggested that the potentially most productive forest areas were to be converted to industrial plantations. The forests are lost through clear-felling and also largely because plantations have failed.

In the enthusiasm for plantations, resources were drawn from what appear to be worthwhile programmes. For instance, from 1981-4, 35 per cent of funds for industrial plantations were drawn from Rs. 9 crores allotted for the rural Drought Prone Area Programme. Two out of the six plantation sites are in NEPA's catchment area, viz., Betul and Khargaon.

Social Forestry and Farm Forestry

NEPA set about producing newsprint (NP) from unsuitable raw material. Perhaps anticipating problems of shortage, NEPA's own plantation activity began as early as in 1948, nearly 8 years before production started. In 1954 the Forest Department started another 11,400 acres of *Salai* plantation specifically for NEPA's consumption. Since the urge for dynamic forestry had not been felt, both plantations were presumably in forest areas which had already been degraded by the British for their war needs. British war needs created intense fuel famines. The term Social Forestry was then unknown.

FAO was aware of the fuel famine, it also noted that bamboo forests were dying; its mandate was to create pulpwood plantations for PPI. The National Commission of Agriculture (NCA 1976: 57) mentions that in 1965 FAO sponsored a project to draw up an indicative plan for forestry in the Ministry of Agriculture. Perhaps it was a blueprint for the NCA, the final version in vol. IX on forestry was released in 1976.

The concept of Social Forestry (SF) was formally announced in the mid-1970s apparently to provide fuel, fodder, fibre, food, etc., to village communities who were deprived of their basic rights to forest produce. But the notion of Social Forestry coincides with the acute forest raw material crisis expressed by PPI. NCA which recommended SF had no illusions about the real use to which the tree crop would be put. It states that the shortfall in industrial wood was to be met by SF programmes (NCA vol. IX, 1976: 65). The species selected for SF are of little use to villagers unless one is willing to believe that eucalyptus, *subabul*, casuarina—all high calorific but expensive wood—would be bought by village folk for fuel when there is a high commercial and industrial demand for them. Scientists at FRI in 1985 were also busy testing out these species for pulping qualities and for the manufacture of chip boards. Although NCA states that the main purpose of SF is to meet the needs of the people, it truthfully adds that it is also meant to lighten the burden of productive forestry.

Although the Central Ministry of Industry knew by 1981 that the cellulosic content of *E. tereticornis* is worthless for paper, the Forest Department and NABARD were kept in the dark.

Social Forestry in forest-rich MP began only in 1981-2 with financial assistance from USAID. Work began in 24 out of the total 45 districts of MP. Four of these are in Khargone, Khandwa, Hoshangabad and Betul which form NEPA's catchment area. The total area it has undertaken to afforest was 63,000 ha but the project ended in disaster, and the scheme was called off in 1986. Arun Kumar Singh of Eklavya (A voluntary social organization for the education of rural children), lists the causes of its failure, starting with unmanageably vast areas, where 'maintenance and upkeep were highly neglected' resulting in 'low survival rate of barely five to ten per cent' of the seedlings. By 1972, FAO was aware that *Eucalyptus tereticornis* does not yield in arid areas in India (FAO, personal communication, October 1995). NEPA records indicate that its Plantation Cell has afforested the hills surrounding Nepanagar, and that NEPA is also involved in farm forestry projects on both a bipartite as well as tripartite basis (see Section 3 for details). Besides these programmes, research at FRI has been towards augmenting the resource base for newsprint production. Newsprint grade pulp can be made not only with hemp and kenaf but also obtained from inexpensive jute sticks for which laboratory experiments have been carried out. Standard pulp sheets were made from a mixture of 60 per cent mechanical pulp from jute sticks and 40 per cent leached sulphate pulp. The strength, properties and brightness of the standard sheets made under suitable conditions have shown that jute sticks are an ideal raw material for production of newsprint (Mukherjee 1971: 101).

M.A. Waheed Khan (1972: 359) suggests that, besides laboratory work, attempts need to be directed more rigorously in the field. The bamboo yield per acre per year does not exceed one ton of *Dendrocalamus strictus* from natural forests. Plantations of *D. strictus*, too, lead to problems and thus plantations of *Bambusa vulgaris* should be carried out as the cost of extraction is lower than in *D. strictus*, also as it has high productivity. *B. vulgaris* is known to grow throughout the state.

Industries play a critical role in forest resource utilization. It has long been known that PPI have been extractive in nature rather than supportive. Their role has been purchase of produce against payment. This practice works well when one deals in a smaller quantity of materials used, but when excessively large amounts, e.g. 1,00,000 S.U. of bamboo for NEPA, are utilized, this line of thinking is unjustified. The units concerned, should put in more effort in the field (and also in their respective R and D units in their factories) to augment resources. One of the reasons for the lower rate of regeneration of bamboo and other forest produce from the exploited forest areas is the damage to the regeneration caused due to excessive felling which cuts into the natural food chain. But the foresters in India, by and large, however, like their mentor, continue to put it down to fire and biotic pressures such as grazing which is very high in MP (Shrivastav 1961: 465). This could be one of the major areas where the units concerned could channelize their efforts by employing workers on their payroll to guard the coupes against these particular factors.

SECTION 3

NEPA Limited – (Formerly, National Newsprint and Paper Limited)

Even before the actual transfer of power in 1947, the Tata-Birla Plan in 1945 prescribed mixed economy for the country, but also implied that private enterprises should get support through government policies. Unfortunately they did, to the detriment of public sector enterprises.

A decade earlier, in the thirties of this century, the up and coming Indian entrepreneurs in the eastern sector, i.e. Calcutta, set up small forest-based pulp and paper units in far-flung, bamboo-rich, backward areas of Bihar, Orissa and Uttar Pradesh, and one unit in north-west UP where soft coniferous wood was readily available.

But forest-rich MP remained unexplored for its pulpable raw material. It was in the sleepy little village of Chandani in East Nimar, in the south-western corner of MP, far from the old established paper manufacturing centres, that a little known managing agency, the Nair Syndicate of Bombay ventured to set up its National Newsprint and Paper Mill in January 1947 as a public limited company with an authorized capital of Rs. 5 crore (*Stock Exchange Directory*, 1988), to manufacture paper and straw board. This was to become the first newsprint plant in India.

Why Newsprint?—Government's Interest in NEPA

The Newsprint Control Order of 1941 under the Defence of India Rules (DIR) had been lifted after World War II. The Industrial Policy Resolution of 1948 had recognized the need to manufacture both paper and newsprint in India, and placed them under category C. Any manufacturer of newsprint in India could anticipate that under the dual policy of planned development and proposed universal education, the demand for newsprint would rise rapidly. Interestingly, although seven units including BILT in 1963 and West Coast Paper Mills were approved companies for the manufacture of Newsprint (NP), none of them took it up. Investment in newsprint is heavier than in paper; it al so implies a long gestation period. The scale of capital involvement at that time was beyond the private sectors' capacity, the returns being uncertain. In view of the lack of interest by the private sector, the government had to come forward to produce newsprint and check the recurring outflow of foreign exchange for import. The raw material to be used for the manufacture of NP was *Salai*, *Boswellia serrata*, much despised by the knowledgeable British foresters as economically worthless (Champion and Osmaston 1962: 258), though available in plenty. The Parliamentary Committee of 1973 on NEPA stated that 'with the possibility of manufacturing newsprint from *Salai* wood, the project was converted into a 30,000 tpa NP mill in which the Government of MP took an active part', not only in guaranteeing and subscribing to the share capital but also in promising free use of water from Tapti river (Joshi 1973: 1, 60). It also promised the mill a continuous supply of *Salai* from its reserve forest at a very low cost of Rs. 40 per tonne (ESRE 1969: 100). But the managing agency had neither adequate resources nor the business acumen to run the mill. MP government took over the mill in October 1949, and included it as its major project in the First Five Year Plan (Joshi 1973: 1). The cost of the project worked out to be Rs. 5.5 crore, financed by the loans from the central and the state governments. Finally, on 29 December 1958, the Government of India acquired 51 per cent of the shares and took control of the 14,000 tpa capacity NP Mill giving NEPA the status of a public sector undertaking (Dagli 1979: 306).

According to the memorandum and articles of the Mills Association, the main objective of taking over NEPA Mills by the government in 1958 was to increase NP requirement of the country and save valuable foreign exchange. This is confirmed by Dagli (1979).

At present the government holds 97 per cent of the shares and the public around 3 per cent (*44th Annual Report*, NEPA, 1991-2: 26). The mills' production did rise after the takeover by the government from a mere 13,567 in 1955 to its target 33,000 tpa in 1963-4. Its quality is however poor because of *Salai*, the raw material that was allotted to it. FAO formally entered India in 1952 with the ostensible purpose to help in the development of PPI. It had noted that paper production was adequate but not NP, it is not known if FAO recommended *Salai*, the untested species but it is certainly known that it did not caution the Government of India against the use of untried wood (FAO 1954). FAO's mandate was to find pulpable species and help to set up either pulp or PPI units in the tropics.

Salai as Raw Material

It is not clear why the government chose *Salai* as the main raw material for its precious core industry or, rather, persisted in its use without putting it to initial or even later pilot plant tests when it failed to produce good results. It would be salutary to remember that experiments in bamboo as pulpable material began in the then British-controlled FRI in 1905, and laboratory experiments proved to be successful in 1912 (Pearson 1913: 159-207). The pilot plant studies interrupted by war, had to be postponed, and a mill fabricated for the use of bamboo was commissioned only in 1924 at Nanbita, where the raw material was found to be good. This was called the Indian Paper Pulp Company. World War I protectionist policies and granting of fiscal autonomy gave a fillip to the British controlled paper industry in India (Ahuja 1980: 36). Into this British-held monopoly, as British power was beginning to wane, some inroads were being made by the up and coming Indian business represented by Birla, Dalmia and Bagoria as well as by the Mysore state. The Indian government's approach to building its industry stands in strong contrast to that of the British. *Salai* had remained unknown and unexplored as pulpable wood, by the British rulers who were deeply interested in various industrial uses of forest material.

The FRI had investigated *Salai* only for its essential oils and other products (Pearson and Puran Singh 1923: 289-306 and Simonson 1918: 259-302). After 1947, the Sirpur Paper Mills had commissioned the FRI to evaluate *Lannea grandis* as a pulpable material. But the FRI scientists Bhat and Guha (1951: 568-75) took on two other species *Boswellia serrata* and *Gargua pinnata* as well, as all the three are found as mixed stands in dry deciduous forests. Their findings show that these three species yield chemical pulp only for wrapping paper. Guha's experiments in 1980 also confirm these results (Guha et al., 1980: 490-3). It must be mentioned that Guha is an over-enthusiastic scientist who seems to have explored every possible pulpable material and published his findings in FRI's journal, *Indian Forester*, from the early 1950s to 1976. In other words, there is no evidence of laboratory investigation, leave alone pilot plan tests on *Salai*. After four and a half decades of NP production, the NEPA was still saddled with poor quality raw material till the early 1990s. The nagging question remains—why is *Salai* the major raw material especially when MP is rich in bamboo resources? But we know that investment in the forest-based paper industry was controlled only by five industrial houses at that point of time and that they had chosen bamboo as their raw material. *Salai* had another major drawback as pulpable wood—it was not renewable either in the disturbed forest areas or even in well-maintained plantations. The government had these reports even before it invested its resources and finally gave NEPA the status of public sector industry at the end of 1958, as the commission reports we quote will show.

The Ranganathan Committee in July 1953 warned that though *Salai* was plentiful, the available reserves on which NEPA banked as its main raw material would last only for a period of 20 years. It recommended that the company's plantations which had been started in 1948-9 should become the exclusive responsibility of the Forest Department of MP as the growth and survival of saplings were found to be unsatisfactory. The Committee went even further to say that a separate NEPA Mill Forest Division should be established. It assumed the average yield of *Salai* to be 1.1 or 1.2 tonnes per acre. The NEPA plantation had to be abandoned in 1954-5 after incurring an expenditure of Rs. 1.10 lakh. This was well before the government took over NEPA. In 1955 the Forest Department started another 11,400 acres plantation of *Salai*, the results again being unsatisfactory (quoted in Joshi 1973: 54) By 1972 investigations had revealed that *B. serreta* seedlings are highly succulent with an abundance of parenchymatous tissue and are thus especially susceptible to physical, physiological, climatic and parasitic damage under field conditions and 'all conceivable methods of soil working, raising of seedlings, planting devices, sowing, affording protection against fires and grazing, planting of branch cutting, etc., were tried but all proved in vain . . . ' (Waheed Khan 1972: 437-8).

Waheed Khan a sensitive scholar forester, was reported to have been humiliated and punished for not falling in line with GOI's stand (A senior Conservator of Forest, Jabalpore, personal communication, March 1994). But NEPA persisted in using *Salai.* The question is why. What is surprising is that *Sabai* grass which grows abundantly in *Salai* forests, known and used as pulpable material even today, is not harvested for use in NEPA.

Newsprint

Newsprint is a special type of paper used for printing newspaper and periodicals and should possess certain qualities: (a) Strength to withstand the speed of printing press machines, (b) capacity to take print on both sides, (c) absorbency so that impressions are not blurred by the mineral oils present in printing ink, (d) printability, (e) pliability, (f) low grammage so that the newspaper does not become too bulky and (g) low manufacturing cost so that newspapers are available at reasonable prices.

Earlier, the newsprint was a mixture of 70 per cent ground wood mechanical pulp and 30 per cent chemical pulp. This combination gave the required properties of strength. More recently chemico-mechanical (cold soda processed) pulp is used to partially replace mechanical pulp (ICICI 1992: 49). NEPA used all the three methods in this proportion, but *Salai*-based newsprint refuses to yield the desired quality of whiteness. Its R and D had raised it to 50 per cent whiteness by 1992. Incidentally elsewhere, for instance in the Karnataka and Tamil Nadu Newsprint Mills, where good quality NP is produced even though bagasse has to some extent replaced wood, newsprint is produced by a modern chemi-thermo-mechanical technology.

NP Production 1955-64: First Phase

A brief discussion on various issues related to production is in order. On 16 June 1955, the first paper machine was put into operation with imported chemical pulp and unbleached indigenous ground *Salai* wood. Regular commercial production began in March 1956 after the commissioning of the chemical pulp mill. Initial production was only 13,567 tpa which was gradually stepped up to reach the target capacity of 30,000 tpa in 1963-4 (Joshi, p. 2).

Forest Raw Material

Salai: In 1956, as soon as commercial production began, 4,70,000 acres of *Salai*-bearing forests in South Khandwa division were leased on a 90-year basis. By 1973, one-third of the leased area had been worked. When the parliamentary report of 1973 declared that *Salai* plantations were a failure, NEPA made a request for allotment of additional *Salai* bearing forests in South Khandwa, North Khandwa and Khargone divisions covering 6,31,000 acres, but received no response (Joshi 1973: 56).

Bamboo: NEPA had also entered into a supplementary agreement on 9 February 1956, under which *inter alia* forest supply of bamboo 'not exceeding 30,000 tpa but which in reality never exceeded 28,000 tonnes annually would be made available'. The royalty for extraction of bamboo was Rs. 5 per tonne, higher than the rate offered to Birlas OPM. NEPA's 90-year agreement for bamboo was from 26 felling circles in Nimar, Hoshangabad and Betul which would yield 26,000 tpa of air-dry bamboo. In other words, the initial catchment area for the small supply of bamboo stretched beyond Nimar into Betul and Hoshangabad. NEPA's production of NP had reached its target capacity of 30,000 tpa by 1963. By 1963 NEPA put in its claim for more bamboo. Initially the MP government agreed on 12 September 1963, to allot on an *ad hoc* basis, seven felling circles in south Betul, north Betul and Harda. It also agreed to release some more bamboo, but at Rs. 12 per tonne, which in reality was twice the rate charged to the private paper mill in Vindhya Pradesh, viz. Orient at Amlai (Joshi 1973: 58, 60). NEPA was taken to task by the 1973 Parliamentary Committee on Public Undertakings (p. 63) for agreeing to pay such a high rate, and it urged the Ministry of Industry to take up the issue of differential royalty rates with the MP government. Incidentally, the royalty issue points out the government's bias in favour of the private sector, especially that of Birla. A similar bias has been noted in Kerala and Karnataka too, much to the annoyance of other private sector PPI units.

The Second Phase of Expansion: The Newsprint Control Order of 1962

The Newsprint Control Order of 1962 (see Notification dated 17 Jan. 1962), only created problems for NEPA, primarily by the government imposing dual authority and control. On the one hand, the administrative control and control over prices were assigned to the Ministry of Industry, while allocation and distribution as well as import of newsprint was given to the Ministry of Information and Broadcasting (Dagli 1979: 305-7). These worked at cross purposes resulting in tremendous loss in production and profit. In 1963 NEPA decided to expand its capacity from 30,000 tpa to 75,000 tpa and obtained a license to install the second NP machine in 1964 which would produce an additional 37,500 tpa. But due to bureaucratic delays and distant authority, the second machine was not commissioned till 1969. It still faced teething problems for some time. The plan to renovate the old machine was given up, for fear it too, would break down and bring production to a halt. But the major blunder was that a new cold soda pulping machine which was necessary to feed the second NP machine was commissioned only in 1975, which meant that both mechanical and chemical pulp had to be imported involving a loss of Rs. 4.7 crore for NEPA, particularly as the rupee had since been devalued (Joshi 1973: 27, 31, 33, 36, 37). But the pulping machine was defective and even after it was repaired and reinstalled in 1977 it continued to be plagued by problems. It is not clear when it began to function properly.

The growth of NEPA was uneven, both production and sales registering a decrease. In 1969-70, though the turnover increased, the working resulted in a loss while the selling price of newsprint remained static. This is not surprising. For one, the price of NEPA NP was pegged much below the imported NP resulting in a loss, till it was corrected in 1974 (Dagli 1979: 306). The 1973 Parliamentary Committee (p. 86) had also remarked that there was hardly any justification compelling the NEPA mills to continue to sell the NP at less than the imported prices, when it was actually running at a loss. It also reported that the demand for NP was inflated by 'upstarts' by putting in 'benami' demands for non-existent presses to import NP which was superior to NEPA and sold quietly to the large presses. The IB Ministry appears to have connived with these agents. NEPA had to repeatedly approach the government for revision of selling prices when its cost of production was higher than the sales prices, and the Ministry of Industry's tardy response came only after it was raised through the Audit Report of July 1971 (Joshi 1973: 86). The 1962 control order made it impossible for NEPA to sell even its rejects and cuttings, while the Ministry of Information and Broadcasting was not interested in the economics of NEPA (Dagli 1979: 306). Despite these hindrances, production began to rise fitfully by gearing up the company's old mechanical and chemical pulp plants, and by purchasing some quantity of pulp from the Central Pulp Mill. NEPA's production soon reached 40,040 tonnes in 1970-1. In spite for increased production of NP, the news-papers were not willing to lift NEPA newsprint when better quality imported newsprint was available. The government failed to support its fledgling industry through tariffs, or give it the necessary support and protection.

As the Dagli Committee (p. 307) points out, NP was one of the few items that was imported without duty, whereas NEPA would have paid duty on its imported machinery and imported chemicals. But then some of the largest units of the newspaper industry are controlled by the powerful industrial houses to which the government has always been sympathetic. True, under the 1962 Control Act the newspapers had to lift a minimal quantity of NP from NEPA, but the newspapers resented it because of the inferior quality of the *Salai*-based NP. *Salai* was not NEPA's choice but was imposed on it. Yet the newspaper industry scrambled to get it whenever there was a shortage of newsprint (Dagli: 306-7). The price of imported NP varied with market conditions at home; with growing politicization, the number of local newspapers was increasing. Unfortunately the restrictive formula of allocation of NP by the Ministry of Broadcasting and Information under the 1962 Control Act favoured the existing newspapers in starting new periodicals, but it made it difficult for the new local ones to get newsprint on the same terms, despite a provision being made (Dagli: 307). Production slowly climbed up to 57,690 tonnes by 1976-7. Incidentally, in 1978, the caustic soda chlorine plant was installed by NEPA which also acquired the Chandani Power Station from the MP Electricity Board.

Till 1980, NEPA was the only newsprint manufacturing mill in India. The private sector was not interested in producing either printing, writing paper, or NP. This study is concerned here with the last. The central government set up HNL in Kerala, while the Karnataka government produces NP in its Mysore Paper Mill. TNPL came in a little later but also essentially as a joint sector enterprise. These are post-1980 entrants and modern and NEPA compares poorly with them. In the public and joint sector NP units the latest pulping machinery was installed as soon as FAO and the Indian team together researched and found a way to save the fast depleting raw material, and yet cope with the troublesome hardwood eucalyptus component imposed by FAO. The research was a mandate given by GOI to FAO in 1970. The results were available in 1980. NEPA was hamstrung with short-fibred *Salai* which yields pulp with a pronounced colour. Due to the bureaucratic indifference of the Ministry of Industry there were other operational imbalances and deficiencies. If the second paper machine was commissioned in 1969, the soda pulp plant went into operation only in 1980, perhaps only after the new information was made available.

Its bamboo chemical pulp plant did not reach its designed production of 90 tpa. It stayed at 70 tpd. Similarly the cold soda pulp plant designed for *Salai* never reached 90 tpd but stayed at 50 tpa. This is understandable, FAO-Indian teams work was on E-teriticornis not on *Salai.* The ground wood pulp performance remained 60 tpa and not 90 tpa. As a result the paper making capacity stood at 67,500 tpa after the addition of a new paper machine with a capacity of 37,500 tpa. In 1978 an attempt was made to lower the grammage from 57 gm to 55 gm and to improve the quality, brought down further to 52 by 1984-5 (Rao 1988: 31).

Augmenting Raw Material for the Second Phase

Salai was showing signs of exhaustion while bamboo was being depleted, so active regeneration had to be carried on.

Plantations

Even before the second phase took off, the Public Accounts Committee 1964 (Para 108) has emphasized the need to assess the existing reserves of *Salai* wood and to take measures to increase the raw material by growing alternative woods through State afforestation programmes. The Seetharamiah Committee (1965) has also recommended plantations of fast-growing pulpwood species, at an annual rate of 6,600 acres on a 15-year rotation basis by a specialized agency. The company's 1966 plan to start a 78,600 acres plantation was scotched as NEPA found the state government's demand of Rs. 2,33,600 too exorbitant; the company had already incurred an expenditure of Rs. 58,172 for its 1961-2 eucalyptus plantation for which repayment would go on till 1970-1. Experiments in growing eucalyptus in Nepanagar plantations in 1972 had yielded poor resultsas in most of this country (discussed in Ch. 8). It had therefore to turn to the Forest Department.

Bamboo from Forest Circles

At NEPA's request the MP government allotted 11 felling circles in Harda, south Betul and west Betul divisions on an *ad hoc* basis for the year 1970-1. These areas yielded 8,000 tonnes of bamboo. In 1971-2, the catchment area was stretched further east as the MP government allotted another 11 felling circles in the south Balaghat division which would yield 15,000 tonnes air-dry bamboo. It is found that 30,000 tonnes of bamboo is inadequate to produce 30,000 tonnes of NP although NEPA uses wood, thereby saving 70 per cent mechanical and chemimechanical pulping processes and only 30 per cent chemical pulp. NEPA's request for the *Salai* forests of Raipur, Bilaspur, Raigarh and Sarjunagar was not granted by the forest divisions. These are now the catchment areas of the Orient Paper Mills at Amlai. *Business Standard* of 23 November 1977 reported that NEPA wanted to enter into an agreement with Himachal Pradesh for supply of a combination of bamboo and pine to make pulp.

But according to an environmental scientist working in MP who requested anonymity, the local contractors resisted and made counter moves, forcing the Chief Executive of NEPA to resign in the accounting year 1977-8.

1980s—The Period of Expansion and Modernization

In an unstarred question on 21 August 1990, Ajit Singh, the Minister of Industry stated that NEPA had undergone renovation, modernization and expansion, and to remove imbalances, Rs. 80.37 crore had been approved for the scheme, which would be carried out by BHPV, BHEL, Jessop, and Richardson and Cruddas. But what really happened was that prices were revised more regularly and production did increase substantially in 1983-4 reaching 67,650 tonnes in 1983 which was 100.2 per cent of the rated capacity; in 1991-2 it jumped to 75,575 tonnes (*44th Annual Report*, 1991-2).

The concern of this study is not to trace the growth of NEPA *per se* but to analyse it in terms of its relationship with forest produce and its impact on the eco-system; also to understand why the government policy is tilted in favour of the large private paper corporations to the utter neglect of its own units. We hold that growth of NEPA and production of NP is for the development of society as a whole. The Minister of Industry claims that the quality of newsprint has improved in terms of its strength, brightness, smoothness, surface finish and more metreage per tonne of newsprint (answer to the unstarred question on 21 August 1990). But its stock has not been lifted.

Stock-piling of Newsprint at NEPA

The major question then is, why is the stock of newsprint piling at NEPA mills and not being lifted by newspapers? Apart from the fact that the 1962 Control Act failed to give NEPA protection from competition from imported NP, it also failed to suggest that indigenously produced NP should be consumed first, while imports be treated as supplementary, as the Kelkar Committee recommended in 1988 (ICICI: 53). The government indulged in what looks like an act of treachery, in the very same year in which Kelkar's report was published. When the large wood-based paper industry was facing a recession and a glut, on 1 April 1988 the Government of India allowed paper mills to supply their creamwove paper to newspapers, printers with excise duty concessions—they only had to obtain the requisite allocation from the Registrar of News Paper Industry (ICICI: 12), which was no problem at all for the powerful paper corporations. Thanks to this act of generosity of the government, the forest-based paper industry recovered from the recession. On the other hand, the stocks of *Salai* based newsprint piled up higher and higher in the NEPA mills as it compared poorly with cream wove paper. The brightness of NEPA's NP was only 42 per cent, though with its R and D efforts it has now reached 49 per cent. The demand for newsprint was rising, with growing literacy and political consciousness, as well as the use of newspapers in the advertising media by the business community. The government was willing to let the paper industry diversify into NP, but stipulated the condition that a paper mill proposing to diversify and supply its writing and printing paper as newsprint, should commit itself to supplying at least 20,000 tpa on a regular basis to newspapers (ICICI: 12). The paper industry found this unpalatable. Besides, as

TABLE 3.5: PRODUCTION AND PRICE OF NEWSPRINT

Year	Price of newsprint (in Rs.)	Year	Production of newsprint (in tonnes)
1980	3,600	1980-1	n.a.
1981	4,700	1982	n.a.
1982	5,200	1983	67,650
1985	7,860	1984	n.a.
1988	8,334	1985	n.a.
n.a.	n.a.	1986	n.a.
n.a.	n.a.	1987	67,650
n.a.	n.a.	1988	n.a.
n.a.	n.a.	1989	n.a.
n.a.	n.a.	1990	n.a.
n.a.	n.a.	1991	75,575

Source: *Annual Reports*, NEPA.

the market for writing and printing as well as speciality paper improved and prices began to rise, the paper industry did not pursue the matter. Incidentally, the Finance Ministry, too, on its part was afraid of a loss of revenue to the government and did not approve the scheme. Nisha Sinha (1992) reports that only two small paper mills, one in Wadwadi and the other in Aurangabad district produced newsprint. Only a mill having less than 33,000 tpa capacity was given this option to diversify (PTI, *Economic Service*, 15 September 1988). The government set up three other newsprint factories during the 1980s. The Newsprint Control Act of 1962 has been amended, and from 1 April 1991, large newspaper establishments are now compelled to lift indigenous newsprint in the ration of 2:1 with the imported NP. The quota lifted from NEPA was around 14 per cent in contrast to 16 to 18 per cent from three other non *Salai* based mills which is definitely superior.

The NP from the Tamil Nadu newsprint, with Seshasayee as its consultants and the backing of the World Bank for instance, produces far superior newsprint from bagasse and imported pulp by initially using the latest thermo-mechanical process and then modifying it to a chemi-thermo-mechanical process, due to the addition of eucalyptus. This is the case with the Mysore Paper Mill in Karnataka too. Hindustan News Print (HNL) in Kerala uses chemi-mechanical technology for pulping. NEPA saddled with it sub-standard raw material stands at a disadvantage. In spite of all the odds that NEPA has faced, it has earned profits in the last five years; Rs. 232 lakh in 1985-6, Rs. 446 lakh in 1987-8, Rs. 331 lakh in 1988-9, and Rs. 650 lakh in 1990-1. This can be explained by the fact that the price of the imported NP being higher, the Indian newspaper industry was more than willing to by newsprint from NEPA. What matters is the government's policy a nd the manipulation of the price of imported NP (see Ch. 7—HNL for details). Encouraged by its performance, NEPA was entrusted with the setting up of another unit, a bagasse-based one in Aligunj to produce 88,900 tpa of newsprint (Corporate Service, *Times of India*, 18 May 1991).

In 1991-2, when the total production of newsprint in India was still inadequate, the policy of permitting the large paper mills to lift NP in the ratio of 2:1 of indigenous to imported was fixed. But production of NP by the Indian mills—all in the public sector—had gone up from 3 lakh tonnes to 5 lakh tonnes by 1993. One would have expected a new policy which would have changed the ratio from 2:1 to 5:1, as any responsible government would have done. But the Indian Government in its new phase of liberalization and its eagerness to please and attract large foreign investment has chosen to permit entry of foreign NP without imposing any customs duty on it. It has also lifted the only regulatory mechanism of canalizing newsprint through the State Trading Corporation (STC), which has hit NEPA hardest because it is saddled with the wrong raw material and ageing machinery.

It is an established fact that NP and the paper industry in the west was facing a severe recession in 1993-4. It was not too difficult for the committee which has in it, representatives of the powerful newspaper industry, to persuade these NP mills to reduce their price to the lowest possible and dump it in India rather than suffer a total loss. It would be interesting to note that the price of imported NP was pulled down to $300-400 per tonne in 1994 and compared poorly to the 1988-9 price of $7,000-8,000. This would save the foreign manufacturer from a colossal loss, and would simultaneously be a positive gain for the Indian newspaper industry to be able to buy a somewhat better quality of NP for a price which is lower by Rs.1,000-2,000 per tonne. Incidentally the heavy customs duty on imported paper has not been lifted. Paper has always been manufactured by powerful industrial houses in India.

For yet another reason NEPA stands in strong contract to the private mills. Its township is not isolated. There is a good interaction between this company and the local community. Apart from engaging local labour, the company runs Hindi, Marathi and English medium schools which are open to outsiders. The company's hospital service is also open to the public for a payment of a small fee. The company also provides other services to the villagers.

Environmental Consequences of Newsprint Production

NP is only a specialized type of paper, and paper is known to be one of the highly polluting industries with its destructive impact on air, water and depletion of natural forests. Some of the waste paper is re-used for pulp thereby reducing some degree of pollution. Chipping of wood causes considerable dust. Since the mid-1980s NEPA has taken some care to reduce water pollution for at least part of the year and has earned a rebate on water charges. The pulping machines have a dual impact (a) on the water and (b) the quantum of wood it utilizes thereby saving precious non-renewable forest material. NEPA's mechanical pulping plant yields 87 per cent pulp from the wood. The chemico-mechanical plant uses a minimal quantity of chemicals and the yield of pulp is around 85 per cent. But there is no attempt at chemical recovery from this plant. The third chemical pulp plant yields only 40 per cent and has some sort of chemical recovery process for its soda contents, but the black colour liquor continues to pollute the Tapti. The Ministry of Environment and Forests has asked NEPA to set up a monitoring station at different locations in consultation with the State Pollution Control Board. The black liquor is neither evaporable nor combustible due to the eucalyptus component, and FAO is responsible for introducing it as pulpwood in India. In 1986-7 a plant for removal of mercury pollution was commissioned. The company has been asked to monitor the mercury cell room exhaust system and report it every three months (Ministry of Environment and Forest (MEF) Office Memorandum no. J-11011/5/89/IA.II/dated 2 January 1990). It is indeed sad that MEF does not demand the same standard from the private sector units as will be evident from the report on the next unit within the same state.

The mechanical and cold soda process has a lot of organic waste. Waste water from these two plants is mixed and piped to the effluent treatment plant which is 3 km away. It is run through a clapper, and the water, cleared of residues of bark, lignin, wood debris and fibres, goes to the aerobic lagoon. In the aerobic lagoon the BOD is reduced to 50 from 60 per cent. The treated water from the aerobic is run through a anerobic lagoon. The wet residue which forms into a sludge, is pumped into open pits to dry and cake into useful fuel, which is either sold or given away to the villagers. But during the rains it runs down to the river. Water from the chemical plant goes directly into the anerobic plant and at the suggestion of NEERI it is now led out into nearby farms where it is used for one-year old *Sebina grandiflora* plantations (*38th Annual Report*, NEPA, 1984-5) see Chapter 4 for NEERI experiments with use of effluents for irrigation. NEPA has been advised by the Ministry of Environment and Forests to monitor the emission of sulphur dioxide and to shut down all coal fired power boilers, which should be kept as standby (Ch. 4, NEERI). Today the air pollution is controlled through electrostatic precipitators. It has also been asked to set up an environmental management cell (MEF, 1990).

The major environmental issue is the depletion of natural tropical forests—mainly *Salai* forests in MP and to some extent bamboo forests. It is presumed that paper has a basic advantage in that it uses a renewable raw material, but this is not at all the case with *Salai*. *Boswellia serrata* theoretically responds to vegetative propagation through cuttings, it coppices, and the seed fertility in dry tropical regions is reported to be good (Krishnamurthy 1973: 466) and it also regenerates through root suckers. But in practice it has proved to be resistant to regeneration. In 1983-4 the 350 ha captive plantations of mixed pulp wood trees raised jointly with the MP Forest Development Corporation (FDC) failed (*37th Annual Report and Accounts*, NEPA 1983/4). FDC everywhere have been authorized by the National Council of Agriculture (NCA 1976: 71) to clear-fell the most productive forest areas to grow industrial wood. Failure of forest plantations implies disturbance of the soil structure as well as erosion and reduction in the natural forest area. NCA (1976) also recommends that state departments tend plantations and industries pay the economic price for the raw material raised in plantations. About the same time, the forestry cell of the company carried out experimental plantations in the lease-in area of the mill and have been able to successfully reclothe the barren hills around the township by cultivating various fast-growing pulpwood species (*37th Annual Report*, NEPA 1983/4). But such activities have failed to provide adequate raw material. Through modernization programmes of the 1980s, and by raising the capacity utilization of NEPA, the mill had to make greater demand on the forests of not only MP but also Maharashtra.

The MP government decided to raise the royalty of bamboo from Rs. 87 per notional tonne to Rs. 175 with retrospective effect from 1 July 1981. The effective rate worked out to be Rs. 219 per tonne as the supply of bamboo was regulated on a linear basis under which a notional tonne barely equalled 0.8 mt. On 30 December 1986, the company signed a 12-year agreement with the MP government for the supply of 11 lakh S.U. of bamboos. Other steps indicated that year for an assured supply of bamboo were (a) to take over irrigated lands in Babai near Hoshangabad from the MP government to plant seedlings of *Subabul*, eucalyptus and other pulpwood species, (b) to set up still another joint venture with the MP Rajya Van Vikas Nigam for raising plantations by pumping up treated effluent water through extensive irrigation system, (c) completion of schemes around Nepanagar with the help of farmers (*Stock Exchange Directory*, 1990). In 1987, the company entered into a 2-year agreement with the Government of Maharashtra for felling and extraction of bamboo from Wada division of Maharashtra (*Stock Exchange Directory*, 1990). But in an answer to an unstarred question, no. 1942, 21 August 1990, Ajit Singh, Minister of Industry, stated that NEPA had entered into a contract with the

Government of Maharashtra for 16,178 mt of bamboos till 10 July 1990. Before the commencement of the monsoon, 4,997.62 mt were felled and transported and the remaining quantity could not be taken as the area came under the submergence zone of the Sardar Sarovar, where the Narmada Bachao Andolan activists were agitating.

With the raising of the installed capacity of NEPA to 88,000 tonnes in 1989, the company asked the MP government to increase the allotment of bamboo from 1 lakh to 1.5 lakh S.U. and *Salai* to 50,000 stacks. The company complained that it received only 77,75,350 S.U. of bamboo and 22,708 stacks of *Salai* during the year. The shortfalls were made up by buying eucalyptus from the open market (*44th Annual Report*, NEPA, 1990-1) but there is no information as to where the eucalyptus was being grown. Information at Nepanagar (1992) revealed that most of the private contractors used to bring forest produce from Punjab, Haryana and Assam. The growing unrest in these places made the supply uncertain. As a public sector NEPA has to call for tenders. This prevents the company from appointing any contractor on a long-term basis to obtain a continuous supply of raw material. Contracts for bamboo operations at Taloda and Dhule circles in Maharashtra came to an end in 1990, the supply falling short by 4,996 mt against the contracted quantity of 16,178 mt. NEPA is trying to persuade the MP and Maharashtra governments to release bamboo and *Salai* (*Annual General Meeting Report*, 1990-1: 11). This, is in spite of the Ministry of Environment and Forests asking NEPA to discontinue the use of *Salai* within a year (see Memorandum, Ministry of Environment and Forest, no. J-11011/5/89/IA.II dated 2 January 1990).

Salai was being exhausted and the alternative for NEPA was to step up Social Forestry Programmes in cooperation with local farmers. It had two schemes, (a) direct bipartite agreements between farmers and the company as recommended by the 1988 National Forest Policy, (b) tripartite, i.e. agreement between farmers, the company and NABARD. NABARD provides the resources to the company at a subsidized rate, the company providing the farmer with seedling, pesticides and technical assistance. According to NABARD's rules, farmers are free to sell their produce to anybody and the company which has encouraged the farmer has no legal claim to the produce (ICICI 1992: 59).

At the 43rd Annual General Meeting (27 Sept. 1990) the Chairman complained that the company got inadequate supply of bamboo and less than 50 per cent of *Salai* from MP.

In 1990-1, 370 ha were covered under the social forestry scheme, the company proposing to extend it to 90,000 ha in the next six years, expecting it to yield 90,000 mt of pulpwood per annum from *Subabul* and eucalyptus plantations [*Annual Report*, NEPA (1990-1): 11]. Large social forestry programmes are undertaken but those who initiate the programme fail to study the requirement of sub-regional eco-systems. *Subabul*, *Acacia auriculiformis* and eucalyptus are known for their aggressive growth, and their negative effects on the eco-system are discussed in Ch. 6: Karnataka. It is reported that eucalyptus is not effective in preventing soil erosion [PTI, *Corporate Trends*, vol. III, no. 2 (10 August 1989): 30]. The same report (p. 34) complains that in many cases 90 per cent of the total outlay of forests is provided for social forestry, while natural forests with diverse species suffer due to lack of funds and therefore land under forest is decreasing. In other words, plantations are no substitute for natural mixed forests.

Conclusion

Circulation of newspapers being an index of development, taking NEPA into the public sector was in theory a step in the right direction. In practice, however, instead of propping it with protective tariffs, supplying it with the right type of renewable raw material like bamboo and appropriate technology, it was hamstrung with a complex control order at the crucial point of its growth. The most pernicious aspect of this control order resulted in pegging of prices of newsprint below the cost of production, and even permitting newsprint to be imported duty free. Also there was delay in releasing resources for modernization. The only respite it had was when the price of the imported newsprint rose. NEPA' balance sheets show profits from 1985 to 1991. In the 1980s when production began to pick up, and when large power-ful wood-based paper manufacturing units which had failed to fulfil their social obligation were facing a recession and a glut, they were permitted by the Government of India in 1988 to sell their accumulated stock of cream-wove paper to newspapers. The recalcitrant paper corporations recovered and their business began to boom within a couple of years, while stocks piled up high in NEPA which pro-duced poorer quality newsprint. New newsprint mills in the 1980s which used the latest technology also produced good quality news-print. Although raw material saving technology was introduced, NEPA was still saddled with *Salai*. The powerful Indian newspapers manoeuvred the international newsprint industry which was facing a glut, into dumping their newsprint in India and the government to permit it to be imported without any import duty. If the newsprint industry as a whole suffered, NEPA with its imposed handicaps was the worst hit. The use of untried *Salai* was detrimental not only for the quality of newsprint but to the eco-system as well, *Salai* was not renewable. On the one hand *Salai* forests were destroyed, and on the other, more and more forest lands were brought into plantations which were also detrimental to the forest eco-system.

Recommendations

1. Newsprint is a socially useful commodity and therefore NEPA must be given autonomy to grow, in fact it must be nurtured and protected by tariffs. The dual controls must be lifted, adequate newsprint production saves foreign exchange.
2. No PPI unit has revealed any worthwhile backward or forward linkages. NEPA has extended its statutory welfare schemes such as provision of schools and medical facilities beyond its narrow boundaries, allowing these to be shared by the local population. The government must strengthen such activities which will contribute to regional development. This in a sense is alternate development. When strengthened it can be used as a model for other industries too. Industrial legislation only calls for provision of amenities to its employees which result in industrial townships in the wilderness. These are merely unhealthy oases which drain the country of its natural and social wealth and in return pollute it with a variety of industrial waste.
3. Damaging hardwoods must be withdrawn to protect not only the eco-system and the industry but also the farmer and NABARD. The precise nature of the damage must be made known to all the parties, including of course, the Forest Department to stop a further drain on the exchequer.
4. Bamboo should be revived in its native habitat. Industry and NABARD should think in terms of getting farmers to cultivate hemp and kenaf. *Sisal* a self-perpetuating species can grow on wet, dry, or arid as a raw material for pulp. (This was recommended by Dr. Yellapa Reddy, a much respected senior forester of Karnataka cadre.) It is cultivated by AP Forest Department. It would be an useful alternative to damaging plantations of exotic pulp woods.

NOTES

1. See Governor-General Hardinge's Minutes dated 28 July 1846.
2. At the same time, Forsythe maintains that much teak may be cut and hacked but if the roots are left, it will continue to grow as long poles. This is the timber needed for small housing (1872: 31). In five years a coppice attains a height of 25-30 ft and a girth of 1-2 ft (p. 199).
3. Cattle are an integral part of the life of the rural people. They yield milk, their droppings are used as a second source of fuel energy and partly as fertilizer. The bull is either yoked to a plough or harnessed to cart; and in times of dire need, even sold in the market.
4. The plains people need bamboo for a variety of uses—rafters, and fences being perhaps the most important (see section on Chanda Forests in Ch. 4 for other details).
5. No mention is made of the myriad medicinal plants known to the tribals, but their bio-diversity was not popular with the nineteenth-century imperialists, nor were there any multinational pharmaceutical companies probing into the inner recesses of the forests for medicinal plants.
6. Forsythe (1872: 91) however mentions another useful function of this practice. *Dhya* clearings attracted all sorts of animals, peafowls, partridges, and hares which were trapped in a 'dreadful' trap set in the run and left open for the purpose.
7. Any area with more than 20 per cent teak trees is classified as teak forest.
8. Grog-shop owners, the Kulars, upgraded themselves to forest contractors and became enormously rich. They seduced the forest people with hard liquor and loans, to indiscriminately cut teak and other hardwood for the growing urban market. Enthoven has clearly shown that a shift from an occasional drink of home-brewed *Mauva* to hard distilled liquor contributed to the further undoing of the tribals in the forest regions.

REFERENCES

Albion, R.G. 1926, *Forest and Sea Power*, Harvard University Press, Cambridge, Mass.

Anon. 1977, *Business Standard*, 21 November 1977.

——— 1977, *Business Standard,* 23 November 1977, Bombay.

——— 1973, Committee on Public Undertaking, 1972-3, *27th Report, NEPA*, Lok Sabha Secretariat, 1973, GOI, New Delhi.

——— 1991, *Times of India*, 18 May, Bombay.

——— 1993, *Journal of Industry and Commerce*, vol. XXX.

Chacko, V.J. 1963, 'Growth Behaviour of *D. strictus* in South Raipur', *Indian Forester,* Vol. LXXX, no. 6: 492, Dehradun.

Champion, H. and H. Osmaston 1962, *Stebbing's Forests of India*, vol. IV, OUP, London.

Corbett, G.L. and R.V. Russel 1908, *Central Provinces, District Gazetteer, Hoshangabad*, Calcutta.

ESRE 1969, p. 100.

Forest Research Institute 1961, *100 Years of Indian Forestry*, vol. II, Dehradun.

Foresythe, J. 1872, *The Highlands of Central India*, Champion and Hall, London.

Gamble, J.S. 1922, *A Manual of Indian Timber*, London.

GOI 1976, *National Commission on Agriculture, Ministry of Agriculture and Forests*, vol. IV, GOI, New Delhi.

——— 1990, Office Memorandum no.11011/5/89/IA.II/dated 2 January 1990, Ministry of Environment & Forests, New Delhi.

GOI, Newsprint Policy, *Tenth Lok Sabha Estimates Committee Report*, Lok Sabha Secretariat, New Delhi, Ministry of Information and Broadcasting.

Guha, S.R.D. and Manmohan Singh 1980, 'Laboratory Experiments on AP Hardwoods on behalf of Bhadrachalam Hardboards', *Indian Forester*, vol. CVI, 490-3, Dehradun.

Hewetson, C.E. 1964, in Champion and Osmaston, *Stebbing's Forests of India*, vol. IV, OUP, London.

ICICI 1992, *A Sectoral Study of the Paper Industry in India*, p. 49, Bombay.

Joshi, Subhadra (Chairman) 1972, *Committee on Public Undertakings*, Joint Parliamentary Committee, New Delhi.

Kadambi, K. 1952, 'On the Occurrence of *Bambusa polymorpha munro* in Bori Valley Hoshangabad Division of MP', *Indian Forester*, vol. LXXV, no. 10, Dehradun.

Krishnamurthy, T. and M.P. Shiva 1977, *Salai guggul* from *Boswellia serrata*, *Indian Forester*, vol. CIII, no. 7, p. 466, Dehradun.

Krishnaswamy, V.S. 1956, 'Bamboos, Their Silviculture and Management', *Indian Forester*, vol. LXXXII, no. 6, Dehradun.

MP District Gazetteer, *Chindwara*, Bhopal.

MP District Gazetteer, *Khandwa*, Bhopal.

MP District Gazetteer, *Seoni*, Bhopal.

Mukherjee, V.N., Y.K. Sharma and S.R.D. Guha 1971, 'Newsprint Grade Refined Ground Wood Pulp from Jute Sticks', *Indian Forester*, vol. XCVII, Dehradun.

Pearson, R.S. 1913, 'Note on the Utilisation of Bamboo for Pulp and Paper', *Indian Forest Records*, vol. V, part V, Dehradun.

Pearson, R.S. and Puran Singh 1918, 'Notes on preparation of turpentine, resin from Boswellia serrata', *Indian Forest Records*, vol. VI, Dehradun.

PTI 1988, *Economic Service*, 15 September.

——— 1989, *Corporate Trend*, August 10, vol. III, no. 2, p. 30.

Rao, Y.A. 1988, *Paper Industry in India: Status and Prospects*, OUP, New Delhi.

Russel, R.V. 1907, *Central Provinces District Gazetteer*, Seoni Pioneer Press, Allahabad.

——— 1907, *Central Provinces District Gazetteer, Betul*, Allahabad.

Sagreiya, K.P. 1961, 'Fodder and Grazing Position in MP', *Indian Forester*, vol. LXXXVII, no. 8, Dehradun.

Savur, Manorama 1983, 'Ratnagiri—Underdevelopment or an Area of Reserve Labour Force', *Sociological Bulletin*, vol. XXXI, no. 2, Poona.

Seth, Y.K. and N.D. Jagu 1979, 'A Methodology for Monitoring and Performance Appraisal', *Indian Forester*, vol. CV, no. 9, Dehradun.

Seth, Y.K. 1974, 'The Potential of MP for Pulp, Paper and Newsprint', *Indian Forester*, no. 7, Dehradun.

Shekar, A.C. 1960, 'Anatomy of *Boswellia serreta*', *Indian Forester*, vol. LXXXVI, no. 11, Dehradun.

Shekar, A.C. and R.K. Bharat 1960, 'Newsprint Grade Ground Wood from *Albies pindrous* (silver fir'), *Indian Forester*, vol. LXXXVI, no. 5, Dehradun.

———, 1961, 'Strength of *D. strictus* from MP', *Indian Forester*, vol. LXXXVII, no. 10, Dehradun.

Shridharan, B.T. 1994, *Indigenous Newsprint*, Letter dated January to K.P. Singh Deo, Union Minister for Information and Broadcasting.

Shrivastav, K.V. 1961, *Indian Forester*, vol. LXXXVII, no. 8, Dehradun.

Shrivastav, P.N. 1968, *MP District Gazetteer, Jabalpur*, Bhopal.

——— 1969, *MP District Gazetteer, East Nimar*, Bhopal.

——— 1970, *MP District Gazetteer, West Nimar*, Bhopal.

——— 1971, *MP District Gazetter, Indore*, Bhopal.

——— 1971, *MP District Gazetteer, Betul*, Bhopal.

Simonson, R. 1923, 'Essential Oils from "Gum" of *Boswellia serrata*', *Indian Forest Records*, pp. 259-302.

Singh, P. 1918, *Indian Forest Records*, vol. VI, Note on Preparation of Turpentine, Resin, Gum from *Boswellia serrata*, pp. 259-302, Dehradun.

Sinha, A.M. 1979, *MP District Gazetteer, Hoshangabad*, Bhopal.

Sinha, Nisha 1992, A White Paper on Newsprint, *Financial Express*, 19 Jan.

Stebbing, E.P. 1922, *Forests of India*, vol. II, John Bodley Head, London.

——— 1926, *Forests of India*, vol. III, John Bodley Head, London.

Thorner, D. 1950, *Investment in an Empire*, Philadelphia, University of Pennsylvania Press.

Tomar, N. 1963, 'Growth Behaviour of *D. Strictrus* in Raipur (MP)', *Indian Forester*, vol. LXXXIX, no. 6, Dehradun.

UN FAO 1954, *World Pulp and Paper Resources and Prospects*, New York.

Vadilal Dagli 1979, *The Committee on Controls and Subsidies*, New Delhi.

Waheed Khan, M.A. 1972, 'Propagation of *Bambusa Vulgaris*: its Scope in Forestry', *Indian Forestry*, vol. XCVIII, 1972, Dehradun.

1908, *Imperial Gazetteer of Central India*, vol. X, East Nimar, Government Press, CVII, 1972, Dehradun.

1969, *MP District Gazetteer, Balaghat*, Bhopal.

1994, Open letter to the Hon. Prime Minister of India, dated 17 February.

1994, *Reviews on Newsprint Allocation Policy*, Newsprint Import Policy and its Implications.

PART 2

Orient Paper Mills and the Disorientation of the Eco-system of Shadol

Trees, Soil, Water—if man is to continue
  to live of nature's bounty
Nature herself must be conserved
And her bounty used wisely

E.P. McGee

Does Orient Care?

M.S.

SECTION 4

Social History

Orient Paper Mills chose a perfect setting for its new unit in the sixth decade of the twentieth century. This was the little village Amlai on the banks of the river Sone or Hiranyavaha in Shahdol district in the erstwhile Rewa state in Madhya Pradesh. Shahdol and its surrounding regions, at that point of time was thickly wooded, and bamboos were found in great abundance. Coal was already being mined and good quality limestone was one of the many minerals that the region yielded. This region is no other than the Baghelkhand, the heartland of the 'tribes' of central India.

There was a continuous process of change in Baghelkhand, which quickened from the mid-nineteenth century onwards and reached macabre proportions by the seventh decade of this century after the entry of the paper industry. It need not have been so.

This study highlights only certain aspects of the social and natural history of the region which have a bearing on the question under investigation. It discusses the changing status of the local forest dwellers and the forest itself, mainly the bamboo forests, and the condition of the river Sone before and after the entry of Orient.

A Glimpse into Social History

The social history of the region is tragic. It is a history of invasions and of a continuous process of marginalization of the inhabitants in the area.

The earliest settlers of the Sone valley and its hilly tracts were the Baigas who pre-date the 'Aryans'. They are food-gatherers and even today are known to live in close symbiosis with the forests.

Of equally prehistoric origin are the Agaria, the aboriginal iron smelters, who are quite distinct from the Lohars, the ironsmiths of the Hindu era.

'Aryans', the cattle breeders, came to India in the second millennium BC, but we do not know at what point of time they came into Baghelkhand, nor do we know the nature of their interaction with the local people. The Baigas and Agarias speak Aryan languages, and over a period of time, the Aryans absorbed the phallic totem into their trinity. It is not clear whether the crumbling ruins of Viratnagari and the muddy poor Banganga in Shahdol district are the same as those that figure in Aryan mythology, in the epic *Mahabharata*.

The Kols (known as Korkus in the Satpuras), the Mundari-speaking primitive shifting cultivators, are believed to have come down from the Naga hills sometime in the eighth century BC and reduced the Baigas and Agarias to a lower social order. The Kols in turn were subjugated by the Gonds. In the fifth century BC the Dravidian Gonds, the plough-cultivators from Krishna valley, came via the Godavari over the Maikala range into central India. They brought with them new concepts of cultivation, although the ownership of land remained communal. The Gonds also effectively cut off the Korkus from other Mundari-speaking tribes in the east.

Next to come were the Maurya and Gupta kings from Pataliputra. They travelled up the Sone river and brought in new concepts of royalty which the Gonds eventually picked up, and by the fourth century called themselves Raj Gonds, setting themselves up as petty rajas.

In the sixth century the kings of Malwa from 300 miles away began their domination and travelled up the Narmada valley to claim a share of the cultivated produce of the Gond rajas and the tribal communities. The Malwa empire disintegrated within 300 years, but the feudatory princes divided the possessions amongst themselves. The Kalchuri Rajputs of Mandla, Panna and Banda claimed the western part of Shahdol as their own. Through military campaigns they extended their rule eastwards into the Sone velley and claimed proprietary rights over all the lands, making the Kol and the Gonds tribals tenants on their own lands. The new rulers settled other new immigrants, the Kurmis and Patels, who were expert farmers, on the best lands. Agriculture recorded a further advance. The monarchs and feudatory princes increasingly demanded a larger share in the agricultural produce as tribute. In the thirteenth century, Baghel Rajputs from Gujarat married into Kulchuri clans.

By the end of thirteenth century, the Khilji rulers travelled into the Narmada valley and to their new domains, they gave the name Baghelkhand. The Mughals who were firmly entrenched in the Gangetic plains by 1562 extended their sway over Baghelkhand and under their tutelage, the Baghels introduced the *zamindari* and *jagirdari* systems, by which a new class of 'landlords' emerged, all the cultivators being reduced to the status of tenants. However some land remained with the tribal communities. In the eighteenth century the Bhonsle Marathas invaded Baghelkhand. Up to this stage Gond kingdoms were allowed to exist as captive states. Beginning with the Mughals and ending with the Marathas, they were systematically deprived of their authority until totally destroyed by the system of *tahsildars* and *patwaridars* who were essentially administrators and revenue collectors appointed by the rulers.

Colonial Intervention

The British merchants coveted the forests. By the middle of the eighteenth century the East India Company began to build up its territorial power in India.

In 1803 after the treaty of Bassein, overtures for an alliance were made to the Rewa chief who rejected them. His scholar son, Raja Jai Singh, was compelled to accede to a treaty in 1813[1] in which he had to accept the protection of the British government, referring all disputes with neighbouring chiefs to colonial arbitration and allowing British troops to march through or be cantoned in his territories. His son, Maharaja Raghuraj Singh, offered troops to the British during the 1857 'Mutiny'. For his good services, Shogenpur and Amarkantak parganas which were held by the Marathas were restored to Raghuraj Singh. His son Venkat Raman Singh was permitted to bear the title of 'His Highness' and 'Maharaja' and receive a salute of 17 guns. He was granted first class power including that of life and death over his subjects.

But he was assisted by two British Commissioners, one for revenue matters and one for judicial. His departments of administration were revenue and general executive, judicial, customs and excise, police, public works, medical education and forests. This brief interlude into social history is necessary to an understanding of the pattern of forest administration (McCridle 1907: 282, 287) and why the pattern of forest administration in a 'native' state is identical to that of British-held India.

It is important to remember that early in the nineteenth century the British concluded a series of treaties with the Baghel rulers for their services during the 'Mutiny'. But every time a treaty was broken the British levied heavy penalties. By 1875 the Maharaja became bankrupt and the administration passed into the hands of the British. Subsequently, the British Tenancy Act took away all the communally-owned land which had still remained with the tribals and gave, sold or auctioned it to landlords (Roy et al. 1982: 105).

The Rajas of Bundelkhand came under the tutelage of the British in 1863, the Maharaja having ceded the land required for the Katni-Bilaspur railway link with all sovereign rights; in 1883, he even agreed to cede the land required for the Bengal-Nagpur Railway.

In 1885, the Rewa state handed over to the British government in India, with other necessary concessions, surface mining rights in the Umria coalfields. Coal mines were also opened in Johilla and Sohagpur. While coal mines were explored and handed over to indigenous capitalists for exploitation (Roy 1982: 42-3).

Political rule was 'stablized' by the British operating through the Maharajas. Stability does not always imply change for the better. The colonizers systematized the method of forest exploitation causing much distress to forest dwellers. In 1901 the forest dwellers or tribes constituted 36 per cent of the population of Sohagpur tahsil in which Shahdol was situated (*State Gazetteer, Rewa*, 1907: 85).

Consequences of Forest Administration in Rewa State

For administrative purposes the forests of Rewa state were divided into two classes, Reserve and Ordinary. 14 Reserve Forests, covering an area of 642 sq. miles, were situated at Bandhogarh, Majholi, Bardi and Shagpur. South of Kaimur range was reserved specially for *sal* and bamboos. These could not be cut without permission of the Darbar. Until 1875 no restriction was imposed on the people if they required these for household, agricultural or other local purposes, but not for export. Prohibition was imposed on 15 species, which was waived on payment of duty or on obtaining a free pass from the forest superintendent. These 15 were considered the most valuable timber trees in the state and their reservation was considered as desirable by the colonial rulers, as a source of revenue and 'as a means of extending protection to the forest'. Such rules, confessed the administrator (Lourd 1907: 33), caused a great deal of discontent and had to be modified first in 1885, and once again in 1902. After its modification people were allowed to cut *sal*, bamboo and other timber trees[2] which were under a span of 5 ft in girth, for household and agricultural purposes and also for *dhya* or shifting cultivation. What needs to be carefully noted is that in 1907 when Lourd's report was written, the forest remained unharmed. This is understandable as the quantum of wood that was cut was only for household purposes and therefore was low.

The situation changed as soon as the purpose became commercial or industrial when outturn was more than the natural regenerative capacity of the forests. Local people lop the lower branches only in a specific season. We do not know whether there is any vital difference in seasonal lopping or the method of lopping by the local people and by interlopers. Nevertheless, some investigation is necessary. The *District Gazetteers* written by the British complain that as a consequence of local pressure even cattle were permitted to graze in forests within the boundaries of villages, while for forests outside this limit, permits were required. Forests harbour both carnivorous and her bivorous animals in large hordes. Bamboo is elephant food, but forests and bamboo did not vanish until they came under mercantile and industrial pressure.

The ban on collection of *mahua* flowers and some minor forest produce like the fruits of *harra*, *chironji*, *anola*, *tendu*, *baichandi* and *bachera*, and gums was lifted, but maintained on the high-revenue yielding lac, *catechu* and *ral* (dhup) the resin of *sal*. These were leased to contractors on a yearly basis (Lourd 1907: 107).

The revenue derived from sale of the forest produce was considerable between 1840 and 1845, averaging Rs. 1.7 lakh. Between 1895 to 1990 it rose to Rs. 2 lakh yearly, the expenditure being Rs. 80,000 despite the fact that the forests were opened out under the pressure of local people. In 1903-4 the revenue rose steeply to Rs. 7.7 lakh while expenditure was only Rs.1 lakh indicating intensified exploitation. In 1902 further concessions had been granted to local people to pick forest produce, compelling us to further confirm that only commercialization of forest produce exhausts forests. The forests of Shahdol produced many types of grasses, the more important being *rusa Andropogon muricatum, bagai Eschamaemum augustifoliom, munj Saccharum arundinaceum, khaskhas Andropogon squrrosus*, spear grass *A contortus, darabh Eragrostis cynosuroides, dab Cynodon dactylon,* etc. In addition there were four unirrigated natural grass reserves at Makundpur, Barapahar, Bahelilabhat and Bir Rampur in the neighbourhood of Rewa and Satna which used to supply fodder for the state cavalry (Lourd 1907: 32-3). There is no information about the status of the forests after 1907. It is not known whether they were plundered to supply wood for both World Wars. What we do know is that Shahdol was rich in forest when Orient entered in 1965 (Roy 1981: 18).

People are part of a forest eco-system. The forests of Rewa/Shahdol is the homeland of 18 tribes, 10 of the tribes mentioned earlier in the social history of the region were employed as forest labour and in the first decade of this century the rate of wages for men was two annas per day and one-and-half annas per day for women. Baigas, Agarias, Kols and Gonds were some of the forest people reduced to the status of daily labourers (Roy 1981: 33).

At certain points natural history becomes entwined with social. One important social technique of preserving vital natural resources, whether forests or rivers, was to weave myths and then sanctify them. This is an effective pre-scientific method, highlighted by this study in the case of the river Sone.

River Sone: The Sacred and the Beautiful

The river Sone (from Sona) 'the red gold', cuts through the district of Shahdol. It rises close to Narmada at Amarkantak on the Maikala range. The legend is it that the two rivers are two tears dropped by Brahma, one on either side of Amarkantak range. The origin of the name Sone or Hiranyavaha or gold-bearing, comes from the red gold colour given to it by the quantity of sand it holds in suspension, during the rains or possibly from the existence of gold in the alluvium. Its water is believed to possess great sanctity, the performance of *sandhya* on its banks ensuring ablution and attainment of heaven, even to the slayer of a Brahmin.

The river is of some historical importance, being the Erranoboas (Hiranyavaraha) of Arian and Megasthenes and the Soa of Ptolemy (McCridle 1907: 94). This river is mentioned in both Valmiki's as well as Tulsidas' *Ramayana*, in the *Sri Bhagvat* and in other literary works (Lourd 1907: 3).

The river, after leaving its course, cascades over the edge to the Amarkantak plateau in the most picturesque surroundings, flowing through Central Provinces into Rewa through a maze of valleys and hills. It does so for the most part in rock channels that expand in favourable spots into magnificent deep broad reaches locally called Dahar, the favourite resorts for the fisher castes. It flows first in a northerly course for about 145 miles and at Sarsi is joind by its tributary Mahanadi. When it encounters the Kaimur range it takes a north-easterly direction. Besides the Mahanadi there are three other affluents of importance. On the left bank is the Johilla which also rises at Amarkantak and the Banas and Gopat which join it on the right bank, near Bardi. In the early part of this century the river was little used for irrigation, but where the banks were low, a rich deposit of silt was annually deposited by the overflow of the river in which excellent crops grew. Earlier records indicate that the river was ordinarily of a moderate dimension and had a steady flow, turning in a few hours into a roaring torrent (Lourd 1907: 4). Today the picture of Sone is totally different in the lower reaches of Amlai where it is dismal, damned and dead—thanks to the effluents of the Orient Paper Mills.

This study has touched on the social history of Shahdol and Rewa, the immediate environment of the mills. It is necessary to go somewhat beyond this to understand the forest ecology of the extended region which provides Orient Paper Mills (Amalai) with its basic raw material.

SECTION 5

Forests in Orients Catchment Area

Status of Forests of Eastern MP (District Gazetteers Old & New Series)

Orient Paper Mills at Amlai was first granted lease areas in its immediate vicinity, viz., Shahdol. With the depletion of bamboos in these forests, ten other forest circles in eastern MP were granted to it. Those that lie due south are Bilaspur, Raipur, Durg, Kanker and

Jagdalpur in that order; Baghelkhand and Bundelkhand in the north and Jabalpur and Sagar to its west (see map).

These are well-watered highlands with soils that yield bountiful crops in valleys and forest clearings. In Shahdol for instance, the only system of wet cultivation was from small embankments of earth raised at the lower end of the sloping fields, so as to retain water for some time after the monsoon has ceased. In the land thus moistened, seed was sown in October, produding a yield three to four times as great as that obtained from the same area of equally good 'dry' soil *Imperial Gazetteer*, 1907: 285).

The focus here, however, is on the forests over the vast region. A closer look at the physiography indicates:

Drainage

Over the northern scarp of the Kaimur range the river Tons fall in a series of magnificent cascades. A similar effect on a smaller scale is produced by streams which precipitate themselves into the valley of the Son river. In the immediate south of Shahdol is Bilaspur which is drained by Mahanadi and so too Raipur and Durg by the upper Mahanadi and its tributaries while Indravati and its tributaries flow through Kanker and Jagdalpur.

Soils

The soils fall into natural divisions, fertile, deep alluvial soil in the north of Kaimur, while in the hilly tract, cultivation is productive only in the valleys where detritus has collected.

Sagar

Sagar region consists of red and brown sandy soil on sandstone and shales in the south-eàstern side; in the south-western side, black cotton soil of the Deccan trap; while in the northern part red and brown forest soil occurs (Shrivastav 1961: 465). The Chattisgarh plateau, the Mailkala range and Chotanagpur plateau are regions which have thick mixed red and sandy type of soil. The black soil is the debris of traps. In Sagar as well as in eastern MP, a wide variety of forests grows on the soil which are mixed black, red and yellow soil on Archaeans and Dharwarian; red and brown forest soil on the

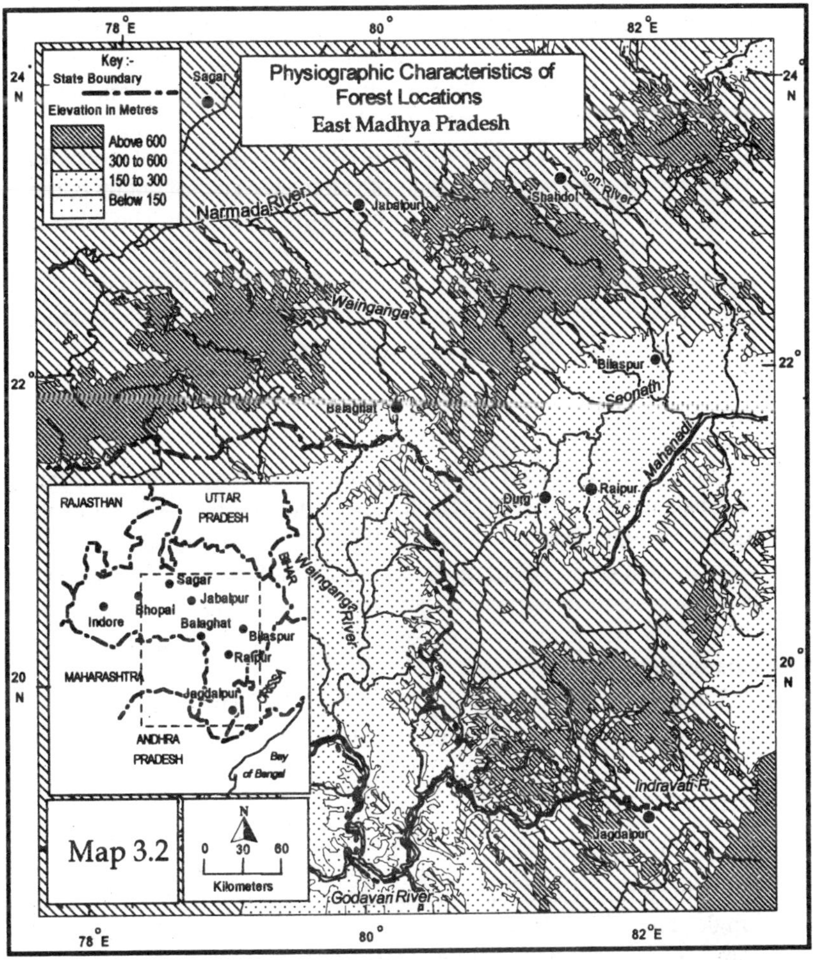
Physiographic Characteristics of Forest Locations
East Madhya Pradesh
Key :-
State Boundary
Elevation in Metres
Above 600
300 to 600
150 to 300
Below 150
78°E
80°
82°E
24° N
22°
20° N
Sagar
Jabalpur
Shahdol
Son River
Narmada River
Wainganga
Bilaspur
Seonath
Balaghat
Mahanadi
Durg
Raipur
Wainganga River
Indravati R.
Jagdalpur
Godavari River
RAJASTHAN
UTTAR PRADESH
BIHAR
Bhopal
Indore
MAHARASHTRA
ORISSA
ANDHRA PRADESH
Bay of Bengal
N
0 30 60
Kilometers
Map 3.2

southern side of the region occur; and red and brown sandy soils on sandstones and shales occur in the remaining region (computed from the map on the soil structure of MP in *Techno Economic Survey of MP*, 1969: 4). The high quality mixed forests with bamboo occur on the schists of Dharwar in the Sokhan range, and sandstone and gneiss in the south loam range and a few *Nala* valleys.

The geology of Rewa is unusually interesting. Almost every formation in the state yields products of value—the gneiss contain corundum, while mica and galena also occur in this formation. The Bijawars contain rich iron ores, valuable lime stone and bright red-banded jaspers. The Lamentas contain ceramic clays of excellent quality. The Umaria coal mines were a source of considerable income to the colonial government, with sandstone as good building material.

Relief

Shahdol was carved out of the erstwhile Rewa state. Rewa state falls into two natural divisions which were separated by the scarp of the Kaimur range. North of the range lay a wide elevated alluvial plain with an area of 3,778 sq. miles; to the south the country is traversed by a succession of parallel ridges the Maikala range enclosing deep valleys, the whole area was covered with dense forests and to east in Chattisgarh plateau.

Vindhya in 1902-3 yielded a revenue of Rs. 7 lakh (IIP, 1907: 280-8), sandstone yielding building material of unsurpassed excellence.

Status of Forests in Shahdol in the Nineteenth Century

Shahdol: Shahdol forests were good and extensive till as late as 1907, covering about 4,600 sq. miles apart from 600 sq. miles of scrub jungle. This was despite *dhya* or shifting cultivation, still practised to some extent by the 'jungle tribes' (McCridle 1907: 285). The colonizers reported that there was little or no real forest on the north of Kaimur except the Makundpur teak reserve, south Kaimpur range which was covered with valuable trees. The predominant tree in the southern forests is *sal Shorea robusta* intermixed in places with large amount of bamboo *Dendrocalamus strictus* 'of first rate quality'. On the higher reaches *sal* is replaced by other species.

The forests are classified into three convenient belts. The deciduous forest belt was a good 200 miles in length, but very unequal in width varying from two to 20 miles (McCridle 1907: 31-2). The deciduous tract filled nearly all the small valleys and low lands. It was estimated to cover an area of nearly 1,000 sq. miles. The most important trees in this belt are *sal*, bamboo, *tendu drospyros* and *Dalbergias*.

The mixed forest belt is situated on the northern slope of the Kaimur range on the Rewa plateau, a few hills on the north-west border adjoining Panna state and on the south-east of the deciduous belt. The mixed forest combines some of the features of both deciduous and dry forests.

The third is the riparian forests which owe their special character to the periodical overflow of rivers and *nalas*. These are situated in the south-west corner of the state, south of Chandia and Singhwara along the banks of Sindhuli and Ghorchatar rivers. The inundated lands produce quick growing forest of teak, *babul* (*Acacia arabica*), *khair* (*Acacia catechu*) and *dhawa* (*Buchanamia latifolia*). Reproduction is rapid in this area Lourd (1907: 31-5) mentions species of common trees and shrubs some of which provided fruit, others small timber and fuel. Lourd also identified the medicinal value of most species and the uses to which they were put by the local inhabitants.

Bilaspur Forest Circle: The best forest exists on Maikala range, the Chattisgarh plain, the Mailkala range. These soil types, along with drainage by Mahanadi and Seonath and their tributaries, support the growth of forests in certain areas. The forests can be classified under *sal* and mixed forests. The chief associates of *sal* in the top storey are *bija*, *dhaora*, etc. Bamboos are common in the *sal* forests and they soon cover opened up areas.

Bamboos (*D. strictus*) are common on hill slopes and on areas above the Ghat. Their growth is luxuriant on hill slopes and tends to become pure. Bamboos are absent from the interior pure sal areas and the only species met with are *D. strictus* and *B. arundinacea*. In Baloda range they are restricted to fewer localities and are absent on the plains in the mixed forests, but lie in abundance on the hill slopes and along the *nalas* above the ghats in Lormi, Lammi and Kota ranges.

Raipur: The forest wealth of this region is concentrated in the southern and eastern hilly areas. The *sal* the predominant species give way to teak as one moves southwards. Today they are classified very broadly into five types:

(1) Northern Tropical Moist *Sal* Forests.
(2) Northern Tropical Dry *Sal*.
(3) Northern Tropical Dry Deciduous Mixed Forests
(4) South Indian Dry Deciduous Teak Forests
(5) Southern Tropical Dry Teak Forests (20 to 90 per cent of the over wood is teak)

Of these, the first is recognized by the presence of teak in 20 to 90 per cent of the overwood. The last two types are predominated by *sal*, whereas the remaining mixed types are so named for the nature of their crop composition, at places giving way to local types of forests formed by the concentration of other species. Bamboo, *Dendrocalamus strictus*, forms an overlapping type of forest throughout the region occurring mostly in teak and mixed forests (Verma 1973; 19).

Bamboo forms the understorey in the northern tropical dry deciduous mixed forests at many places excluding the plateau. The bamboo crop is mostly open and is said to have suffered from the fires and clearance from slash-and-burn cultivation and heavy exploitation for *nistar* demands (Verma 1973: 23) and last but not the least, due to heavy exploitation by and for the OP Mills.

Salai predominates on the top of the ridges, bamboo of poor quality forming the understorey of mixed forests, plains have *Salai* in common with *sal* forests. Most forests of this class are on sandy or clayey loams underlain by granite, gneiss and quartz. The better quality can be seen on the alluvial deposits of Paire, the Sondul and the Udanti where bamboo (*D. strictus*) and Katang bamboo (*Bambusa arundinacea*) form the understorey.

Durg: On the northern plateau of the Durg circle, which lies in the south-western part of the Chattisgarh plain, the chains and large patches of forests extend along the south-western and southern boundaries. The most compact forest belt extends on the Maikala range along the north-western and western boundaries. More compact forests are marked as either Reserved or Protected Forests and managed by the state FD.

Under Champion's classification most of the forests are classed as tropical dry deciduous teak and mixed. Small areas in the north-west around Chilpi and Rengakhar bear moist deciduous forests to be further classified into *sal*, moist teak and moist mixed forests.

In the moist treak forests, teak forms from 30 to 40 per cent of tree crop, bamboo forming a thick understorey. The moist mixed forests occur where the soil is not well drained. The species are similar to those in the moist teak forests but with a higher proportion of *sal, bija, dhaora* and *tendu*. The crops are of good quality and are associated with a dense understorey of bamboo.

The dry mixed forest type is the most extensive in the region. Good quality mixed forests are common in the southern region on the sandstone and also extending on to the gneiss. No species is conspicuous in the overwood. However, among the species in the understorey, *gargari* and bamboo are most conspicuous.

The poor quality forests occupy a bulk of the area in the northern range and northern half of Balod range along the Chandrapur sandstone. In the northern range the density of the crop is 0.5 and the characteristic feature is the presence of *Salai* in varying proportions.

Bamboo occurs in all the dry as well as moist forests, except in the *sal* forests. In the northern range its quality is fairly good. It is usually associated with *gargari* in the understorey. Katang bamboo of planted origin is found in very limited places.

Bastar: The mills catchment area also includes the Bastar region (with Jagdalpur and Kanker forest circles) which is also the most important tribal area in the state. The Bastar region, houses one of the very best forests that exist in Madhya Pradesh today. However, reports indicate that bamboo is not abundant in the Bastar region. The same can be said of Jagdalpur, where it is said to occur only in the periphery (Tiwari 1955: 411). The reasons attributed to this are dense *sal*, bad drainage and over-exploitation in the plateau (Tiwari).

The state of the catchment areas have direct bearing on past management practices (see Part 1).

Sagar Forest Circle: Another catchment area of OP Mills is located in the north central region of Madhya Pradesh. The forests, restricted to the hilly areas, are generally not suitable for permanent cultivation. The forests belong to the northern tropical dry deciduous type according to Champion's classification. Within this broad-based main type, considerable variations occur due to geographical, geological and other factors, the most important of which are grazing and fires (Verma 1973). Bamboo occurs in moist sheltered valleys.

The Jabalpur Forest Circle: The Jabalpur Forest Circle is an important area for the OP Mill. The forests in this region occur on the slopes of the hills and all over the rugged country in the southern, eastern and south-eastern parts. The lowlands, the inter-mountain valleys and the flat tops of plateaus where the soil cover is thick, have long been rather indiscriminately cleared of forests. The forests can be divided into two classes: (a) northern tropical dry deciduous forests and (b) southern tropical dry deciduous forests. Important sub-classes are: (a) bamboo forests and (b) the grass lands.

Bamboo forests and grass lands occur in all the above types. Bamboos mainly occupy areas where the top canopy has light. Bamboos have also been exploited unsystematically except on the hill slopes where they are luxuriant. They generally form an understorey, the prevailing species being *D. strictus*. Occasionally *B. arundinacea* is found on alluvial flats along the rivers.

Forest Management in the Catchment Area of OA

Forest management, through regular working plans was absent. However, the erstwhile 'native states' of which Vindhya Pradesh was composed were content with annual plans of operations which gave them the maximum financial returns. Most of the forests were consequently subjected to heavy felling. But such is the recuperative power of the soil that the forests have reappeared and the regions which had been completely devastated are green once again.

In the case of Jabalpur which was under British control, forest management was first taken up in 1862 with the promulgation of the Waste-land Rules, under which felling of teak, *sal*, *bija*, *saj* and *shisham*, was forbideen without the sanction of the Deputy Commissioner. These forests were first demarcated and formally preserved in 1867 under the Indian Forest Act, 1865. However, heavy felling continued. In 1871 *sal* forests were closed to cutting and grazing. All other forests were treated as unreserved. The unreserved forests were also notified as Reserved in 1879 under Section 34 of Indian Forest Act, 1878. Shifting cultivation was stopped with effect from 20 February 1879.

The forests came under Departmental management in 1887. Working Plans (WP) were thought essential and the first WP of Jabalpur Forest Division was prepared by some forest rangers in 1894-5. However, it proved to be a failure. The plan prepared by E.E. Fernandez was introduced from 1 November 1899. This was revised as the second WP in 1929 by Gurudial Singh. The third WP was made in 1939, it being a revision of Gurudial Singh's earlier WP. In these three overlapping working circles of bamboo, *khair* and *kulu* were also prescribed for felling. Bambooo in twenty-three felling series were worked in cycle of four years. The fourth WP was made by F.A. Lahri for a period of 15 years from 1952-3 to 1966-7. In this too, the three overlapping circles of bamboo, *khair*, and *kulu* were implemented.

As for Bilaspur, the records indicate that forests have been worked irregularly by contractors since 1961. Most RFs came under reservation in 1878 up to 1893-4, forests were open to extraction of timber, no restrictions being imposed as to the locality from which materials such as bamboo, firewood, grasses and all other minor forest produce was obtained. In 1893-4 another system of localized felling was introduced and the felling of all green wood was prohibited except from the coupes specially marked off for this purpose. WP was introduced for Chitta Perdaria Reserve in 1898 and for the rest of the reserves in 1910. Under the J.W. Bests WP of 1912, the situation is reported to have improved (Shrivastav 1978: 35). The vegetation, once probably very rich, had been carelessly hacked, silvicultural principles absolutely disregarded and in nearly every case the forests had been over-cut, neglected and over-grazed. In 1950 the *zamindari* system was abolished.

In 1962 the forest came under a different type of assault; this time by the Government of India-sponsored scheme for 'plantations of fast growing species' after clear-felling the natural forests. Table 3.6 not only shows the colossal expenditure, but the trend away from the local species to eucalyptus hybrid.

Interestingly, the schemes for plantations of fast-growing species such as eucalyptus hybrid were taken up when the OP Mill at Amlai was just taking off. Is it a coincidence that these 'fast growing' speices are also pulpable? However, under these conditions too, the Government of MP leases out the produce to these mills, which play a lead role in creating the mounting crisis of the forest and its resources in the state and in the country.

Around 1965, Orient Paper Mills, entered into a 20-year bamboo lease agreement with the MP government at the rate of Re. 0.37 per ton (see Section 6 on the question of fixation of royalty rate). Under this agreement the mill could extract bamboo from the areas allotted to it. However, with nationalization of bamboo in 1972-3, the lease agreement became redundant.

Shahdol and Rewa, which fall in the first catchment area of MP Mills, forms a special case as regards its present state. The Rewa forest circle is in a totally depleted state. Bamboo resources in Rewa, have be-come negligible. The main causes attributed are industrial exploitation at a very high rate, pressures of grazing, and fire (retd.

TABLE 3.6: AREA PLANTED IN BILASPUR DISTRICT AND EXPENDITURE INCURRED IN THE FOREST DEPARTMENT 1962-3 TO 1967-8

Year	Species	Area under each species	Total area	Expenditure (in Rs.)
1962-3	Teak, Semal, Maharukh, Salai			
	Baranga	280	280	1,86,345.40
1963-4	Eucalyptus hybrid	312	1,122	4,39,616.98
	Bamboo	10		
	Teak	800		
1964-5	Eu-hybrid	1,112	1,224	7,38335.72
	Teak	110		
	Bamboo	2		
1965-6	Eu-hybrid	1,002	1,200	7,59,681.77
	Teak	30		
	Bamboo	168		
1966-7	Eu-hybrid	1,332.20		7,09,094.31
1967-8	Eu-hybrid	1,400	1,400	4,39,361.56 (up to August 1967)

Source: Bilaspur Forest Division DFO.
Ref.: P.N. Shrivastav, *Gazetteer of India, Madhya Pradesh, Bilaspur District*, Bhopal, 1978: 38.

IFS: personal communication, 1992). There is a total moratorium on bamboo felling in Rewa. Destruction of bamboo forests in Shahdol is put down to OPM (Roy et al. 1981: 19). The Government of MP however extended Orient's catchment area to 10 new circles (see OPA's agreement with the government dated 14 November 1986). These circles have a varied quantity of exploitable raw materials (some have good stock while others are depleted). Nevertheless, irrespective of their present status they continue to be the targets of high exploitation by OPA. OPA has been exploiting forest resources from these circles at a rapid rate, so much so that now Raipur and Sagar too have been depleted (see Fig. 3.4). One of the new controversial catchmet areas of OPM is Raipur circle.

Harvesting of Bamboo

Although the Forest Department was most kind and cooperative no data on the annual outturn of bamboo to OPM from different circles was available. The total outturn is given in Table 3.7. But it had no records and desperately sought for data from the OA Mills in 1991,

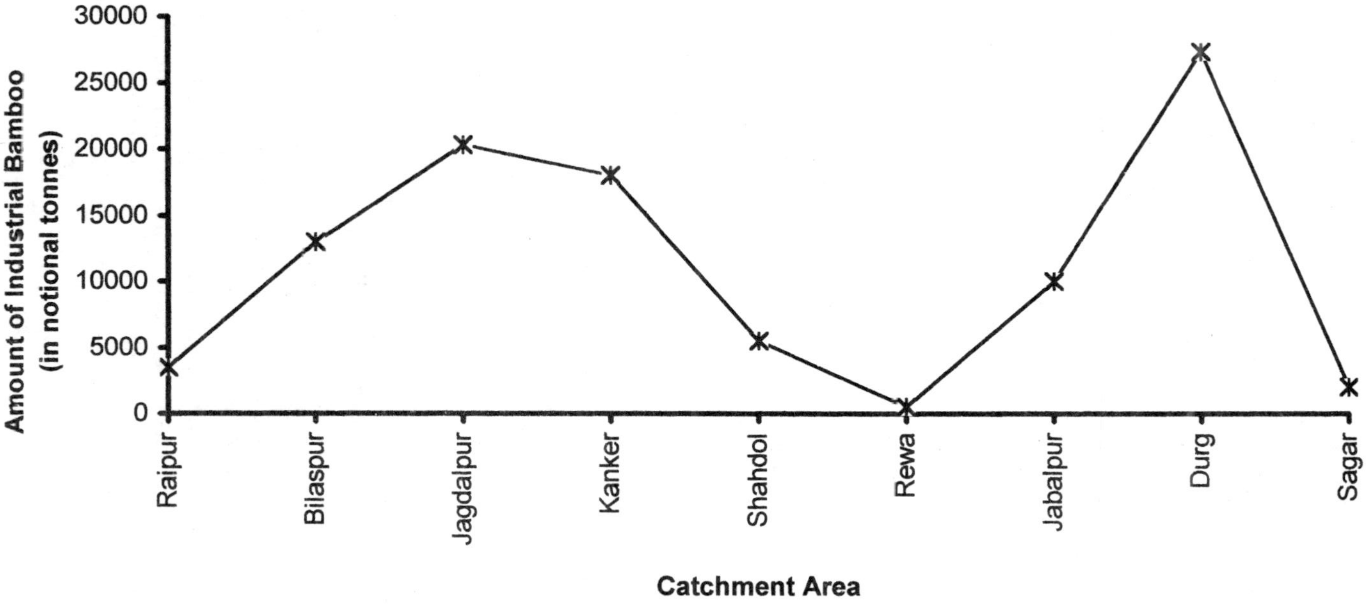

Source: MP Forest Department

Figure 3.4: The Depleted State of Bamboo Forests Comparative Trend for 1988-9

TABLE 3.7: TOTAL QUANTITY OF BAMBOO SUPPLIED ANNUALLY TO OPA AND THE ROYALTY RATE OF BAMBOO

Year	Quantity of bamboo supplied (in notional tonnes)	Price paid against the supply (in Rs./mt or S.U.)
1973-4	1,00,000	175
1974-5	1,01,000	175
1975-6	1,00,000	215
1976-7	1,00,000	215
1977-8	93,500	260
1978-9	94,500	260
1979-80	91,000	285
1980-1	91,000	290
1981-2	n.a.	n.a.
1982-3	1,02,000	365
1983-4	1,00,000	455
1984-5	95,000	510
1985-6	n.a.	550 (from agreement)
1986-7	n.a.	n.a.
1987-8	1,00,000	640
1988-9	1,06,000	690
1989-90	1,00,000	775
1990-1	1,00,000	838

Source: Forest Department, Bhopal.

1992, and 1993, when we asked. Raw material for OPM is weighed at the FD office at Shahdol. There was no worthwhile bamboo area in that circle for our immediate observation. Even the nearby Jabalpur circle drew a blank. We were asked to visit Balaghat, OA's late catchment area, the only area that was worth observation for both its growth as well as its felling practices. We were told by the FD in Jabalpur that the bamboo growth in Balaghat is tangled. This is surprising for the *D. strictus* which is a non-thorny, straight, solid (but which has now become hollow), and of average size, often bordering on stunted, is now tangled growth. M.G. Gogte IFS, a silviculturist from the Maharashtra cadre, was able to explain the strange phenomenon of tangling in this species. A bamboo needs nourishment; its carbohydrate cycle is from the atmosphere. Frequent felling reduces the nutrient cycle and weakens the bamboo culms and their ability to stand straight, firm and tall. This results in a thin, tangled and apparently crowded clump. Weakened clumps and trees are also open to insect and other pathogens.

The bamboo forests were largely destroyed by careless felling by the contractors. OA is believed to have engaged contractors to fell bamboo from its assigned coupes. The nationalization of forests and banning of private felling led to some revival for bamboo, as has happened for instance in Karnataka. The 'forest area' in MP is vast, much too large for the FD with inadequate funds and inadequate staff to cope with efficiently. The tendency is to repeatedly go back to the same easy areas for harvesting, and to neglect the tangled re-growing which is time and energy consuming. The senior officials know about this, but the tendency is to 'protect the supervisory staff of the bamboo workers, rather than fault them' said one forest officer. Misplaced sympathy has contributed to the degradation of the vast bamboo forests of MP.

Prior to nationalization, the mill contractors/agents did work the allotted areas. However, since 1973, when bamboo was nationalized, these contractors/agents have been prohibited from entering the forests and eventually only FD of MP was authroized to fell.

In Madhya Pradesh the working of bamboos follows a 4-year cycle. The silvicultural rule, which is to be strictly adhered to (but is not followed too often), states that 'no *karlas* (first season culms), no *mahilas* (second season culms), are to be cut'. It also states that 'twice the number equal to *karlas*, live or mature culms, are to be left untouched in the clump, subject to a minimum of 10 culms in a clump'.

However, it was observed in South Raipur Division that clumps of various sizes, when worked under a particular combination of felling cycle and felling intensity, tend to attain a definite size in course of time and these would often give a sustained yield (in terms of the number of bamboo culms). Tomar argues 'If the interval of gregarious flowering cycle is 24 years or more, the best combination would be 3 years felling cycle and 4n felling intensity (retention of culms 4 times the number of current seasons culms)' (Tomar 1963: 410). Tomar's observation has been question by V.J. Chacko (1963: 492).

Under the silvicultural prescription, harvesting of bamboo involves converting the cut culms into bundles, stump dressing and transporting the bundles to a truckable spot. The formation of bundles requires that 'all billetted bamboos shall be tied with a *mahul* or *sabai* string and will consist of 20 pieces of 2 m length. No bamboo piece shall be less than 4 cm in girth at the thin end. Each full bundle shall consist of 2 m long pieces and each half bundle will consist of 75-125 cm long pieces. 60 full bundles shall be reckoned as one sale unit and 125 half bundles shall be reckoned as one sale unit.'

Labour Charges

The labour charges are on a piece-rate basis, i.e. per bundle made. This piece-rate is worked out by the conservator, based on two factors, viz., (a) the minimum daily wages fixed by the collector and (b) the average number of bundles that a worker can prepare in a day.

For instance, the minimum daily wages for Shahdol district fixed by the collector is Rs. 21 per day. But the charges paid to the labourers are Rs. 1.80 a bundle. This means that on an average each worker would be able to prepare around 12 bundles a day.

However, this rate can vary according to the working conditions, i.e. if the terrain is undulating and difficult, the charges per bundle might be higher.

Nationalization of Forest Produce

The process of nationalization was initiated in 1964, with tendu being the first product to be nationalized. Subsequently, nationalization of timber commenced in 1971 in a phased manner and complete nationalization of timber was achieved by 1975. In the meantime, during the year 1972-3, bamboo too came under the purview of nationalization. It meant that the earlier lease agreement would stand redundant as the working would be carried out by only one agency, the Forest Department of Madhya Pradesh. Thus the OPM agents or contractors were now prohibited from entering the forests. The government's policy of accepting such private industries became its own liability. The pressing demand by the perpetually hungry mill led to another agreement between the Governor of the state of Madhya Pradesh and OPM in 1974.

The New Agreement

Consequent to nationalization in 1972-3, an indenture was made between the Governor of Madhya Pradesh acting through the Secretary to Government, Forest Department, Madhya Pradesh and the M/s Orient Paper Mills Ltd. on the 19th day of February 1974. The validity of this agreement was for 12 years, i.e. from 1 July 1973 to 30 June

1985 (both days inclusive). The quantity of forest produce to be sold under the agreement was fixed at 1,00,000 mt of bamboo per year. However, the agreement stated, 'the actual requirement of bamboos by the company will be worked out on the basis of the company's present capacity, and subject to the provision of clause five of this agreement'.

Clause five of the agreement deals with the proportion of bamboos in the manufacture of paper and states that 'The company shall use bamboo with other pulping material for the manufacture of paper in the proportion of 70 per cent bamboos and 30 per cent of other material. The total quantity of bamboo supplied shall not exceed 70 per cent of its requirements of pulping material.'

The forest division from which bamboo was supplied to OPM between 1973 and 1985 were Changbhakar, Raigarh, North Raipur, Mahasand, Kawardha, Rajnandgaon, South Balaghat, North Balaghat, North Bilaspur, Bilaspur and Kanker.

However it is interesting to note that in the latest agreement for which an indenture was prepared on the 14th day of November 1986, between the Governor of Madhya Pradesh and M/s Orient Paper Mills, valid for a period of 12 years, i.e. from 1 July 1985 to 30 June 1997, there has been an apparently drastic change in the price payable by the company for the purchase of industrial bamboo and also the area of supply. The price payable under this agreement is apparently higher [at Rs. 550 per S.U., where one S.U. = 2,400 running metres]. The unit now is not in terms of weight but length; and comparison becomes difficult. The price includes the expenses incurred by the Forest Department in making supplies at the depots too.

The change in the areas of supply is due to the fact that in the previous agreement they were just Forest Divisions, whereas in this agreement the term 'territorial circles' is used. The latter encompasses a much wider area; a circle may comprise of three-four forest divisions. In other words more new areas have been opened out for Orient Paper Mills.

It is clear that as the areas allotted get further and further depleted, new areas are opened out to meet the increasing demand by the mill. The degree of 'hunger' can be realized from a communication by the mill to the Forest Department, viz., 'Mill facing closure, supply bamboos immediately' (Dec. 1991, Shahdol).

Bamboo, after all, is a fast-growing species and quickly occupies space. Repeated and excessive harvesting on the one hand disturbs photosynthesis for the carbon cycle and also drains out its nutrients. The rhizome that is responsible for the annual vegetative growth is starved, leading to loss of vigour and even to the death of a clump. The chain reaction is erosion of soil and destruction of micro-organisms. The British laid silvicultural rule of at least four years is inadequate for a sustained healthy bamboo forest (Chapekar: personal communications, Feb. 1994).

Royalty Rate

The price payable to the Forest Department by the mill against the supply of industrial bamboo was fixed during the signing of the indenture. Under the agreement from 1 July 1973 to 30 June 1985, the price payable was at Rs. 55 per notional tonne of bamboo. The determination revision of rates is done in the following manner:

> An increase of 25 per cent over the rate of bamboo fixed for the previous slab of four years or a state equivalent to the price of bamboos worked out from sale price of a standard tonne of paper on the basis of the weighted average according to the company's product mix, whichever is higher for the period under review.

Besides depositing the amount stated which worked out to Rs. 55 per notional tonne of bamboo, the company has to deposit the expenditure on labour charges to be incurred by the Forest Department while working the bamboo.

However, the latest agreement valid from 1 July 1985 to 30 June 1997, states that 'the company shall pay Rs. 550 per sale unit for the first year. The price so fixed includes the expenses incurred by the Forest Department in making the supplies at the depots. For the subsequent years in blocks of 4 years, the price payable by the company would be fixed at 8 per cent above the price fixed for the preceding year, and the figure would be rounded off to the nearest rupee.

Therefore, though the price paid (Rs. 550/S.U.) seems to be apparently higher, it is not really so because it includes the price incurred during the working. Thus the payable price is equal to actual rate payable plus expenses incurred by FD in working. The rough estimate provided by the forester indicates that the working expense is nearly always more than 50 per cent of the price payable by the mill.

This payment is made in annual instalments, the dates for which are fixed in the indenture. They are as follows:

	Instalment	Date of instalment	Quantity for which payment has to be made (S.U.)
A	1st	16 October	20,000
N	2nd	1 December	16,000
N	3rd	1 January	16,000
U	4th	1 February	16,000
A	5th	1 March	16,000
L	6th	1 April	16,000
Total			1,00,000

In case of delayed payment the company has to pay an interest of 8 per cent per annum on the amount of the total price due.

Figure 3.5 and Table 3.7 show annual amount of bamboo provided to OA.

Augmenting Natural Resources

The PPI's demand for forest-based raw material seems insatiable so that there is always a raw material crisis. But the Indian Government is always willing to take it on. For instance, i.e. some time in late 1960s it 'asked' the National Commission of Agriculture to recommend ways and means of meeting the mill's demand. In response, in 1976 NCA recommended the setting up of Forest Development Corporations, to set up social forestry, agro-forestry and farm forestry. The nomenclatures are misleading, but all boil down to industrial plantation whether it be in prime-forested area or plain agricultural land, or common panchayat land or road sides, or along the railway tracks, or in fact wherever space permits. In NCA's words 'the requirement of the industry should be the major reason for initiating forestry activities (NCA 1976: 4) and in the management of production forest, the site potential should be determined for realization of maximum production with the shortest period (NCA: 25)'. The best 'potential sites' are of course the well-preserved natural mixed forests, 1 per cent of which, as Von Monroy of FAO recommended, should be clear felled and planted with fast-growing species (Von Monroy 1961). This was recommended in the third Five Year Plan. The percentage of clear-felling to make room for industrial plantations was raised to 2 per cent in the Fifth Plan and was emphatically referred to as 'dynamic forestry'. NCA also recommended pro-

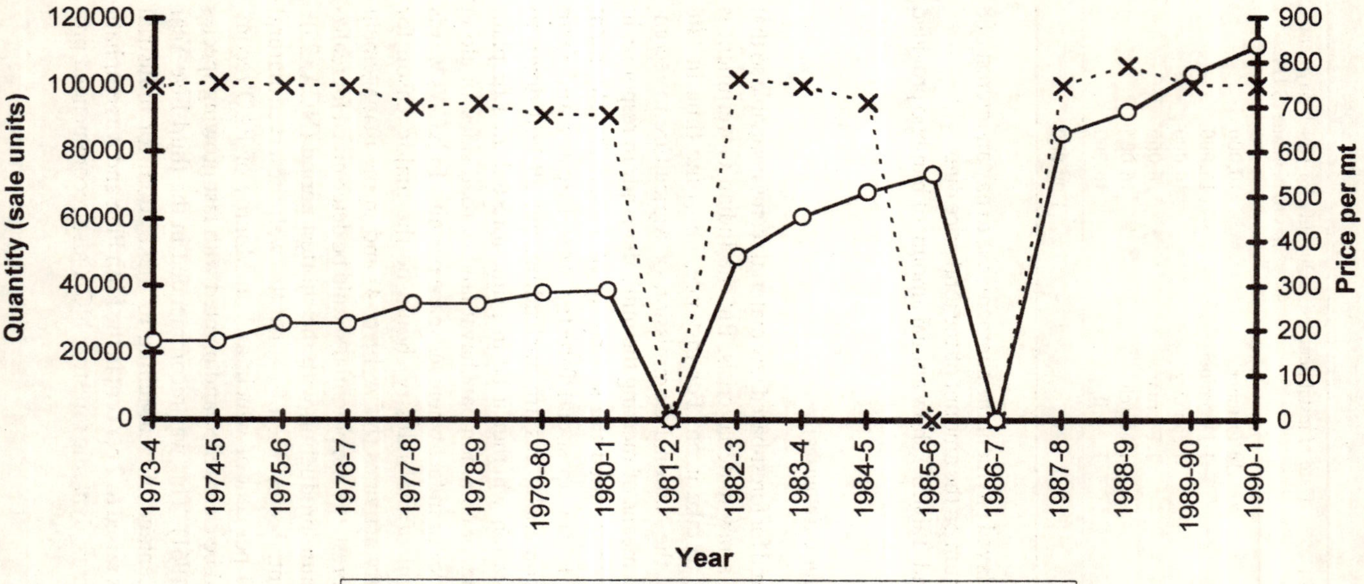

Source: Forest Department, Bhopal

Figure 3.5: Exploitation Pattern of Bamboo by Orient Paper Mills (Amlai) 1973-4 to 1990-1

ductive forestry to be initiated even in inaccessible forest regions (NCA 1976: 56). These are the last remnants of natural forests which had so far been free from both biotic and industrial pressures. In order to reach the inaccessible forests NCA strongly suggested 'a high tempo of road construction' to 'speed up the process' (NCA: 56). It wanted each state to have one or more FDC (NCA: 71). For MP it suggested that plantation timber from the whole of the Bastar forests could be used as industrial wood, as pulp wood and fuel wood are interchangeable (NCA: 67, also see Bastar Project, September 1973).

MP Forest Department Plantation Forestry

NCA (1976) also recommended that industry should be charged a price for plantation produce so as to pay the FDC's cost of clear felling, planting maintenance and still leave a profit (p. 37), and ARC (now NABARD) should amend its statutes to provide long-term finance to FDC (p. 72) with which NABARD readily complied. But the industry continues to buy the raw material at the arbitrarily-set, low royalty rate (Kothare 1971: 297). NCA is aware that raw material for PPI is heavily subsidized. MPs FDC known as Madhya Pradesh Raj Van Vikas Nigam Limited (MPRVVNL) and, henceforth referred to as Nigam or MPRVVNL, was set up around 1975-6. This information is gleaned from its Annual Reports, the Nigam authorities being unwilling to provide any information.

Commercial plantations of teak and bamboo were undertaken in Project I. The concern here is only with bamboos. This project was divided into 12 sub-projects covering a total area of 2.67 lakh ha of Reserve Forests which were handed over to Nigam by the state FD on a lease rent. The lease rent is worked out as the net surplus revenue after deducting revenue expenditure. The capital expenditure in respect of both phases of Project I were met partly from a term loan of Rs. 600 lakh secured from commercial banks and NABARD. For Phase III, NABARD sanctioned a term loan of Rs.1,224.24 lakh for commercial plantation of both teak and bamboo in the year 1987. The exact share for bamboo plantation activity is not available (MPRVVNL 1981-2: 1). Details of its activity and achievements are given in Table 3.8.

Nigam needed more resources than were available from financial institutions. Funds from other sources were directed by the state

TABLE 3.8: NIGAM'S BAMBOO PLANTATIONS 1979-80 TO 1981-2

	1979-80	1980-1	1981-2	% increase over last year
Area Planted with Bamboo (in ha)	1,928.00	2,360.00	2,445.00	+ 3.60
Actual Number of Culms Planted	12.05	14.75	18.40	+ 25.08
Industrial Bamboo (NT)	100.34	9,917.00	8,640.00	– 12.87
Commercial Bamboo (in lakh)	85.35	37.81	41.76	+ 10.83

Source: Madhya Pradesh Raj Van Vikas Nigam Limited (MPRVVNL) *Seventh Annual Report and Account for 1981-2*, 1981.

government. Starting with 1980 and for the next three consecutive years, 35 per cent of Rs. 900 crores earmarked for Drought Prone Area Programme (DPAP) was diverted to work Nigam's first in 4 districts and in 1984 stretched to 2 more districts. Dhar and Shahdol, the catchment areas of Orient Paper Mills (Amlai) fall within the 6 selected districts. The funds were meant to carry out afforestation and pasture development (see Table 3.9).

No information is available on the species used for afforestation. Since the funding is to intensify productive forestry it is logical to assume that industrial species were cultivated.

In 1982, International Development Association (IDA) sanctioned $4 million as Forestry Technical Assistance to study the feasibility of setting up large scale tropical plantations. This was for Project II (MPRVVNL 1981-2: 2). The purpose of the technical assistance was to convert 1,586 ha of natural forests into plantations of tropical pine. Giving in to popular pressure the government was compelled to call

TABLE 3.9: YEAR-WISE COVERAGE OF AFFORESTATION UNDER DPAP

July 1980	4,710 ha
1981	4,653 ha
1982	5,776 ha
1983	5,528 ha
1984	6,819 ha
1985	6,176 ha
1986	5,393 ha
1987	235 ha

Source: MPRVVNL, *Seventh Annual Report and Account for 1981-2*, 1981.

off the project in December 1982, and assets created in the project were transferred back to the FD (MPRVVNL, 1983-4: 4).

Dr. Ram Prasad, Director of the State Forest Research Institute at Jabalpur, is a great votary of eucalyptus. He notes that between 1961-81, MP the once surplus state in forest produce, lost more forests than in any other state in the country. Denudation is most noticeable in the western region especially in Bundelkhand and Rewa (the original lease areas of Orient Paper Mills). The districts are facing a shortage of firewood, timber, bamboo and other forest produce, and eucalyptus is an answer to overcome the shortage (Ram Prasad 1987: 16). The focus of research in the two Forest Research Institutes (FRI), one run by the state and other by the Centre or rather FRI Dehradun, is to increase the biomass of eucalyptus (B.N. Gupta, Director, Tropical FRI and P.K. Shukla, Director, State FRI: personal communications, Feb. 1994). They were not concerned with the effect of the species with the eco-system and their relationship with the soils. Large scale trials of *E. tereticornis*, 45 provenances (p) *E. tereticornis*, 40 provenances of *E. camaldulensis* and 80 provenances of *E. citrodora* in the State FRI were carried out (Ram Prasad 1987: 8). This despite the fact Mysore gum a provenance of *E. tereticornis* plantations division that were set up in Bilaspur, Shahdol and Bastar between 1956-74, covering an area of 45,000 ha had to be stopped in 1974 due to poor growth rate and high mortality (Ram Prasad 1987: 2). Incidentally, Mysore gum is also referred to as a hybrid and a provenance of *E. tereticornis*. These plantations were set up to provide raw material for the paper industry (Ram Prasad 1987: 2) and failed.

These researchers and heads of prestigious institutes had not cared to keep in touch with Indian findings in the Kerala Forest Research Institute at Peechi or with the findings of the international organization—its mentor, FAO. FAO in its Forestry Research Paper 19/1 and 19/2 had written that *E. citradora* had failed to acclimatize in India, it requires deep soils and does not tolerate prolonged dry season (19/2, 1980: 79). *E. camaldulensis* has failed to acclimatize in many tropical countries and in India. Yield of the sulphate pulp is average (19/2, 1980: 133). *E. teriticornis* has shown inferior growth in semi-arid zones and low resistance to drought. Sulphate pulping requires relatively low alkali charges and gives low pulp yield compared to average hardwood (19/1: 237; 19/2: 11). FAO's results were available by 1971 but were not published till 1980. FAO had not cared to communicate its findings to MP Forest Departments.

Despite failure of eucalyptus plantations in MP and its official withdrawal, Ram Prasad insists that eucalyptus controversy raised in Bangalore in 1983 was based on misguided hostility towards that species (p. 11). He has failed to read FAO's rejoinder published in Forestry Paper no. 59 (1985: 18), on the question of water consumption and recharge of aqueducts. The paper states (a) Lateral spreading and depth of penetration of root systems of the eucalyptus vary with species and this has to do with the intensity of water uptake; (b) In the regions of deeper soils and higher rainfall regimes, soil water deficit created by eucalyptus forest, seems to be in the region of 250 mm/year. In comparison with crop or pasture land this means eucalyptus forested land would yield approximately 70 mm/year less steam flow on ground water recharge; (c) The effect of soil moisture reserves of eucalyptus plantation apparently starts to appear at the age of approximately 4 to 6 years; (d) Transpiration rates differ among eucalyptus species, varying apparently between 20 litres/tree/day to 40 litres; (e) Some eucalyptus species have not developed mechanism for controlling higher rates of transpiration. But he was compelled to concede that the negative reaction of eucalyptus to the soil or to the undergrowth must be judged on what it is replacing. In his words 'when it replaces natural stratified mixed forests of *sal*, teak or dense miscellaneous forests, all the arguments (against eucalyptus) may be valid. However, when planted on mine overburdens devoid of vegetation and micro organisms' or on barren hard laterites, in deep ravines, in water-logged areas or along roadsides, railway tracts, on community waste lands, canal banks and degraded forest 'or as farm forests' then eucalyptus plantations are more beneficial than any other indigenous or exotic species' (Ram Prasad: 16). The effect of uncropped eucalyptus on soil depends on the state of soil in which they are planted. There is evidence that eucalyptus litter decomposes less well because of low activity of soil microbes (1985: 54). The FAO Forestry Report, no. 59 also states cropping of eucalyptus on short rotation, especially if when the whole biomass is taken, it leads to rapid depletion of nutrients in the soil (p. 55).

But we differ, especially on the last three sites suggested by Ram Prasad. Also, as farm forestry programme, if eucalyptus is introduced as monoculture or as a dominant species in any degraded land, it will prevent colonization of other native flora and fauna including micro-organisms.

Eucalyptus is an exotic from Australia. The Australian continent was cut off from the rest of the land mass millions of years ago. Evolution of Australia's flora and fauna is strikingly different from that of other continents. The breakdown products of metabolism of the flora have a dual effect, a negative one on other species and are therefore referred to as allelopaths, but a positive one for itself for its survival as the fittest. In the long-term evolutionary process other local species, both flora and fauna have also evolved to co-exist peacefully but not necessarily if transposed to other continents.

Ram Prasad as a forester and an ex-director of an FRI must know that forests are not static, and eucalyptus plantation, at our present state of knowledge, has not evolved into higher forms of forest in India to form stable forest eco-systems. Empirical evidence for almost two centuries of its existence in India seems to point towards its inability to allow most other species to co-exist.

Lands of poor quality, whether belonging to the poor, or dubbed as degraded waste lands, do yield millets or coarse grains, the staple food of many poor people in rain-fed areas. To deprive even sheer survival for those on the brink is inhuman, and to invest state resources to grow a species which does not suit the site, only to feed the powerful industry, is therefore nothing short of a crime. Farm forestry in a region of former established feudal relations and where no land legislation has come into operation, is genocide. It is often on the leased-in land of the poor that a rich farmer experiments with damaging crop like eucalyptus, tobacco, etc. (see Chapter 5 on Andhra).

We do not know at whose instance eucalyptus was introduced as a plantation crop as early as in 1955 in MP to feed the pulp and paper industry. However, we do know that eucalyptus is the brain child of FAO's forestry section, prescribed as an alternate pulpable hardwood for the tropics. Now, FAO Forestry Papers, nos. 19/1 and 19/2 (1980) indicate that in 1955, FAO initiated trials of tropical hardwoods in Latin America, Asia (including India), and Africa, which included several varieties of eucalyptus species. The Working Paper, Feb. 1980, prepared by FAO's Pulp and Paper Industries Division, Rome (1980: 17), puts down India's investment to Rs. 200 crore for establishing plantations between 1955-79 over an area of 3 million ha. Although it was aware of the danger of clear felling as early as (1954: 61), and points out 'contrary to what is often believed the soil of the tropical forests is of low inherent fertility and humus cover, not normally thick it is very rapidly destroyed by exposure; the ground is sensitive to heavy rainfall and drought and once damaged recovers slowly. Clear felling followed by replanting is therefore not recommended'. This is perhaps the viewpoint expressed by a Forester in FAO (1954: 62) but a dominant voice in the same organization and the same paper, (1954: 5) thought 'Forests constitute a renewable resource which with proper care can be relied upon to give an undiminished yield. . . . Unless raw material is suitably located the price of paper becomes prohibitive' and this voice carried the day. Experiments were conducted in Bilaspur, and Dr. Ram Prasad approved them. There is also evidence that FAO's first entry in the east was in 1949, when a conference was held at Mysore where the Austrian forester Von Monroy and FAO's advisor for the region was present (*Australian Forestry*, vol. XV, no. 1, 1951. In 1951 funds became available to FAO for its expanded programme of technical assistance in India, Burma and Ceylon where forest services were well established (*Journal of Australian Forestry*, vol. XLIX, no. 8, August 1951).

A review of the fourth meeting of the *Technical Committee on Wood Chemistry* in FAO held at Brussels in 1949 reported that in Tasmania, *Eucalyptus gigantea* was being utilized as high grade pulp for paper product. In western Australia low grade deformed and faulty eucalyptus was used by clear felling the area. One of the participants, Cohen, states that eucalyptus had much more in common with tropical hardwoods than with softwoods of the north temperate zone particularly in respect of the extractives which not only affect the pulp yield but consume excessive amounts for chemicals in pulping. While at Rome, PPI experts, after two weeks of consultations, had concluded that a pulp industry based on tropical and subtropical woods and agricultural discards is a feasible commercial venture and such an industry would more than offset the rising demand for paper and newsprint (*FAO Release*, FAO/52/12/8196, 15 December 1952). It also reported that 'a variety of well-established pulping processes are available for the conversion of the hitherto non-conventional raw materials' (reported in *FAO Release*, 15 December 1952 I/R/Pres 36).

Another report at the same time disclosed that a 30-member international meeting of chemists and foresters agreed that if facilities 'were available, there could arise pulp and paper industry in tropics of world importance'. Though it was agreed that chemicals, power, transportation and available money are more expensive in underdeveloped countries than in industrial areas, it would require far more capital to establish a pulp and paper industry in these countries than in Scandinavia. They argued however that the additional cost element is frequently offset by the existence of cheaper wood and labour. Their calculations showed that 'production cost for pulp and paper from tropical woods and agricultural residue can be higher, equal in some cases and even lower than average production cost in Europe or in North America according to local circumstance'. There is no information on the actual composition of this 30-member committee, but the chairperson of the consultation committee was Dr. G.H. Cidester, chief of Pulp and Paper at the Forest Product Library at Madison, Wisconsin (FAO/52/1281/8196 reported in *FAO Release,* 15 December 1952). Dr. R.V. Bhat of FRI's Pulp and Paper Division was India's representative reports FAO (Paper no. 6, 1953: 1).

However, a year earlier on 15 September 1951, there was one Indian among the 13 experts who met at FAO's' fifth meeting of Technical Wood Chemistry at Appleton, Wisconsin, to review the scientific and technical exploration of tropical wood species and agricultural residues (FRI 1951, File on FAO documents, Dehradun).

It was only as late as in 1961 that Von Monroy of FAO made his formal recommendation for transforming one per cent of India's most potentially productive forests into fast-growing eucalyptus plantations. This suggestion was accepted and translated as a recommendation in the Third Five Year Plan and by the National Commission of Agriculture (vol. IX, 1976). Eucalyptus as a plantation crop has been finally discredited by the Indian Council of Forestry Research and Education [see *Eucalyptus* (ICFRI, undated: 434)]. Most foresters in the states covered in this project regard it as a folly while bamboo is a worthwhile species to be cultivated and protected.

The technical bulletin no. 22, *Bamboo Plantations*, 1987, by SFRI Jabalpur shows that except for a feeble attempt in Bastar in 1954 and by Nigam, there was no programme undertaken to plant bamboo. The colonial government however had experimented in 13 of its forest divisions from 1870 onwards. From the 1930s there were no experiments, Balaghat in 1940 was an exception. Interestingly, this is the first document that does not hold local people and their cattle responsible as the destroyers of young bamboo shoots. It states 'pigs are the greatest menace to a bamboos plantation especially in the year of their formation. They relish the succulent rhizomes and uproot the entire plants where the pig population is already a dangerous level . . . it continues throughout the rains and it becomes really severe after the winter rains, if normally received in the area and persists till the soil hardens. The damage is practically negligible once the plantation is established.' The question is why FAO failed to explore the full potentiality of bamboo.

FAO on Bamboo

FAO discovered that India was self-sufficient in paper by 1952 and its raw material was bamboo (FAO 1953: 110). Forests are a state subject and the Indian foresters were angry at the gifting away of their forest produce at throwaway royalty rate (see Section 6). Both Sagreiya who protested and Tomar who was critical, were from the MP cadre. FAO had established a rapport with the central government and its Ministry of Industry and was dismayed that the Forest Department was totally indifferent to this. This was a constant refrain by several senior FAO officials at Rome (personal communication, October 1995). There seems to have been little interaction between the FAO officials and the India foresters on the bamboo question, nor was there any independent research by FAO on it, although much literature on bamboo was generated by the Forest Department in India. This is evident from FAO's misreading of the species and its total reliance on its empirical observation. It was the starting point of FAO's folly and its damage to the forest eco-system in India.

FAO's mandate was to identify tropical woods that are pulpable. In 1952 FAO was impressed by the excellent quality of bamboo and even pointed out its advantage over conifers 'as a three-year culm is ready for pulping' and therefore suggested that far more consideration should be given to it as a source of chemical pulp. Yet it cast doubts on it as a sole source for any paper project (FAO 1952: 110). FAO's scientists had not cared to understand the physiology of this species and its dual system of reproduction. It therefore erroneously noted, 'bamboo gives rise to one or more shoots in the second year and ceases to produce any more culms in the second year' and therefore 'it is not felled before it is two years old' (FAO 1952: 107). Evidently this was an observation made in the catchment area of a PPI unit and that 1952-3 were the years when bamboos, *D. strictus* had flowered gregariously in several states. Perhaps it also explains why in a two-year later report FAO (1954: 64) considered bamboo as a rapidly renewable resource whose yield is 3 to 4 tonnes per ha in India and 7.5 tonnes in East Bengal.

This differential yield calls for an explanation. East Bengal in undivided India used to feed British Paper Mills situated in West Bengal. After the Partition of the country, East Bengal had no mills on its territory and the bamboo undisturbed by mill's heavy demand must have begun to flourish. Bamboos go into hybernation and revive to their old glory once overfelling ceases. This has happened in Karnataka after a ban was imposed by the State Forest Department on felling by PPI units in 1992.

It is important to remember that several forest based PPI units were installed in 1951-2 and earlier ones had increased their production capacity and the royalty rate was by and large Re. 1 per tonne. Reckless harvesting had resulted in gregarious flowering. This appears to be nature's defensive technique of survival. To throw up the seeds, disperse them and perpetuate the species.

International organizations by and large, offer exotics. The Indian government sometimes was a willing collaborator and sometimes succumbed to pressures to open the country's fragile forests. In MP, for instance, under the technical assistance project of the World Bank, the MP Forest Development Corporation was aiming to plant 3,100 ha of clear-felled forest with *Pinus caribaca*. The FAO, SIDA combination estimated 40,000 ha as suitable for this species in Bastar. It also estimated that 50,000 ha in Bastar region has red, loamy soil with gritty sand which ensures good aeration and drainage [FRI Documentation Centre, File 5-12 MP (Part I) not dated]. The same document states that replacement of dense well-grown moist deciduous high quality forests with 100-150 m^3ha stock of tropical pine is questioned by many. Finally, it was put to a halt by a popular resistance movement under the leadership of some scientists and environmental-journalists, media support, giving the movement a further crucial thrust. This episode is important. It led to setting up of the powerful Department of Environment to keep a surveillance on the pliable Forest Department. But, possibly under pressure from the forest-based industry Jagdalpur circle or Bastar as it commonly referred to, was assigned as catchment area to OA. It is a poaching area of PPI. The Departments of Environment and Forests were clubbed together under a single ministry in 1985. A senior official of the Environment Department believed that under the patronage of the incumbent minister, the environment wing in the Joint Ministry would be unable to take a correct, independent position (private communications, March 1994).

SECTION 6

Orient Paper Mills–Amlai Division (OA)

The Amlai unit of OPM is an outgrowth of its parent company in Brajrajnagar, Orissa (OB). It was set up in the late 1950s and its history cannot be understood without referring to the older OB. Although today the Amlai Unit (OA) has outgrown its parent unit in productive capacity by a substantial 9,000 tpa, its umbilical cord has not still been cut. Its figures of production and consumption of raw material, are jointly maintained. Both deal with the government as one. Their problems as well as achievements are remarkably identical and therefore constant reference would be made to OB, the parent unit. This is done in order to avoid repetition.

OB graciously gave the study team the necessary data and arranged for an observational visit to its erstwhile lease area. However OA was on the defensive. Right from 1991 it repeatedly postponed our visit and finally stalled it in February 1994. It is the only unit out of the 16 forest-based paper units in operation right from 1991 to 1994 which the team was not permitted to visit. This report on OA is therefore necessarily based on information from secondary sources.

Choise of Amlai and the Royalty Question

When OB decided to expand the Madhya Pradesh government was anxious to attract industry, it offered bamboo at a throwaway rate of 37 paise per metric tonne in the late 1950s (Agrawal 1985: 72). It is reported that the minister in charge of forests opposed this deal because of the ridiculously low royalty rate for which he was quietly given marching orders by the authorities (Keswani 1981: 103). The protest perhaps explains why the Governor and not the Chief Conservator of Forests signed the agreement. The question arises as to who then could have fixed this ridiculously low royalty rate.

How is Royalty Determined?

Foresters are angry that there are no set rules or regulations for determining the royalty rate. According to K.P. Sagreiya, a former President[3] of FRI and a retired IFS member of MP cadre, the question of computing royalty on pulpwood, whether growing naturally and therefore needing to be protected and tended; or created artificially as in the case of eucalyptus plantations, has to be based on incurred cost along with a reasonable profit to the grower. He writes angrily that 'as soon as the royalty rate is established by the Forest Department (FD) the information is conveyed to the powers-that-be, they immediately start negotiation for a lease for raw material and actually finalize the deal or recommend giving a licence to a particular party and after this, settle the royalty'. The FD is formally consulted but the figure mentioned by it is dubbed too high and a much lower rate is fixed. 'It is under such circumstances that we have bamboo licence at Re. 1 with other conces-sions thrown in.' Sagreiya goes on to say 'there have been worse cases where the lessee has been allowed to extract raw material on the understanding that he will pay any rate fixed by the government. Yet facts do not confirm this assertion. In 1970 it was stated in the legislative assembly that 9 lakh tonnes of raw material had been removed from the forests. Even at a nominal rate of Rs. 10 per tonne the royalty would have worked out to Rs. 90 lakh and the annual interest at 8 per cent to Rs. 7.2 lakh which the government is losing. Rational methods of fixing royalty have been ignored' (Sagreiya 1971).

The Development Council of Paper and Pulp and Allied Industries had arbitrarily recommended that the royalty should be a flat Re. 1 of the selling price of paper. Even if this formula is applied, the royalty collected would have been considerable, rather than the nominal rates at which certain mills have secured under the *modus operandi* mentioned earlier. A working group was set up in the Planning Commission at Sagreiya's recommendation in 1965 to consider the fair royalty payable. Its recommendation was 'The royalty on bamboo, wood or other raw material removed from the government forest should be so fixed that there is a fair margin of profit to the forest department after allowing full allowance on growing it (or replacing it) including development overheads.' This recommendation has not been adopted (Sagreiya 1971: 557-8).

It is time that all concerned realized, says Sagreiya rightly, that forests are national assets which the Forest Department holds in trust for the people as a whole, and as such the government would be fully justified in working out royalty charges on a commercial basis. Sagreiya forgets the injustice done to the forest dwellers by the FD, however, that is not the issue under consideration here.

Why Amlai?

Birlas chose Amlai with care. It was situated in the thickly wooded bamboo forests of Shahdol district. Its neighbouring districts were also densely forested to yield what then seemed as a perennial source of bamboo. A paper mill needs an enormous quantity of water, and the beautiful Sone river flowed through Amlai. Besides, the railway track on the Bilaspur-Katni route of the south-eastern railway passes through Amlai. All that it needed was a railway siding. Amlai is happily situated in a major coal belt and coal is absolutely necessary to generate the enormous quantity of energy needed for production of paper. It acquired a coal mine to supply coal to its power station. Following nationalization of coal mines in 1973, it purchased coal from the government's Western Coalfields (Roy 1982: 65). OA's needs would of cource, be supplemented by the MP State Electricity Board. Whether by chance or design, at that point of time a thermal power station was being built at Chachi within Shahdol district. Limestone was available from nearby Katni and 3,000 acres of prime forests in its vicinity were leased out to OA to provide bamboo. What more could an ambitious paper mill ask for, especially when the obliging government continues to open out 10 other forest circles with the exhaustion of the older catchment area.

Paper Production and the Raw Material Question

OA was initially given a licence to produce 100 tonnes a day, its licensed capacity increasing to 150 tpd in 1959-60. The gestation period is long. In 1965, a new German-made paper machine together with power and some ancillary units went into trial production using imported pulp and paper trimmings. In 1965-6 the pulp section was fully commissioned.

From 1966-7 onwards, there was a progressive improvement in the working of the plant, contributing to its overall production. True, in 1968-9 the unit had to be temporarily closed for a month to overhaul the plant and a fire in the Brok conveyor system delayed production by another eight days. During this period the MP State Electricity Board continued to supply additional power. OA however, moaned that it had to pay 'heavy electrical charges'. Perpetual complaining seems to be OPM's hallmark. In 1975-6 it was acute shortage of water in the Sone river combined with the 'go-slow' of the workers; in 1977-8 there were acts of intimidation, in 1978-9 labour unrest and

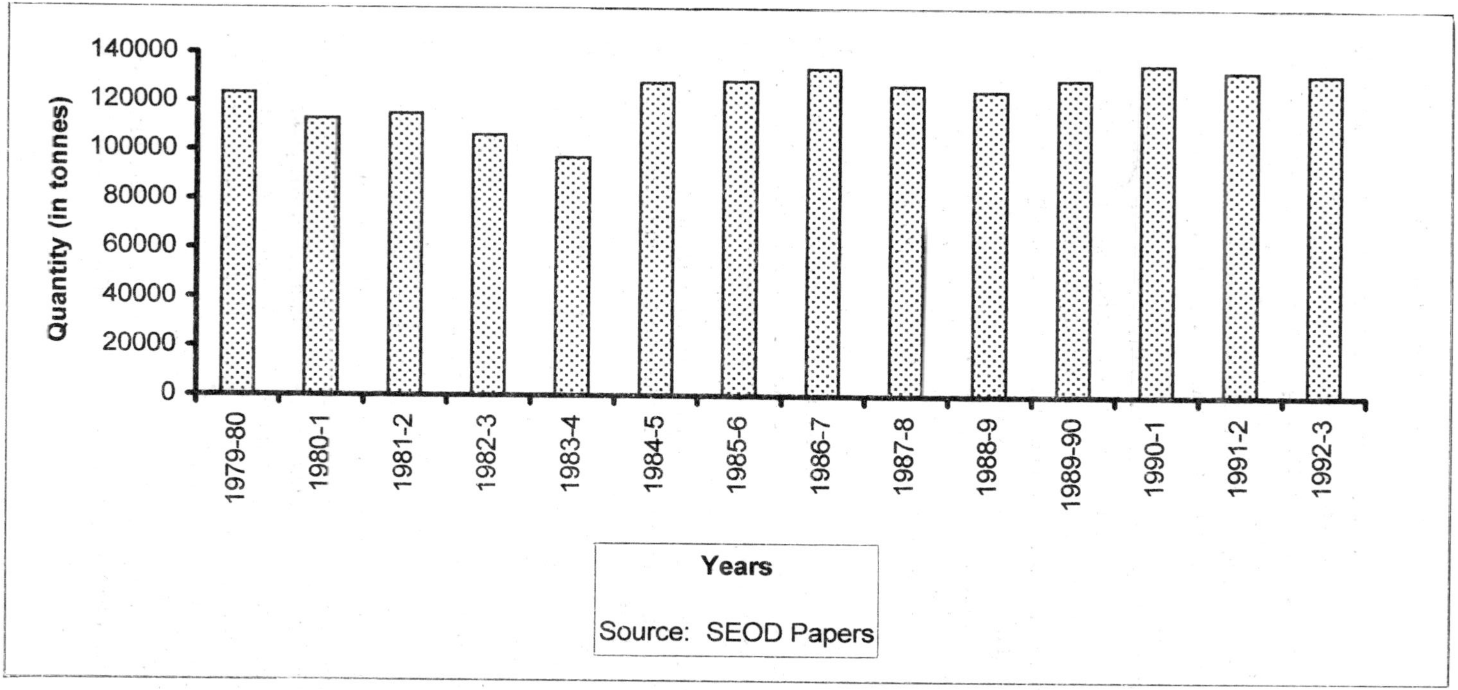

Figure 3.6: Production of Orient Paper Mills Brajrajnagar and Amlai

Note: * Fig. Include production of OPM Brajrajnagar and OPM (Amlai). Separate production figures are given for OPM Brajrajnagar in chapter on Orissa. Production figures for Amlai not made available.

strike, 1980-1 again a water shortage in Sone, 1981-2 a strike in the company; 1983 production was affected due to suction-press roles getting damaged and 'necessitating slow-down of paper machine'. Finally, production began to increase after 1984-5 apparently in anticipation of lifting of the Paper Control Order in 1987, although the market for paper was sluggish and the price of raw material was on the increase.

No attempt is made here to say why the control order was lifted nor why the 1974, 1978 and 1979 orders were brought in, nor the game that Orient and the industry as a whole played with the government, nor even how it made its monies nor how it was finally absolved of its social obligations. The entire drama of the apparent contradictions between the industry and the government have been dealt with in some detail while discussing the Orient Mills at Brajrajnagar. This needs no repetition as the Amlai unit is not independent of its parent company.

In 1968 the Price Control Order had been lifted but the company claims that it made only 'moderate' increase and adjustment in the selling price and 'the working results were satisfactory'. That year it exported Rs.76 lakh worth of paper and claims it could have been more 'but for the difficulties faced in shipment' with the numerous problems in the ports. Meanwhile, to use Roy's apt phrase, the bamboo forest in its first catchment area was rapidly being converted into paper, and bamboo forests in and around Shahdol were becoming depleted. Simultaneously, new forests were being prospected. OA for the first few years used to get its total requirements from the neighbouring forests and the forest of Vindhya Pradesh. Soon, negotiations were on for fixing royalty rates for new forest leases.

Forests and minerals are a major source of wealth for MP. The government had to call a halt to squandering all its forest wealth. On 8 November 1973, an ordinance was promulgated and OA was forbidden to fell or remove bamboos and other pulpable wood like *Salai* from MP forests. OA immediately reacted by filing a writ petition in the MP High Court challenging the ordinance. The MP government showed no signs of being offended. On the contrary, records indicate that OA immediately negotiated with the government, the latter agreeing to sell all the available stock of bamboo to it when it was dwindling (*Stock Exchange Directory*, 1983).

Initially, when production started in 1965, OA's machines were fed only with bamboo, the best possible pulpable raw material from tropical forests. With depletion of bamboos, select local hardwoods were introduced and later hardwoods were even imported from Haryana. Haryana farmers had turned from cereals to eucalyptus farming. So the composition of the raw material changed from 100 per cent bamboo to 25-40 per cent hardwood and about 70 per cent bamboo. Its production process was further modified to take in a variety of mixed wood and bamboo. Even at the risk of repetition, it is important to remember, that bamboo when well nurtured and carefully harvested is an inexhaustible raw material, while hardwoods are far from easily renewable and can even cease to reproduce themselves as is known to have happened in many prime tropical forests in India in the nineteenth century, forests soon being reduced to scrub and even degraded wastelands.

Relationship between Orient and the Government

How OPM flouts government control orders has been dealt in detail in the OB section. Perhaps the government regards OP as its prodigal child and therefore pampers it. Despite nationalization of forests, bamboo continued to be supplied at subsidised rates. OPM's *Annual Report* (March 1977), states that the supply of bamboo from MP was satisfactory. It cannot but be so as that catchment area was widened to cover 10 forest circles. In 1978-9 OA received as much as 1,03,833 tonnes of bamboo (Roy 1982: 58). Simultaneously, the Forest Department set up plantations of eucalyptus as an alternate source for pulp and the MP Forest Department Corporation also set up its plantations in Bilaspur, Bastar and Balaghat specially to supply raw material for the paper industry (Roy 1982: 19, 63). (For details, see section on augmenting resources in Part 2, Section 5.)

In 1985-6 a joint-sector company under the name of MP Plantation was incorporated to ensure long-term supply of bamboo and hardwoods from plantations to be raised by it 'in due course' that is, after allotment of suitable areas in the forests. Thc plan did not materialize as the government's decision was to not allow either the private or joint sector in forest areas (*Stock Exchange Directory*). This is understandable. Dr. A.S. Maslekar a much respected Conservator of Forests and a one-time Managing Director of FDC of Maharashtra, fears that a powerful wood based industry might usurp forest lands. This fear is shared by many foresters in the Ministry of Environment and Forests (personal communication, March 1994). Maslekar suggests that for commercial success what is needed is forest farms, but he is afraid that this is not feasible in wastelands (*Times of India*, 4 April 1987). Maslekar's fears are fully justified, and are probably based on the failure of the Forest Development Corporation of Maharashtra's experiments to raise plantations in the potentially best forest lands which were clear felled for the purpose of raising pulp wood and other plantations at the dictates of the National Commission of Agriculture (vol. IX, 1976). OA had to meet the shortfall of forest raw material from long distances. The company was unhappy for it meant high transport costs. Among OPM's perpetual complaints one that stood out was 'rising cost of raw material' and it claimed that this was one of the grounds for its failure to supply levy paper for vulnerable sections in the educational sector.

Social Commitment vs. *Expansionist Programmes*

OPM had no valid reasons for forgetting its social obligations to the nation. It instead diverted its resources to set up paper units abroad; at first in 1970, it entered into an arrangement with Birla Brothers Pvt. Ltd., for participation in a paper manufacturing project of Nigeria Paper Mills Ltd., at Tebba, Nigeria. Before the end of the decade it granted technical assistance and entered into financial collaboration with Pan African Paper Mills (East Africa) Ltd., for setting up a pulp and paper mill in Kenya.

A research centre was set up in Amlai to provide technical know-how to its joint ventures abroad. The centre also started serious research on various agricultural and forest-based fibrous materials to find suitable pulpable material for the manufacture of paper. With wasteful use of raw material, and the drying up of the nearby resources, the institute also started research on plantations to solve the problem of shortage of raw material. But no attempt was made to introduce raw material saving pulping technology.

Initially OA had two generators of 30 mw each imported from Germany and installed in 1964. Another 240 mw capacity generator was added in two expansion phases. By 1978 the installed capacity was 300 mw. OA also established backward linkages, the first as early as 1965. It set up Hukumchand Jute Mills (HJM)—a misnomer. A major part of the plant was imported from Japan to produce 60 tonnes of caustic soda, 30 tonnes of chlorine (liquid), 15 tonnes of hydrochloric acid and 12 tonnes of sodium thiosulphate per day on a three-shift basis. Most of the products are used for the manufacture of paper. HJM draws all its power from Chachi for electrolysis of brine. But to date, villages around Chachi have not been electrified.

For a commercial firm what matters primarily is profits. OA diversified into cement in 1977-8, while in 1985, land was purchased in Haryana to set up a unit to produce a variety of air pollution-control equipment not necessarily related to pollution created by paper production, and also to produce heat and mass transfer and tele-communication equipment. The company also issued debentures at various points of time starting from 1982-3 (*Stock Exchange Directory*, 1988).

Industrial Relations

A 1982 report states that 400 workers were engaged in the HJM plant. Though industrial regulations demand that a canteen be set up for units of over 250 workers, HJM has not cared to do so. Most of the workers are casual labourers employed through contractors.

In 1978, HJM labour struck work to abolish casual labour. Thirty were made permanent, nine were converted to substitute workers but contract workers, however, remained.

OA has three saw mills to chop wooden logs into smaller pieces and a 20 tonnes a day lime kiln. We do not know if the wood is from local forests and used for pulping (see Ch. 5, SPM—also owned by Birla). Both units are sub-contracted to labour contractors. Roy reported (1982: 69) that in 1982 the OA Trade Union did not take up the case of these workers because small units are not covered by industrial legislation.

Social Consequences of Obsolete Technology

Orient is one of the largest paper mills and the oldest too, but the Amlai division was set up only in 1965 and had the tremendous advantage of getting its raw material at throwaway prices. The resources thus saved ought to have been used for modernizing the technique of pulping, which would have given it a sustained yield of raw material. It still uses the sulphate process, a wasteful one that utilizes only 35-40 per cent of the raw material. The rest is thrown out as waste, polluting the once-beautiful and sacred Sone river on whose banks the industry is situated.

Effluents: Sone—the Dismal and the Damned

Orient Mills draws 65,000 cubic metres of water daily from Sone river out of which nearly 85 per cent flows back as effluents into the river. An 1986 assessment states that Grade I effluent is relatively clean water and is recycled for use. Grade II and Grade III flow back into the river and pollute it. Grade II is a yellowish brown water flowing from paper machines, outflows from the chlorination tower and hypo-section of the bleach plant, supernatant from causticizers plus wash water from chipper house and grit collector. But the worst effluent is Grade III which comes out mainly from the digester house, liquors and overflows from the pulp mill, plus caustic extraction effluent from the bleaching plant. It has a dark brown or brown black colour with a foul odour. The spent chemicals such as bleaching powder and caustic compounds are highly toxic and the liquid from the hardwood or bamboo is not degradable.

During the peak discharge months of the river, viz., during the monsoon, the flow of the Sone is about 2,00,000 cusec; it drops down to 14 to 12 cusec. This strange arithmetic with nature is achieved by Orient, through using primitive engineering techniques. Every year around November OA constructs two earthen bunds or dams up-stream to store fresh river water for its use till the next monsoon arrives. In other words, the entire flow of the river is consumed by OA. During the lean months, nothing but the dark poisonous effluents flow in the downstream till the monsoon torrents wash away the dam or the bunds every year to flush out the area (Keswani 1986: 1301).

The government is not unaware that a paper industry can pollute a river dangerously. It was alarmed that the OPM plant was to be set up at Amlai. But OPM belongs to Birla and Birla has tremendous clout. OPM's licence was not cancelled. Instead, on 28 June 1960, the government set up a committee to find out (a) how to purify the water of the river which was bound to be polluted by OA's effluents once production began, (b) to suggest ways and means to keep a check, so that the polluted water could be properly purified and could not adversely affect the health of the villagers (Keswani). There is no information what precisely the committee suggested to forestall pollution.

OPM's trial production began in 1965. In the same year 165 villages were affected, 30 very severely by the effluent discharge in the river. So the government set up a second committee in October 1965. The committee only recommended that 'the company should make alternate drinking-water arrangements for the affected villages'. OA has not cared, or even if it had, it is important to remember that mere provision of small quantities of drinking water to the villagers would not solve the problem. Clean water is basic to life and living. The eco-system has been badly damaged. People who cross the river barefoot, contract skin diseases. There is no aquatic life in the river, not a single fish can be traced, even 10 km from the factory. The estimated catch prior to setting up of OPM was 277 quintals each season. Fishing which was a traditional occupation of some local communities (vide Section 4) had to be totally given up. Cattle too die in the large numbers after drinking the polluted water and for agriculturists it is almost impossible to farm without cattle. A second pair of bullocks cost money which the people do not have (Keswani: 1302) and even if they raised money to buy a second pair there is till no clear water for them to drink. The report of the Vidhushakh Kharkhana survey 1984 put down the death rate for cattle due to pollution of the river Sone to 79.2 per 1,000 per year.

Watermelon which grew in abundance on the river banks does not grow there any more. The soil on the river banks has been polluted. For the same reason grass does not grow and there is shortage of fodder for the cattle. The air has been polluted by noxious sulphur fumes which has affected the growth of pulses. The only option for the fisher folk and the peasants deprived of their livelihood is to seek employment in OA—but employment opportunities are limited and sometimes dangerous, even life-threatening. OA's safety measures are primitive and inadequate, particularly for those who work in the caustic recovery section where hazardous chemicals like chlorine and sulphuric acid have to be handled by workers who are not provided with basic protective gear like aprons, gloves, goggles, etc. The workers who have suffered from accidents are chided and asked to be more careful (Roy 1981: 56).

The perceptions of the management to the water pollution differ. According to J.M. Bhandari, Vice President 'Our effluent is not toxic, just slightly coloured with a little higher level of BOD. And it is not harmful; villagers' complaints are all merely politically motivated'. Plant Manager P.N. Agrawal admits the problem of pollution 'We are trying to minimize it' he comments a little more modestly. An inspection by MP Pradooshan Nirvan Mandal in June 1983 revealed that a 25 cm thick layer of lime had deposited on the river bed at a distance of 200 m. A huge amount of coal ash was found deposited on the banks of the rivers. In 1986 the Collector of Shahdol issued a notice to the company under Section 133 of CPC for polluting the source of drinking water.

How does OPM get away with such misdeeds? Are there no reactions. Yes, there are, but as one villager puts it: 'How can we fight such a big man? The barking of a little dog cannot affect the clouds.' An attempt was made in 1970 for concerted action. About 100 villagers from 26 villages launched an agitation. Meetings were held, scores of representations and memorandum submitted to district administrators, ministers, management of the factory, etc., highlighting the grievances of the people. They fell on deaf ears and brought no relief. The villagers decided to resort to direct action which included breaching of the two bunds constructed by the factory. But the management of the factory created a rift among the leaders giving them cash and other gifts and the movement died down. When the company anticipates joint struggles, it distributes booklets of hymns (bhajans) or harmoniums, sweets to children and seeds to farmers; 'We distribute all these things as part of our welfare work' says Vice-President Bhandari. 'So many people come to us for donations and various other things and we do help them.' But the company is all-powerful, it does not take kindly to those who do not concede the company's demands. A farmer who rejected OA's demand to make a by-pass for a bund through his land was refused employment even during the months of acute drought (Keswani 1986: 1302-3).

The company may deny that its water pollution treatment is inadequate. But the government agencies do not find it is so. The level of pollution in 1983 and 1986 remained unchanged. The tragedy is that the government cannot compel OA to conform to the standard regulation that has been laid down. An even greater tragedy is that OPM's words and action do not match and it can get away with whatever it does.

MP Pradooshan Nivaran Mandal in 1986 found that only four out of the six aerators were working in the aerator tank. Again, it found the 'treated affluents of grade III also exceeds in three parameters out of its four' confirming that the treatment plant was not working satisfactorily. It estimated that 0.5 to 0.55 tonnes of lime sludge is generated per tonne of paper. The mills produce 200 tonnes of paper per day. The lime sludge discharged into Sone is therefore a substantial 100 tonnes per day.

An all-India government body, The National Environmental

Engineering Research Institute (NEERI), clearly forbids the supernatant from being discharged into the river directly. OA has plainly ignored this and the untreated supernatant continues to be discharged into river Sone.

Even the waste water from the OA pilot plant which generates effluents, containing suspended solids creating high BOD and COD is run untreated into Sone along with domestic waste water. OA treats only limited trade effluents, and even in this case, the treatment is not to the prescribed standard. Whatever arrangements have been made are not properly continued. The government agencies are helpless to compel OA to maintain standard regulations. However, a more recent report by NEERI on the anerobically treated effluent on crops and soil at the end of three years showed that on a coarse textured soil, the yield of salt tolerant crops was successful. But the rise in the soil-exchangeable sodium per cent (ESP) was 30.0. The waste water needs to be diluted by chlorination and hypochlorite waste water, in either 4:1 or 3:1 proportion, even for coarse or medium textured soils. The soil pH increased by 2.3 units with undiluted waste water irrigation and in the case of diluted waste water, by 1.6 and 2.1 units. The experiment shows that there was a linear increase only in the first year from *kharif* (Jul.-Oct.) and *rabi* (Nov.-Feb.) cropping seasons. The organic matter doubled at the end of three years and there was an increase in the total dissolved salt particularly in the post-harvest soils of the *rabi* season. Juwarkar and Subrahmanium (1987: 693-700), warn that if pulp mill waste water as such is applied to soil, the irrigated soil will become alkaline and may create a problem of sodicity. Even diluted waste water will affect heavy-textured soil. Soils of coarse texture will not show any severe deleterious effect but they will certainly need soil reclamation.

This is not surprising. The Indian government is closely linked with the business class and dependent on it, particularly if it happens to be as powerful as Birla is. There is no evidence that the Indian government has exercised any relative independence in this case. On the contrary, it has colluded by supplying bamboo even from the dwindling bamboo resources.

Conclusion

If OA failed to fulfil its social obligation, it also defied the government control orders. Its behaviour is identical with its parent organization and therefore the information available in the OB report is applicable to OA too. This study has only highlighted some aspects of its destruction of the ecosystem, the forests, the river and a section of the local people which it has regarded with such careless unconcern. We have also raised the question of why eucalyptus and why not bamboo, as well as revealed FAO's folly.

Recommendations

Government, FRI scientists and OA should fulfil their social obligations. The government must apply the same stringent rules of effluent treatment on OA as it did on NEPA and OA should carry it out.

Scientists of both Forest Research Institute at Jabbalpur must be permitted to carry out their research dispassionately. Resources of these institutions should not be tied up with production of the larger biomass of pulpable material. The scientists too must be willing to rigorously investigate the problem that is assigned to them and be willing to stand by their findings even if it goes against the prevalent notion (e.g. 'the notion' that eucalyptus spp is a panacea for PPI's rising demand for pulpwood). The scientists should keep in touch with national and international findings on the subject that they are investigating. OA should modernize its pulping technology to save the raw material famine and use a type of raw material that does not pollute the rivers. Its corporate resources should be used to repay its debt to society.

GOI should beware of international organizations and trust the judgement of its own foresters.

NOTES

1. Although the East India Company continued to be formally in charge till 1855, in reality the sovereignty of the British state as ruler of the new conquered territories was established, since Lord North's Regulation Act of 1773 consolidated it with Pitt's Act of 1984. The distinctive economic role of the Company was brought to an end with the ending of its monopoly in 1813.
2. An exception was made in the case of *mahua, kasam, harra, khair, chiula, sagwan, tendu* tin and *shisham*. These are a combination of hardwood timber trees and high-revenue yielding 'Minor Forest Products' trees.
3. One has to remember that the office of President of the Forest Research Institute was the highest in the IFS cadre next only to the Inspector

General of Forests till 1992. Now this position has been replaced by that of the Director, Indian Council of Forest Research and Education at Dehradun.

REFERENCES

Agrawal, Anil 1985, *Citizens Report: State of India's Environment*, Centre for Science and Environment, New Delhi.

Anon. 1951, *Journal of Forestry*, vol. XLIX, no. 8, August.

———, 1951, *Journal of Australian Forestry*, vol. XV, no. 1.

Bombay Stock Exhange Directory on Orient Paper Mills, Bombay.

FAO 1952, Dec. Release I/R/Pres/36 (FRI Documentation Library), Dehradun.

FAO, 1952, Dec. FAO/52/12/8196 (FRI Documentation Library), Dehradun.

FAO, *Sales Publication Bulletin*, no. 9 (FRI Documentation Library), Dehradun.

FAO *Release* (FRI Documentation Library), Dehradun.

FAO 1954, *World Pulp and Paper Resources and Prospects,* New York.

FAO 1980, *Pulp and Paper Making Properties of Fast Growing Plantation Wood Species*, vols. I and II.

FAO 1985, *The Ecological Effects of Eucalyptus*, Rome.

FRI 1985, *File on FAO Documents* (FRI Documentation Library), Dehradun.

GOI 1976, *Report of National Commission of Agriculture*, vol. IX, New Delhi.

ICFRE (n.d.), *Bamboos*, Dehradun.

——— (n.d.), *Eucalyptus*, Dehradun.

Juwarkar, A.S. and P.V.R. Subrahmanium 1987, 'Impact of Pulp and Paper Waste Water on Crop and Soil', *Water Science Technology*, vol. 19, RIO.

Keswani, R. 1986, 'Tragedy of River Sone', *Economic and Political Weekly,* 26 July 1986.

Khan, Waheed 1971, 'Determination of Royalty on Pulpwood', *Indian Forester*, vol. XCVII, no. 9, Dehradun.

Kothare, P.S. 1971, 'Royalty on Pulp Wood', *Indian Forester.*

Lourd, C.E. 1907, *Central India State Gazetteer*, Rewa, Clarendon Press, Lucknow.

Madhya Pradesh Raj Van Vikas Nigam Limited 1981, *Seventh Annual Report and Accounts for 1981-2.*

McClure, F.A. 1966, *The Bamboos: A Fresh Perspective,* Harward University Press, Cambridge, Mass.

McCridle 1907, *Imperial Gazetteer of India, Rewa*, Government Press, Allahabad.

NCER 1960, *Techno Economic Service of Madhya Pradesh*, National Council of Economic Research, New Delhi.

Prasad, Ram (n.d.), 'Bamboo Plantation', *Technical Bulletin*, no. 22, State Forest Research Institute, Jabalpur.

Prasad, Ram 1987, *Eucalyptus Plantations in Madhya Pradesh*, SFRI, Jabalpur.

Roy, Dunnu 1981, *Planning the Environment*, Vidhushak Kharkhana, Annapur.

Saigal, D.D. 1954, 'Forests in Vindhya Pradesh', *Indian Forester*, vol. LXXX, no. 12, FRI, Dehradun.

Sagreiya, K.P. 1971, Letter to the Editor, *Indian Forester*, vol. XCVII, no. 6, Dehradun.

———, 1971, 'Royalty on Pulp Wood', *Indian Forester*, vol. XCVII, no. 9, Dehradun.

———, 1961, 'Fodder and Grazing Position in MP', *Indian Forester*, vol. LXXXVII, no. 8, Dehradun.

SFRI (n.d.), *Bamboo Plantations, Technical Bulletin II*, Bamboo Forest Research Institute, vol. XCVII, no. 6, Jabalpur.

Srivastav, P.N. 1968, *Gazetteer of India, Madhya Pradesh, Jabalpur District*, Bhopal.

———, 1971, *Gazetteer of India, Madhya Pradesh, Indore District*, Bhopal.

———, 1974, *Gazetteer of India, Madhya Pradesh, Damdit District*, Bhopal.

———, 1974, *Gazetteer of India, Madhya Pradesh, Narsipur District*, Bhopal.

———, 1978, *Gazetteer of India, Madhya Pradesh, Bilaspur District*, Bhopal.

———, *Gazetteer of India, Madhya Pradesh, Shahdol District*, Bhopal.

———, *Gazetteer of India, Madhya Pradesh, Surgunja District*, Bhopal.

Tiwari, D.N. 1955, 'Bamboos Do Not Grow', *Indian Forester*, vol. LXXXI, no. 7, Dehradun.

Tomar, M.S. 1963, 'Growth Behaviour and New Culm Production of *D. strictus* in Raipur', *Indian Forester*, vol. LXXXIX, no. 6, Dehradun.

Varma, Rajendra 1973, *Gazetteer of India*, Madhya Pradesh, Raipur district, Bhopal.

———, 1972, *Gazetteer of India, Madhya Pradesh Durg District*, Bhopal.

———, 1973, *Gazetteer of India, Madhya Pradesh, Sagar District*, Bhopal.

Von Monroy, J. 1961, *Report on the Integration of Forests and Forest Industries*, Rome.

CHAPTER 4

Maharashtra

The Bamboo Forests of Chanda and the Growth of BILT

The bamboos swayed, the streams ran full,
when the Tiger on the prowl
came to Chanda to roost.
In Chanda, the Tiger cub grew
but, the keeper held it in check;
so, the bamboos swayed but the river trickled.
Remember,
this was Kipling's country, where the Tiger cub grew,
where the Gonds had been cleared but the
  bamboos swayed.
Years rolled by and then—
In flew a long silent Owl
and it hooted
Toot-Toot, Toot-Toot, Toot-Toot.
Hail wise Owl, said Kipling's son.
Amen, echoed his nephews.
Oh No, It's only a foolish Owl.
Why, its our own old Uloo,
chorused the local beasts.
Beware. Beware the Owl is a bad omen.
Now the hunters might kill us, said the Tiger.
Come, lets put our heads together.
Hurry up wise Jackal, hurry up Grizzly Bear
come fight the big game hunter.
But wait,
can the hunted fight the hunter?
Will the king of the jungle help us
now, that the rule of the jungle is broken?
Perhaps, the hunters would spare us

if we give them a helping hand,
in the land where the bamboos sway.
But, the hunter will tame you,
you, handsome Tiger cub;
and the wild ones confined in cages,
and the little ones trampled.
Alas! Alas! what shall we do?
No—
The king of the jungle won't help you
since you broke the rule of the jungle,
Toot-Toot, Toot-Toot, Toot-Toot,
hooted the silly Uloo.

M.S.

SECTION 1

Social History of Bamboo Forests of Chanda (CP)

The focus of the present study is the relationship between the growth of a large forest-based pulp and paper industrial unit and the bamboo forests in what we may call the central Paper Belt in contemporary India. The social history of Chanda region throws up a few important questions on the paper-forest relationship, especially the bamboo forest relationship. Chanda is a 'tribal belt'. It is the homeland of the Gonds, Bhils, Halbaras, Tolis and others. The forest provided them with all the basic essentials of life and living. Some of them were agriculturists practising slash-and-burn (S & B) agriculture.

Bamboo Forests of Chanda

The first question referes to the prevalence of dense bamboo forests in Chanda. Chanda during the British period was a part of the Central Provinces (CP).

The famed British forester Stebbing observed that the widely prevalent, and what he describes as a pernicious system of S & B agriculture by the Gonds in CP led to the destruction of thick timber forests even before the British entered the area in 1866 (Stebbing 1921: 398). He laments that the fertile pile of burnt-down slash and the remnants of the first crop softened by the rain provided a rich base for bamboo and other thorny bushes to sprout and flourish as thickets . . . 'in which no timber tree can force their way to day-light' (Stebbing 1921: 400).

Stebbing's explanation does not seem to hold true for the thick virgin forests in the eastern section of Chanda which is rich in unharvested bamboo. There is no reason to believe that in this difficult hilly terrain, with impenetrable forests, tribals could have at any point of time carried out S & B agriculture. There is no evidence to support this observation either in relation to the Forest Department, or the pulp and paper industry (PPI) working these forests at any time, even though the latter is perpetually bamboo-hungry.

C.E. Hewetson, Conservator of Forests (CF) in the 1930s gives a more reasonable explanation. According to him, 'forests with heavy rainfall, rich soil and absence of grazing encourage a rank growth of grass and bamboo' which make natural regeneration of teak impossible (Champion and Osmaston 1962: 245). The present PCCF confirms there is no pressure of grazing (personal communication, Feb. 1994).

F.B. Bergbie, Assistant Commissioner of Chanda district at about the same time as Stebbing, reports that Chanda possessed 'vast forests of both high class timber as well as bamboo', and that the size of the forests alone surpassed that of many districts (Bergbie 1910: 1). The Maharashtra *District Gazetteer of Chandrapur District* (1973: 539), confirms that Chanda's forest vegetation is still one of the best in India.

The Question of Slash-and-Burn

The destruction of valuable timber forest by the S & B technique of agriculture was raised by Stebbing. Bergbie did not agree. In fact, he grudgingly reported that the yield of rice was 80-fold and of good quality, and that the cultivated plot which was generally abandoned after one rice crop yielded luxuriant forest growth within a span of 10 years. He also asserted that 'if care is taken to conserve valuable timber it (S & B) is the best method that could be devised in the wilderness'. Many of the hardwoods like teak are resilient to fire. Timber, after all, was the richest source of revenue for the colonial government. Besides revenue, durable timber was needed in large quantities by the British to build infrastructure, i.e. to lay railway lines and erect telegraph poles, for shafts to be sunk in mines, to build offices and residential quarters for its staff in urban areas and in cantonments. In fact, until 1893, indiscriminate felling was allowed, working plans being made only in 1889 (Bergbie: 263). S & B became disreputable only after 1904 in this region. It was the year when 1,307 sq. miles in the north division of Chanda were declared as A class, i.e. permanently Reserved Forests which were henceforth not to be made available for cultivation. It was the administrative interference of the British with the local Madia Gond practice and enclosure of forest as 'Reserved' that upset the land-person ratio in the forests and made S & B destructive to forest growth. As against this, the present Principal Chief Conservator of Forests says that the Chanda forests are very vast and not subject to biotic pressure (Maslekar: personal communication, Feb. 1994).

Bamboo as a Minor Forest Produce (MFP)

It is interesting to note that till 1902-3, bamboo, *D. strictus*, was the most valuable source of revenue for CP. It flowered gregariously that year and died down almost completely (Bergbie 1908: 2) giving rise to the fear that it would not fetch a revenue for a good eight to ten years. That bamboo had fetched the highest revenue in 1900 is not surprising. Apart from being the poor man's timber with multiple uses such as the framework for huts, rafters for roofs, shaft handles for axes and other implements, barge-poles for boats, rods for fishing and fencing material for homes, to name the most important uses, a new urban market had developed. Its lightweight and immense tensile strength reported by both Champion (1954: 41) and V.D. Limale (1952: 558) must have persuaded builders into using bamboo for scaffolding in their massive construction activities in large urban centres.

However, what is puzzling is that bamboo was and continues to be labelled as MF produce, or a non-timber produce although it is used as timber; and again particularly today, when it feeds the pulp and paper industry its major raw material. Technically, timber alone is classed as a major forest product, perhaps because its trunk is sawed at ground level and it is lost to the forest. But not so with bamboo. Technically too, bamboo belongs to the family of grasses (CSIR 1988: 7). But the more plausible reason why bamboo is classed as a MFP is because it 'resurrects' itself after its death through gregarious flowering and seeding. Bamboo has two forms of reproduction, not only the familiar annual vegetative one but also a sexual type which occurs at certain specific intervals. *D. strictus*, the species prevalent in Chanda, appears to have a long 40-year cycle. Available data shows that it had flowered in 1902, then in 1942 and again in 1983. In Bamragad a large stretch flowered in 1994. The physiological cycle was broken. So too, the careful felling practice by BILT, since 1991.

Gregarious flowering in bamboo is an inexplicable phenomenon. Flowers get pollinated and the seeds that are formed, scatter and germinate. The process of growth of a full mature clump from seeds perhaps takes eight to ten years which is much longer than its vegetative method of reproduction. In vegetative or asexual reproduction the rhizome spreads under the ground, and during the monsoon gives rise to several young bamboo shoots, termed as culms, to form the cluster of a bamboo clump. The young and mature culms are easily differentiated and are useful for silvicultural harvesting. A bamboo from vegetative growth is ready to be harvested in just four years.

Unanticipated Contributions of British Rule to the Entry of Paper and Pulp Industry

While building their infrastructure, the British set up a forest department to harvest wood for the industry and laid a railway track in the district in 1877 to transport coal from Warora State Colliery (Bergbie 1909: 251). They then stretched it to Ballarpur. The work on the extended tract was carried out between 1904-9. At Ballarpur a shaft was sunk for new collieries. The railways were meant to carry coal to Bombay and Calcutta, the administrative-cum-power centres. These two factors are mentioned because infrastructure attracts industry.

The coal in this region is not the best quality but nevertheless an important source for generating thermal energy. The paper industry is the largest consumer of energy, second only to aluminium. Ballarpur Paper Industries Limited (BILT) draws its power not only from the nearby Durg grid and the Maharashtra State Electricity Board, but also has its own captive power plant where it uses the local coal. The colonizers had built 2,324 miles of roads between 1924-47 for transporting timber. In 1931 motor lorries were introduced which cut into the bullock-cart transport system of the local Madias, which had till then, provided much needed employment during the slack agricultural season (Champion and Osmaston 1962: 253-5).

Handmade Paper Industry

Chanda has a history of paper making. Paper used to be made by pulping gunny bags, and like all indigenous industries, was snuffed out. It is not known whether gunny bags came from Calcutta with which strong links were maintained through forest contractors, or whether hemp was locally grown. This point is raised because hemp's cousin, *kenaf*, experimented by the West Coast Paper Mills in Karnataka, is now used as a substitute for soft wood in many countries, including USA and Thailand. The study will show later that BILT has interests in the Thai Paper Industry, its contribution being mainly to give technical assistance to the kenaf-based Thai Paper Industry. No attempt is made to cultivate either kenaf, hemp or jute in Chanda although it has black cotton soil on which cotton is grown. Cotton and hemp belong to the same family—*Malvaceae*. Tushu N.K. (IPPTA 1991: 100) refers to use of mesta and hemp as an excellent pulpable material for paper.

Colonial Forest Management

Management of the forests by the colonial foresters was geared almost totally to the exploitation of timber, mainly teak. Silviculture and conservation of the lesser-known species of good timber trees was a part of their enlightened self-interest, so that some day the lesser-known trees would be recognized as valuable. The post-First World War boom and the expanding market led to the creation of new posts like Forest Utilization Officer and Superintendent of Working Plans, introduced in 1925. The first area in which departmental working was introduced was inevitably in Allapalli in south Chanda (Hewetson 1962: 235, 238, 239). During the period of economic depression the government enforced stringent economy in all its activities and the working plan branch was abolished and no new gazetted staff recruited (Champion and Osmaston 1962: 236, 238).

Although there was a gradual transfer of power to the provincial government from 1921 onwards, and more importantly when Congress governments were elected to power in several states including CP in 1937, the Congress ministry did not make any substantial changes in the forest policy (Hewetson 1962: 235). The situation worsened when the Congress ministry resigned in October 1929, when India was unwillingly pulled into the Second World War. The Forest Department then busily engaged itself in supplying timber to the Defence Department not only for use in India but for exports to the War fronts in Egypt and the Middle East. FD's felling activities intensified during 1940-5 as India remained the only source of wood after Malaya and Burma were lost to the Japanese. Wood was largely drawn from the western circle (Hewetson 1962: 256) leading to denudation of the forests of this area. 'The first call was for poles and bamboos.' Then came a demand for 'hutting' timber of any species and 'even the despised salai' was not spared. The other needs were for jetty piles, wharf timber and telegraph poles. *Sal* from other parts of the Central Provinces was used for sleepers. The quantum of forest lost for timber from 1941 to 1946 was 362,606 tonne which measures 18,09,68,860 cu. ft. It is important to remember that bamboos were classified as MFP. The outturn of MFP from 1940-1 to 1944-5 was valued at over Rs. 31 lakh at pre-War prices. This loss, of course, was detrimental to India's ecosystem. For the British government in India, on the other hand, forests brought in a surplus of revenue throughout the period. From CP alone the amount was Rs. 81.7 lakh which jumped to Rs.114 lakh in 1946-7 (Champion and Osmaston 1962: 259). The British knew that they had to quit and had lost interest in Indian forests.

Did the situation improve for the forest dwellers after Independence, were their rights restored to them?

Forest as a Resource Base—Conflicting Claims on Bamboo

The forests of Chanda, once the home of the Gonds, Bhils, Halbaras, Tolis and others, had provided them with all the basic essentials of life and living. The mat and basket makers, 70,000 of them in this region, are today in a precarious situation. The British had introduced a pass system to enter the forests for a fee of Rs. 10. The Maharashtra government in 1960 started a rationing system, where each card-holder was permitted 150 bamboos for Rs. 90. In the early 1980s the price was increased to Rs.175. The government also restricted the right to cut bamboos for repair of huts, allotting only 75 bamboos annually (Agrawal 1985: 76). But now, under the control of FD, they are provided with a modicum of seasonal employment as woodcutters in the felling season and in forest labour cooperatives. Today the forests provide them with some minor forest produce (MFP) such as *avla*, and *tendu* leaves, which they sell to *beedi* labour contractors, as well as some *nistar* bamboo.

The Gonds

The Gonds need a special mention. They are the main woodcutters of Chanda and are listed as 'backward tribals'. It is important to remember that the Raj Gonds ruled Chanda and other parts of the Central Provinces from 1240 to 1751. They left behind impressive monuments, and also built large dams some 30 ft high and 2,600 ft long and practised the ingenious *ghata* irrigation system, impressing the British rulers with their marvellous engineering skills (Bergbie 1909: 171). Their economic base was agricultrue—slash-and-burn agriculture. A century of neglect by the Bhonsle rulers and another century of oppressive colonial forest and agricultural legislation effectively marginalized them to the status of a backward forest tribe, fit only to be woodcutters, earlier for forest contractors, and now for the Government Forest Department and forest-based industries. In other words with the first intervention, the Gonds history was arrested, and with British intervention it actually took a regressive turn. Their status has remained unchanged even for decades after a core industry, The Ballarpur Industries Limited (BILT), entered. It is an irony that BILT stands where Balarsha, the Gond king, proudly built his famous fort and that the Gonds work as casual labourers for BILT. The Gonds were an integral part of the Chanda forest ecosystem. The ratio of small population and extensive forests was so favourable that any clearing made for agriculture quickly recovers as this study will point out later.

Minerals as Resource Base—But no Development

Chanda, with such a wealth of mineral resources, has largely been a host for only extractive industries, for both national and multinational corporations. These are Associated Cement Company (ACC) and Larsen and Toubro which have set up cement factories, while others such as Birla, Small Industries Corporation of Maharashtra (SICOM) and Hyderabad Asbestos were in the process of doing so (CMIE 1985). Interestingly, the raw materials used by BILT are soda bicarb, sand, lime, felspar and borax which are locally available (*Chandrapur Gazetteer*, 1973: 393). So is coal used for BILT's captive power station. It produces its own chlorine. Its caustic soda plant went into production at Ballarpur in 1974.

The one exception to the pure extractive industry is perhaps Maharashtra Electrosmelt which produces Ferro-Manganese steel. The minerals are of course first extracted locally. This study has clubbed two forest-based industries as extractive. The first is Indian Plywood Manufacturing Company (IPMC) which harvests the trees for its trunk in a tropical forest. BILT uses bamboo, soda bicarb, sand, lime, felspar, borax for paper production and coal for its power station. All these are locally available (*Chandrapur Gazetteer*, 1973: 393). Tropical forests are known to have shallow soil which does not permit easy renewal or regeneration. The second is BILT; listed under the core manufacturing sector, it has now turned into a mere extractive industry. True, bamboo is renewable and the corporation had harvested it with great care strictly on the basis of silvicultural rules and regulations. The reason for this will be discussed in the next section.

This will pertain to the question of BILT's failure to its first commitment to produce socially useful paper, meant for the deprived section, in order to generate literacy and universal education. Secondly, it will refer to the industry's responsibility to conserve both forest wealth and the ecosystem, by installing modern wood-sawing but less polluting technology. No other indicators of development are used as it may be unfair to expect a single industrial unit, core or otherwise, to generate much development in a backward area. Its intake of local labour is also poor. CMIE's index of the level of development of Chandrapur is a mere 59 in comparison to the state's 169 [All India 100 (CMIE, 1985)].

SECTION 2

The Forests of Chanda

The reconstruction of the natural and social history in the nineteenth and the first half of the twentieth century in the earlier section was meant to indicate the degree of resilience of the natural forests of Chanda as well as its destruction after the transfer of power in 1947. It is in this setting that one unit of pulp and paper manufacturing company grows and flourishes, in which the state policy plays a major role. It is also necessary to mention that the Indian Plywood Manufacturing Company harvests hardwood trees for their timber.

Under the land-use classification of Chanda today, more than 56 per cent is marked under forests. The percentage of forests to total geographical area in erstwhile Chanda was 71.6 per cent, while the per capita forest area is 1.12 sq. km being the highest in India. According to the latest statistics, the total forest land is 18,420.20 sq. km. Reserve Forests constitute 8,874.73 sq. km, Protected Forests

9,494.10 sq. km, villages cover 51.37 sq. km. Raw material for industry is drawn from the Reserved Forests.

For the purpose of administration, the Indian government placed the subject of forests on the state list and then in 1977 on the Concurrent list, where both the central and state legislations operate.

Chanda forest circles largely coincide with the present districts of Chandrapur and Ghadchiroli which were together the single largest district in Maharashtra till 1982. Undivided Chandrapur was originally a part of the Central Provinces till it was transferred to the bilingual Bombay state in 1956. With the linguistic division of the states, Chandrapur became part of Maharashtra in 1960. In this analysis the more appropriate term, Chanda is used.

In order to understand the complex relationship between the natural resources particularly a renewable bamboo resource and industry, and where the government policies play a major role, the study presents some details regarding the climate, rainfall, drainage, soil structure and topography that influence the growth of forests, together with the available water and mineral resources which make the site particularly attractive for the growth of the industry.

Location and Rainfall

Chanda falls in the tropical area18°-4′ to 20′-50° N and 78°4′ to 80°-55′ E in central India. Its hot wet summer is conducive to a variety of plant growth. The average rainfall in the region is 1,420.2 mm of which 87 per cent occurs during the activity of the south-west monsoon. The rainfall generally increases from the west to the east and with it the tropical dry deciduous forests of the west give away to the moist deciduous forests in the east which abounds in the thickest of bamboo clumps.

Drainage

Chanda is bestowed with an abundant river system, 80 per cent of the river system is perennial. There are five rivers, the Painganga (40 km), the Wardha (160 km), the Wainganga (160 km), the Indravati (120 km) and the Pranahita. Painganga flows into the Wardha river near Ghugus and the combined flow empties into Wainganga near Asti. In addition to these five rivers there are numerous tributaties, most of them originating in the upland and drawing in a larger catchment area.

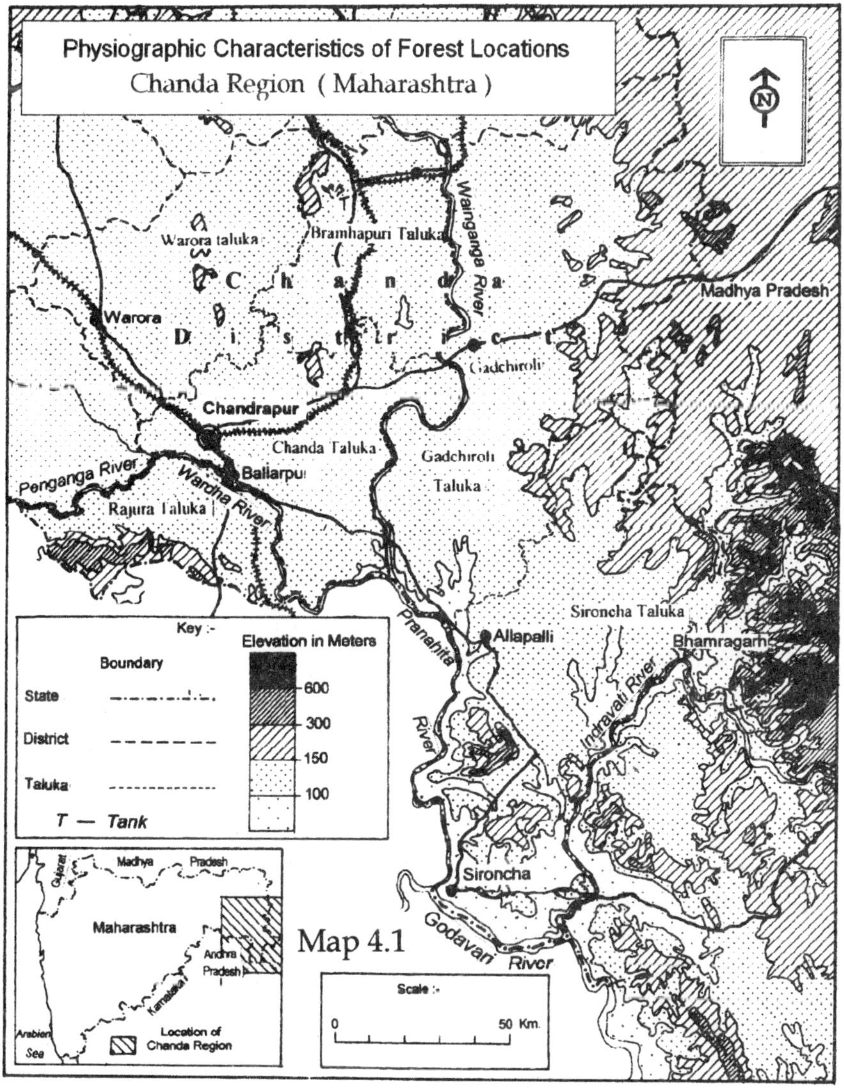

Physiographic Characteristics of Forest Locations
Chanda Region (Maharashtra)
N
Warora taluka
Bramhapuri Taluka
Chanda District
Wainganga River
Madhya Pradesh
Warora
Gadchiroli
Chandrapur
Chanda Taluka
Gadchiroli Taluka
Ballarpur
Penganga River
Wardha River
Rajura Taluka
Sironcha Taluka
Pranahita River
Allapalli
Bhamragarh
Indravati River
Sironcha
Godavari River
Key :-
Boundary
State
District
Taluka
T — Tank
Elevation in Meters
600
300
150
100
Madhya Pradesh
Gujarat
Maharashtra
Andhra Pradesh
Karnataka
Arabian Sea
Location of Chanda Region
Scale :-
0
50 Km.

Map 4.1

Incidentally, BILT draws its water from the river Wardha and its new plant is at Asti.

Soil

The soils can be divided into five physiographical units. Their properties vary according to the geology and topographical situation. The soils on the broken *mesas* and butts on the eastern uplands are predominantly red or dark brown to brown in colour. These are the forest lands where bamboo thrives. The soils on the steep escarpments and hill ranges are shallow to very shallow. The flood plains in the Western uplands have stratified soil of recent origin, whose colour varies from light to dark brown. In the old flood plains in the Wainganga basin soils vary from light sandy to dark brown to grey brown.

Minerals

Chanda has a large reserve of kaolin, barytes, lime, sodium bicarbonate, quartz, sillimanites, felspar, sandstone, mica and coal, borax, talc. Lead, graphite, corundum are also found.

Topography

Topographically the region can be divided into four zones:

1. *The Eastern Uplands*: The physiographic picture of the whole of the eastern and southern portion along the eastern border is made up of hilly and densely forested areas.
2. *The Western Uplands*: The region between Wardha and Wainganga river basins comprises part of Warora, Chandrapur and the major part of the Brahmapuri tahsil where smaller, dense mixed forests with teak and other quality timber thrive.
3. The Wainganga basin comprises the eastern part of Brahmapuri and Chandrapur tahsil and the western part of Gadchiroli tahsil.
4. The plains lie in the western half of Warora, Chandrapur and Rajura tahsil.

Description of the Chanda Forests

The forests of Chanda are of a tropical, dry deciduous type in the west, and moist deciduous in the east. Most of the forests are concentrated in Gadchiroli and Sironcha, and the smaller forests in Rajura, Bramhapuri and Warora regions. The forests are of the mixed type.

The major teak-yielding areas are in the neighbourhood of Chandrapur, Maikanda, Somanpalli and Jimalgatta, with heavy concentration in Allapalli. Allapalli teak is today rated as the best in India second only to the famous Burma teak. It is not surprising that it was in Allapalli in south Chanda that the first departmental working was introduced (Stebbing 1992: 581) and a saw mill erected in its vicinity (Champion and Osmaston 1962: 259).

In the mixed forest grow miscellaneous species which are economically less important—*ain* (*Terminalia tomentosa*), *bija* (*Pterocarpus marsupium*), *tendu* (*Diospyros melanoxylon*), *dhawda* and *awla* (*Embelica officinalis*). Along rivers and canals grow *kusum* (*Schleichera trijerga*) and *anjan* (*Terminalia arjuna*). Besides these common species, other species that are found scattered or in small compact patches are *kekda* (*Garguga pinnata*), *kasai* (*Bridelia squamosa*), *parada* (*Sterresspermum suavebbens*), *karai* (*Saccopetalum tomentosum*), *haldu* (*Adina cordifolia*), *semal* (*Bombax malabaricum*), *salai* (*Bosewellia serrata*) and *tiwas* (*Qugenia dalbergiodes*), depending on the micro-climatic and edaphic factors. *Palas* (*Butea monosperma*) prevails in areas where the moisture conditions are short-lived and the soil gets comparatively hardened. In degraded soil, *khair* (*Accacia catechu*) becomes prominent. *Steculia urensi* is found on low hill tracts. These species are mentioned because many of them are potentially threatened. S.R.D. Guha et al. state that their laboratory results testify to the suitability of many of these species for quality writing and printing and good quality wrapping paper with satisfactory strength. These results have been confirmed by pilot plant tests (Guha et al., 1970: 64). BILT, like other PPI, has progressively shifted away from writing to industrial paper. This was the trend till 1993. BILT in 1994 asked the forest department to supply available hardwoods (see Appendix I).

The understorey of the mixed forest are formed of *ghent* (*Zisiphus jujuba*), *dhaman* (*Grewia tilliiaefolia*), *achar* (*Buchanania latifolia*) and similar plants. Bamboo (*dendrocalamus strictus*) needs special mention as a common associate in the understorey in the teak as well as some of the mixed forests. Bamboo forests in the district occupy an area of 4,258.48 sq. kms.

Bamboo Forests

Guha and Kumar (1977: 212) report that humidity is the determining factor for the growth of bamboo *Dendrocalamus strictus*, the prevalent species. At the same time it is reported that *D. strictus* can thrive in tracts suffering from periodic droughts (Guha, Adarsh Kumar 1977: 212; and Yadav 1963: 177). *D. strictus* flourishes in Chanda in dry deciduous teak forests, with geology of sandstone and a topographically flat plateau, containing lateritic soil.

The quality varies from poor to good. Good quality bamboo is found in remote areas of east Chandrapur division and are also in better areas of Allapalli although it was vigorously suppressed in order to permit teak. It occurs in great abundance in Bhamragad, major parts of west and central Chandrapur divisions and Rajura subdivisions. Bamboos in these two districts have flowered gregariously around 1903, 1943 and again in 1983. In 1993 some areas in Bhamragad have flowered rousing the curiosity of the foresters.

Bamboo (*D. strictus*) continues to be both a minor forest produce as well as the sole pulpable material for BILT, except during 1978-80 when eucalyptus was tried on an experimental basis for pulping and then given up. BILT has been exploiting bamboo at a very rapid rate from the areas earmarked for it by FD. The pristine area of Bhamragad and Sironcha divisions had remained so, owing to its inaccessibility due to a combination of factors such as difficult terrain and almost impenetrable jungles. But how long these would so remain, is a moot question. The question is complicated because of the peculiar gregarious flowering phenomenon of *D. strictus* combined with the highly subsidized price, making bamboo the cheapest pulpable material coveted by every unit of PPI. It is not surprising that BILT officials continue to express deep anxiety, even though *D. strictus* has demonstrated a well-established 40-year cycle and flowered only in 1983. This phenomena is also confirmed by Gupta (1952: 547). The Indian Plywood Manufacturing Company almost had a monopoly control in this area over expensive species like *Tectona grandis*. Its felling rights ceased after the Forests Conservation Act 1980 was passed.

Structure of the Forest Department

The structure of the FD even after the transfer of power continues to be mainly concerned with the management of forest resources. As in the colonial past, so today, the largest area of the forest is marked as Reserve forest and forestry is confounded with excise-generating activity. The Chief Conservator of Forests is in charge of production and not silviculture. The intensity of exploitation has increased with the Government of India's conscious attempt to support the industrial

FOREST DEPARTMENT: CHANDA REGION

Head Office: Nagpur

South Chanda Circle
Chandrapur

- Allapali Division
- Bhamragarh Division
- Sironcha Division
- Central Chanda Division

North Chanda Circle
Chandrapur

- Gadchiroli Division
- Brahmapuri Division
- Wadsa Division
- Chandrapur Division

FOREST DEPARTMENT STRUCTURE

PCCF—Nagpur

CCF Productiuon—Nagpur

FUO Nagpur

CF—SCC Chandrapur
Dy. CF–Allapali Division
Dy. CF–Bhamragarh Division
Dy. CF–Sironcha Division
Dy. CF–Central Chanda Division

CF—NCC Chandrapur
Dy. CF–Gadchiroli Division
Dy. CF–Brahmapuri Division
Dy. CF–Wadsa Division
Dy. CF–Chandrapur Division

1. PCCF–Principal Chief Conservator of Forests
2. CCF–Chief Conservator of Forests
3. Dy. CF–Deputy Conservator of Forests
4. SCC–South Chanda Circle
5. NCC–North Chanda Circle
6. FUO–Forest Utilization Officer

bourgeoisie. In order to bring about rounded development, paper was considered necessary to usher in full literacy and universal education. The National Commission of Agriculture (NCA), looking into this matter, made a policy statement which bluntly states: 'The requirement of the industry should be the major reason for initiating forestry activities' (NCA, 1976, vol. IX, 1976: 4). Pulp and Paper Industry gets primacy. The structure of the FD with the CCF (Production) holding the most important post, next only to the PCCF, is an eloquent testimony to this point.

Working of Bamboo

Bamboo is industrial raw material. Management of forests is carried out according to guidelines laid down in the working plans. Earlier the working plans were revised every 20 years. Now they are to be revised every decade. According to the current plan, the exploitation of bamboo needs to be done in a scientific silvicultural manner. The bamboo areas earmarked for exploitation are broadly divided into three forms of felling series, namely:

(a) The *nistar* bamboo felling series, (b) Commercial bamboo felling series, and (c) Industrial bamboo felling series.

The *nistar* bamboo felling series are marked out for local use. In the commercial bamboo felling series, the outturn is for sale to outsiders. These include poles for building, etc. Most of it is for the urban market, whereas in the industrial felling series certain areas are earmarked only for industrial use in this case only for BILT. There are three main agencies which carry out the working of bamboo—FD, FDCM and BILT.

Here we are concerned only with the industrial bamboo felling series. The areas to be exploited are divided as follows:

BAMBOO FELLING SERIES

	F.S.-1	F.S.-2	F.S.-3	F.S.-4	F.S.-5
1st Year	■ B C	■ B C	■ B C	■ B C	■ B C
2nd Year	A ■ C	A ■ C	A ■ C	A ■ C	A ■ C
3rd Year	A B ■	A B ■	A B ■	A B ■	A B ■
4th Year	■ B C	■ B C	■ B C	■ B C	■ B C

Key: F.S. = Felling Series A, B & C = Coupes ■ = Coupe worked

After categorizing the areas into a number of felling series, each felling series is divided into three coupes, namely A, B and C. The same coupe (A, B or C) from each felling series is worked in a given year. Thus the working of bamboo follows a three-year cycle. Previously a four-year cycle was followed during which an additional coupe D was marked out. This implies that each coupe used to be worked every fourth year.

Reasons for Reduction in the Working Cycle

Officials at the FD were not willing to explain why the working cycles of the coupes have been reduced from the usual four-year to a three-year one. The most logical explanation for this could be sought in the fact that in recent years the increasing demand of bamboo as a raw material for BILT has outrun production, i.e. the yield from the coupes. Thus, in an attempt to augment the raw material, a decision to shorten the working cycle was conceived and implemented. This would give less time for the culms to regenerate. During the team's second visit in February 1994, a District Forest Officer (DFO) explained that reduction of the felling cycle to three years was done to help to thin out the eroded clumps so that they could regenerate. The team's observation in Allapalli did not tally with this rationalized explanation. The present Deputy Inspector General of Forests from the Maharashtra cadre felt that the condition of the bamboo forests was poor because of the three-year felling cycle prescribed to BILT, under its long lease (personal communication, March 1994, New Delhi).

The working of bamboo follows a particular silvicultural rule. It requires that ($n_1 + n_2 + 8$) culms should be left behind and the rest cropped,

where, n_1 = first-year culms (standing shoots)
n_2 = second-year culms
8 = eight mature (three-year and above) bamboo culms evenly distributed in the clump to augment the regenerative process.

It is important to understand the phenotypic differences between first, second and third year, and mature culms for proper implementation of working of bamboo.

First-year bamboo culms are covered with white powder all over them, the sheaths originating at internodes cling tightly on to the culms.

Second-year bamboo culms are still covered with white powder but the sheath usually drops off or loosens up. The third-year bamboo culms have neither a sheath nor the powder covering. The colour of the younger culms is usually tender green whereas mature bamboo culms attain a dark green colour. Culms above the age of four are harvested for pulp and paper making.

In the leased area BILT harvests bamboo under the supervision of the FD. The qualitative assessment of the team in October 1991 of the previous year's harvested area revealed that BILT had followed the silvicultural regulation to the letter, leaving the clumps undamaged. Unfortunately, in 1994 the team's observation of an allotted coupe in Allapalli division showed a marked deterioration in BILT's felling. Our observation was confirmed by the findings of the Working Plan Division (WPD). BILT had been working in an area covering three lakh ha. A new working plan is in the making. Its first step is to assess the status of the forests of the region by the method of sample survey. The tree species in the sampling areas are enumerated. In the bamboo areas where the working plan is being made, particular note is taken of the average size of the culms and the number of culms in each clump. It was found that the number of culms that had been harvested in the clumps far exceeded the permissible number. The girth of the bamboo had shrunk below the average, which is an indication of an abuse of silvicultural practices that tells on the health of the clump. The officer concerned, of the WP Division, stated that the abundance of bamboo in the region has masked the current faulty felling techniques.

True, the observation of the team in 1991 was totally different, but then as the CF, Dr. B. Majumdar, Conservator South Circle observed, it was the FD's close vigilance that had compelled BILT to resort to a perfect harvesting practice. The system of long lease has been terminated. Coupes are now allotted every year. The team is not aware if this has reduced BILT's responsibility.

Ballarpur paper unit had an agreement with Government of Maharshtra for a 40-year lease and another one in 1968 for an extension of the area. This easy accessibility to the raw material has played a major role in the prosperity of BILT. The lease agreement expired on 30 July 1991. Efforts were made for another agreement, but the Government of Maharashtra did not agree.

BILT cleverly linked weighment of bamboo with wages. It hires skilled workers from adjoining states for working of the bamboo from its assigned coupes. It has worked out a rough and ready method to compute wages and to estimate the amount of bamboos cut. Each bamboo is cut into three pieces; each piece is about 2 metres long, 18 to 20 pieces being bundled together. Roughly 6 to 7 culms make one bundle and 60-65 bundles comprise 1 mt. Thus, 360-455 culms make 1 mt. It is no wonder that the State Minister for Forests was grilled for selling bamboo to BILT at an adhoc rate of Rs. 500 per tonne when the market rate was Rs. 1,800 per tonne (*The Times of India*, 16 April 1991). The amount of bamboo harvested from its leased areas is presented in the graphs. BILT also buys from coupes worked by the forest department and those worked by FDCM. In other words it additionally consumes bamboo from outside its allotted lease area which has not been quantified in the graphs.

Observations on Harvesting Bamboo from Leased Area

The data regarding the amount of bamboo exploited by BILT annually from the North Chanda Circle (NCC) and the South Chanda Circle (SCC) according to the Forest Department records, is provided in Figure 4.1. The figures do not tally with data provided by the mills. The discrepancy is brought out in Figure 4.2. From the FD data the general trend that comes to the fore is that there have been more or less similar forms of fluctuations. The available data from 1973 to 1990 shows that there has been an increase and decrease during alternate years respectively, with reference to the previous year. Data prior to 1967 was not made available by FD.

Exploitation of Bamboo from South Chandrapur Circle (SCC)

Thc SCC which comprises central Chandrapur, Bhamragad or Bhamragarh, Sironcha and Allapalli forest divisions, has provided BILT the major component of its raw material requirement.

The initial years, 1968 to 1971, show more or less a similar amount of bamboo exploited. 1972 shows an increase in the yield, whereas in 1973 the amount exploited from these divisions declined. From here on till 1977, it follows an up-and-down trend with the average amount

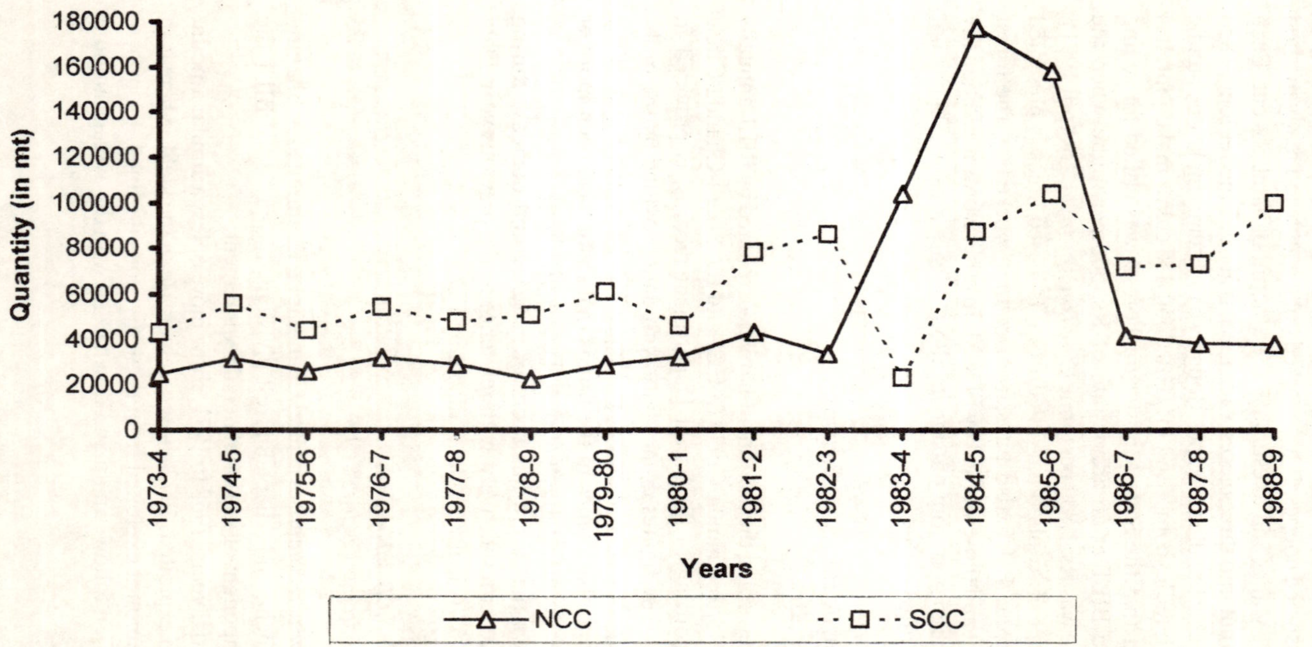

Source: Forest Department, Maharashtra

Figure 4.1: Annual Exploitation of Bamboo by BILT from SCC and NCC (1973-4 to 1988-9)

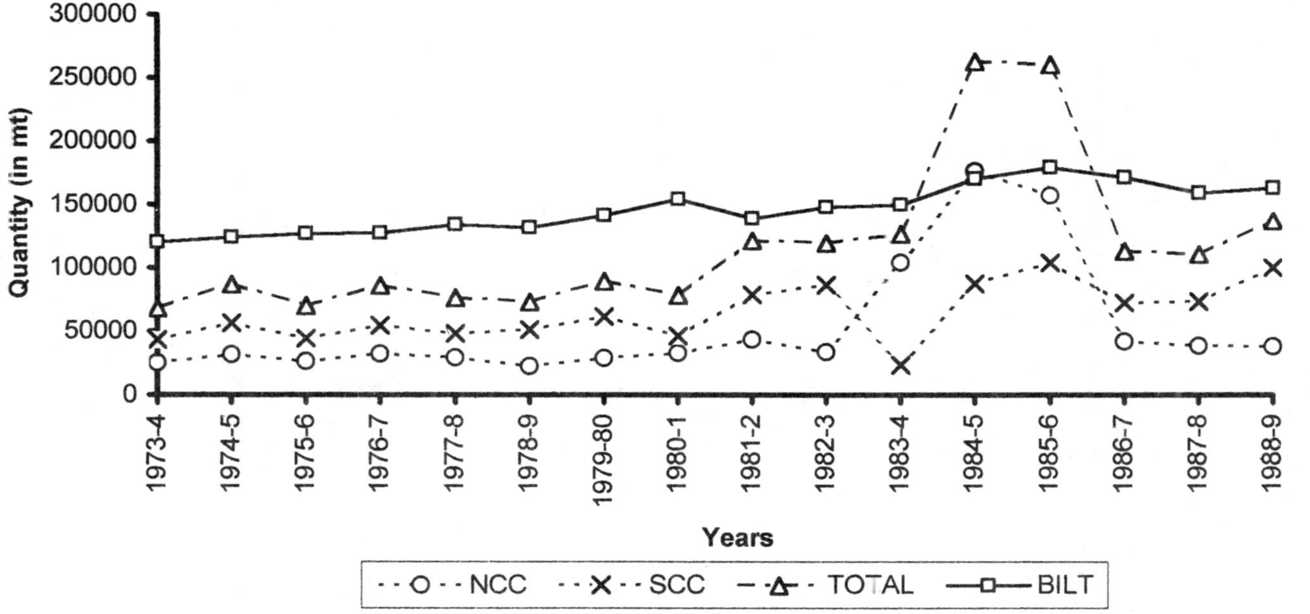

Source: (1) Forest Department, Maharashtra, (2) BILT

Figure 4.2: Differential Information on: Consumption of Bamboo by BILT. Supply from North Chanda Circle (NCC) and South Chanda Circle (SCC)

exploited at about 50,000 mt. From 1977 to 1979 there has been a gradual increase in the amount exploited. The subsequent year shows a lower amount exploited. However, there is a leap from roughly 45,000 mt to about 85,000 mt from 1980 to 1982. We have no information as to whether during these years new alternative areas were allotted for exploitation. The fall in amounts in the subsequent year is inexplicable as the figures drop from 85,000 mt to about 25,000 mt. Probably this, and the boom (85,000 mt in 1984 and 105,000 mt in 1985), could be due to the flowering of bamboo during 1982-4 when all the bamboos had to be harvested and stored, and the clumps died after flowering. 1986 shows a decline to about 72,000 mt where it remains till 1987. This coincides with the non-harvestable regeneration period of the bamboo and the slump in the industry, but for different reasons. In the third year new clumps could account for the steady increase for the next two years. In 1990 the amount exploited shows a dip as low as about 50,000 mt. The paper industry appears to have recovered by 1990, for it continues to use only bamboo as its raw material.

Exploitation from North Chanda Circle (NCC)

The trends in exploitation from North Chanda circle, which includes Gadchiroli, Chandrapur, Wadsa and Brahmapuri forest divisions, show a more or less similar picture as that provided by the SCC data. However, the amount exploited from NCC which has always been less than that from the SCC, shows that during the years 1982 to 1985, a higher amount of exploitable bamboo was harvested for BILT. This was probably due to the gregarious flowering of bamboo in these regions.

The years from 1986 to 1988 correspond with the regeneration period as in 1989 the amount of bamboo harvested fell from 55,000 mt to 45,000 mt. The year 1990 shows a trend which corresponds to that of SCC (see Fig. 4.1).

The amount of utilization of bamboo by BILT and that supplied by North and South Chanda circles do not coincide. See Figures 4.2, 4.3 and 4.4.

Wages to Labourers

Bamboos are worked by both the Forest Department and by BILT, and more recently by FDCM. The working of bamboo requires a

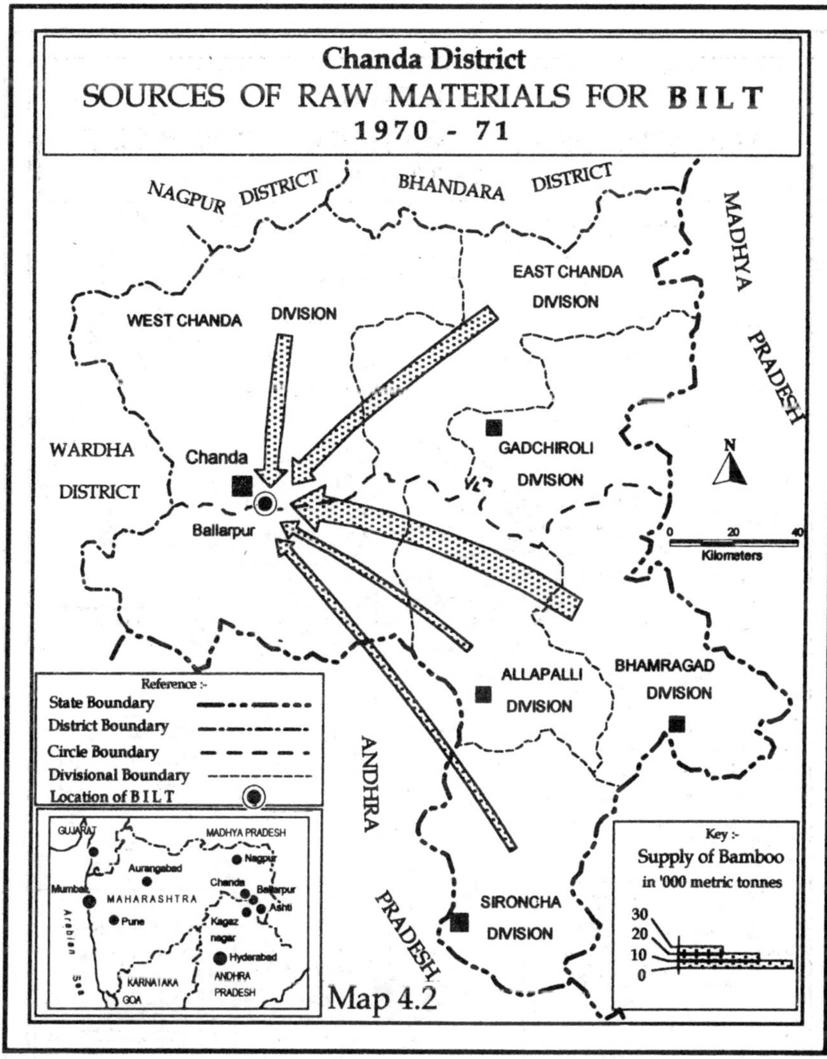
Chanda District
SOURCES OF RAW MATERIALS FOR BILT
1970 - 71
NAGPUR DISTRICT
BHANDARA DISTRICT
MADHYA PRADESH
EAST CHANDA DIVISION
WEST CHANDA DIVISION
GADCHIROLI DIVISION
WARDHA DISTRICT
Chanda
Ballarpur
N
0 20 40
Kilometers
ALLAPALLI DIVISION
BHAMRAGAD DIVISION
ANDHRA PRADESH
SIRONCHA DIVISION
Reference :-
State Boundary
District Boundary
Circle Boundary
Divisional Boundary
Location of BILT
GUJARAT
MADHYA PRADESH
Nagpur
Aurangabad
Mumbai
MAHARASHTRA
Chanda
Ballarpur
Ashti
Pune
Kagaz nagar
Arabian Sea
Hyderabad
KARNATAKA
GOA
ANDHRA PRADESH
Key :-
Supply of Bamboo
in '000 metric tonnes
30
20
10
0
Map 4.2

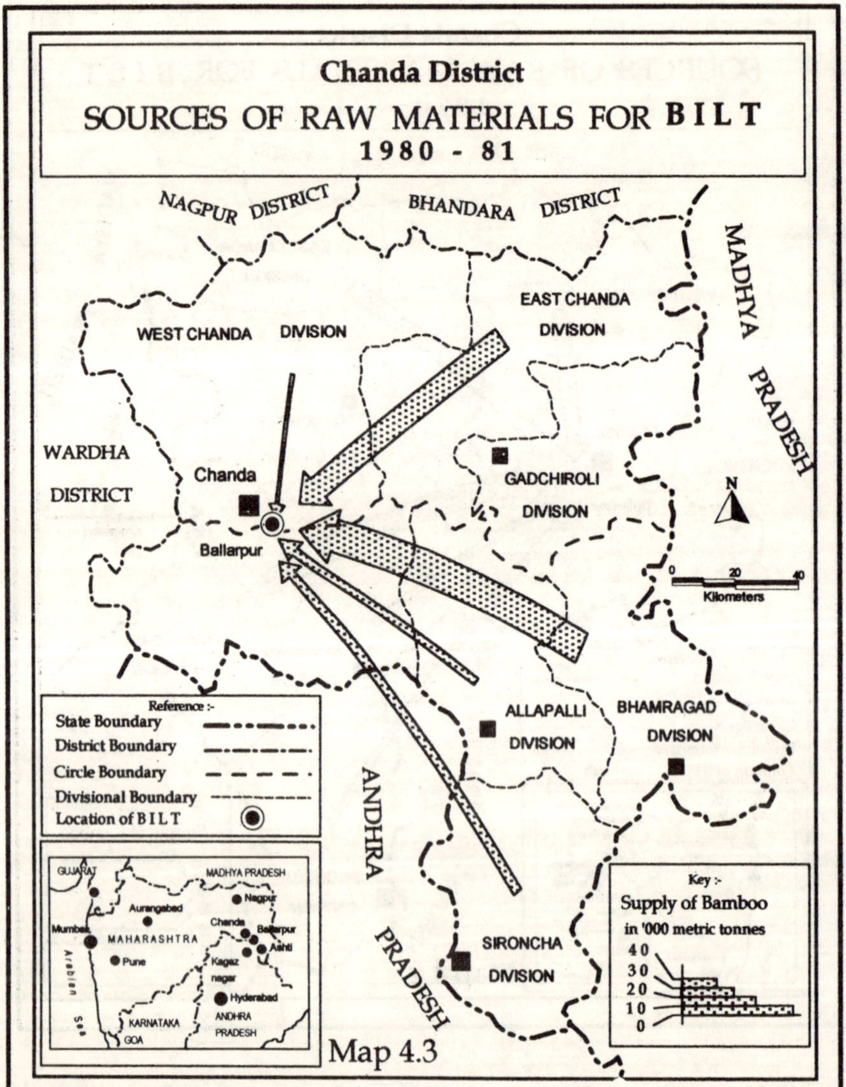

Chanda District
SOURCES OF RAW MATERIALS FOR BILT
1980 - 81
NAGPUR DISTRICT
BHANDARA DISTRICT
MADHYA PRADESH
EAST CHANDA
DIVISION
WEST CHANDA DIVISION
WARDHA
DISTRICT
Chanda
Ballarpur
GADCHIROLI
DIVISION
N
0 20 40
Kilometers
ALLAPALLI
DIVISION
BHAMRAGAD
DIVISION
ANDHRA PRADESH
SIRONCHA
DIVISION
Reference :-
State Boundary
District Boundary
Circle Boundary
Divisional Boundary
Location of BILT
GUJARAT
MADHYA PRADESH
Nagpur
Aurangabad
Chanda
Ballarpur
Mumbai
MAHARASHTRA
Ashti
Arabian Sea
Pune
Kagaz nagar
Hyderabad
KARNATAKA
GOA
ANDHRA PRADESH
Key :-
Supply of Bamboo
in '000 metric tonnes
40
30
20
10
0

Map 4.3

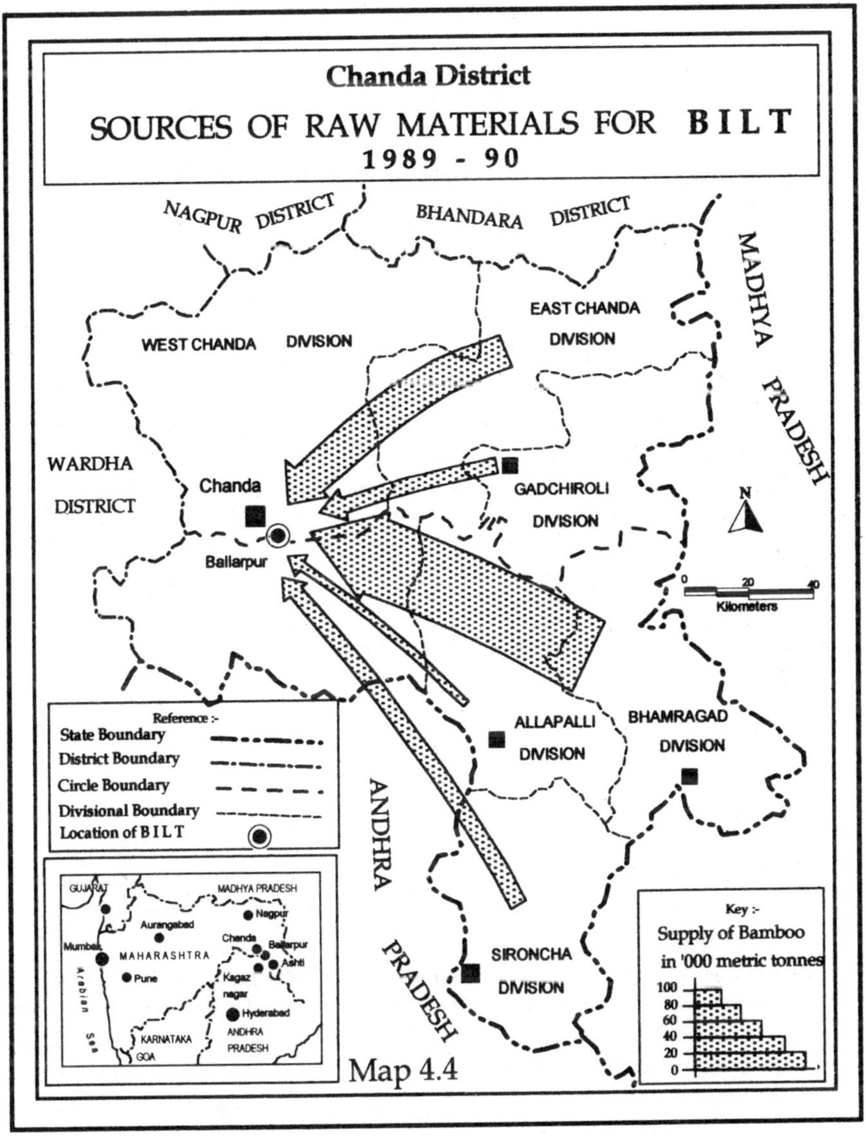
Chanda District
SOURCES OF RAW MATERIALS FOR BILT
1989 - 90
NAGPUR DISTRICT
BHANDARA DISTRICT
MADHYA PRADESH
WEST CHANDA DIVISION
EAST CHANDA DIVISION
WARDHA DISTRICT
Chanda
Ballarpur
GADCHIROLI DIVISION
N
0 20 40
Kilometers
ALLAPALLI DIVISION
BHAMRAGAD DIVISION
ANDHRA PRADESH
SIRONCHA DIVISION
Reference :-
State Boundary
District Boundary
Circle Boundary
Divisional Boundary
Location of BILT
GUJARAT
MADHYA PRADESH
Nagpur
Aurangabad
Mumbai
MAHARASHTRA
Chanda
Ballarpur
Ashti
Pune
Kagaz nagar
Arabian Sea
Hyderabad
KARNATAKA
GOA
ANDHRA PRADESH
Key :-
Supply of Bamboo in '000 metric tonnes
100
80
60
40
20
0

Map 4.4

large labour force. This is not only because the outturn is high but because the technique of harvesting is both delicate and difficult.

There are militant labour organizations in the Chanda Forest area as well as in BILT.

Our concern here is with the differential wages of bamboo workers. Labour engaged by the FD gets higher wages, almost double of that paid by BILT. FD hires local labour-power on wages which work out to approximately Rs. 1.12 per culm. The work involves cutting, dragging to the forest depot, sorting and stacking. Both contractual and bonded labour exist in the region.

FD is sore that in contrast to its own labour BILT usually hires skilled labour from across the border, mainly from Bastar, Bhandara and Mandla. There is no information as to whether this migrant labour is organized or not. But it works at a rate much lower than what FD pays. BILT's mode of payment is different; it is believed that BILT pays around Rs. 3.25 for a bundle made up and cut from six to seven culms which works out to half the rate paid by FD. BILT's workers cut bundles and load them into waiting trucks; this is transported to transit depots located near all-weather roads where the stock is maintained for six to seven months.

According to BILT officials, in 1991 its 10,000 annually employed migrant labourers were paid a piece-wage rate which worked out on an average to Rs. 70 per day. The company provides to-and-fro transport costs, while labourers are permitted to construct their makeshift homes near the site. Medical facilities are provided whenever necessary, but somewhat reluctantly. Malaria, especially life-threatening cerebral malaria, is endemic. No effort is made by BILT to provide prophylactic treatment to migrant labourers who have no resistance to the disease. Every year they die in droves; a few who find their way to a hospital or a social worker are treated. But BILT hesitates to pay even a meagre Rs. 25 per patient (Amte, personal communication, Feb. 1994).

BILT condemns local labour as lazy, given to alcoholism and absenteeism. Dr. Prakash Amte observed that what BILT fails to note is that the Madiya Gonds have physically weakened. Deprived of their rights and old occupation of carting which ceased with the introduction of lorries, their earnings and their food intake have lowered to less than the minimal need. They resist taking up gainful occupation after being repeatedly cheated and underpaid by forest contractors (Amte, personal communication, Feb. 1994).

Outturn of Bamboo

The Forest Department records show that the outturn of bamboo has increased from 60,000 mt in 1968 to 90,000 mt in 1990, from the same bamboo area. This seems questionable but the FD's official explains this away on two grounds, viz., in terms of the changing working plans from a 4-year to a 3-year cycle and secondly in terms of its vegetative growth.

(1) According to an FD official the working plans are not worked every year but only when the working plan is written or revised. It used to be once in every 20 years, but it is now planned to be revised every decade.
(2) The forester pointed out that bamboo has a vegetative form of propagation. Proper thinning or working of culms encourages the growth and spread of the rhizome under the soil throwing up fresh culms every year. Thus bamboo under the same area increases.

Royalty Rate

It is generally believed that royalty rates are fixed by the state government as a contract with the PPI unit. However, the former President of the Forest Research Institute (FRI) K.P. Sagreiya, IFS, retd. (1971: 558) complained that royalty is negotiated and the FD only formally consulted. Rates are subject to revision every five years. The earliest rate was fixed by the government in 1947 at Rs. 5 per tonne of industrial bamboo. Over the years it slowly increased and was pegged at Rs. 230 per mt in 1982-3. BILT promptly challenged this revised rate in the Bombay High Court. An interim order was passed by the court directing BILT to pay Rs. 115 per tonne of industrial bamboo. This rate applies even today, bamboo still remaining a highly subsidized commodity.

It is not surprising that a former IFS officer K.P. Rustogi (1977: 235-8) says that 'forests leased out for a long period at nominal rates deprive the forest department of potential income which would not have occurred if the forests had not been leased out'. Some minor changes have been introduced. BILT has now to pay an additional sum which works out to a mere 5 per cent of the royalty as a Forest Department (FD) tax, to the FD for its maintenance and renewal of forests. But out of this fund, first Rs.15.49 lakh and then Rs.17.9

lakh, were sanctioned for maintenance of naturally regenerated bamboo seedlings for the years 1984-6 (G.R. no. FDM-1804-7234 F, dated 20 June 1984).

In contrast to this dismal royalty rate for the PPI, the rates for commercial bamboo (1991) today vary from Rs. 7 to 25 per culm depending on quality. Even if BILT pays Rs. 7 per culm which is the lower end of the scale, it would still have to pay: Rs. 7 x 6 (least number of culms required to make one bundle) x 60 (least number of bundles required to make 1 mt) = Rs. 7 x 360 = Rs. 2,520 per mt of bamboo.

However even after giving BILT the concession for the labour charges it pays during the working of bamboo, and other forms of direct and indirect benefits that it is supposed to provide in the local area, the state and the country, the astonishingly low price of Rs.115 per mt cannot be justified. Yet it continues to be a part of government policy to provide forest produce at a highly subsidized rate to a core industry.

FD officials are divided on the question of the leasing agreement. One of them strongly felt that the leasing agreement with BILT should be drastically reconsidered. During the team's first assessment in 1991 he even stated that instead of one long 40-year lease, the area should be subdivided into various categories; a part of it could be leased out for instance, on a 20-year long lease while the other areas could be leased for shorter periods, say annually for 5 years or at the most for 10 years. In our second assessment in February 1994 we found that for the last couple of years, FD has annually allotted coupes to BILT. Even before the annual allotment came into force, we feared the possibility that the shorter the lease period, the higher would be the risk of over-exploitation by the industry. Our fears were confirmed in the Allapalli division, on the basis of our own observation and by the findings of the working plan officer. Given the present business culture, the industry has flouted the silvicultural demands of harvesting which it had rigorously followed till 1990 and of which the Forest Department was once proud. But today, it has evidently yielded to the temptation to exploit more than the permissible level since the chances of that specific area being re-allotted for this purpose during subsequent years is minimized. In 1994 we can link the unnatural gregarious flowering of bamboo with the faulty felling which BILT has resorted to since 1991.

This drawback of over-exploitation of areas under short lease, the

FD official explained, could be overcome if only one agency instead of the present system of three agencies, viz., FD, FDCM and BILT are authorized to work bamboo in the earmarked areas. And that agency, he firmly says, should be the FD itself. The risk of over-exploitation would then cease. A bamboo coupe could be worked and the bamboo brought to the forest depot. From there, it is the responsibility of the industry (BILT), to buy and transport the material to the production site.

Regarding the Forest Department lease paid by BILT, he feels that the amount should be increased and a major component should be ploughed back into the relevant areas. This would improve the economic situation and lead to a better management of the area.

Contrary to these suggestions, however, another forest official feels that when an industry is given a long lease, it is in its own interest to keep a check on over-exploitation. This was his opinion in 1991. At that point of time he did not concur with the opinion that only FD should be authorized to work bamboo in the area on two grounds. First, the quality of work done by the BILT was far superior to that done by the FD. However, he refused to divulge the degree of difference. He also said that financial constraints and lack of flexibility, and the large labour force skilled in the art of cropping which the industry acquires by paying a specific amount in advance to workers living across the border, is not possible for the government FD. He therefore believes that the suggestion offered by the other officials is not feasible. What he, however, fails to note was that less than a decade ago BILT used to pay the unorganized forest labour a mere 30 P per bundle, which used to work out to Rs. 6 to 7 per day for seven hours of hard labour. The present wages, under pressure from Naxalite groups, has been raised to Rs. 4.20 per bundle, which on an average works out to Rs. 80 to Rs. 100 per day. BILT is under pressure to raise it further (Amte, personal communication).

Programmes to Augment Forest Raw Material for PPI

PPI was nurtured and supported by the Government of India. Wasteful use of forest raw material through obsolete technology was leading to a raw material famine. Both the Ministry of Agriculture and Irrigation (forestry was under this ministry) and FAO from 1965 onwards had been working out various programmes which included pre-investment surveys to assess the forest stock. Finally, as this study has stated in an earlier chapter, the National Commission of Agriculture (NCA) was instituted in 1972 to look into this matter; it made certain recommendations which were accepted and became part of government policy. The *NCA* (1976), vol. IX recommended three sources for raw materials: *Social Forestry* (SF) : 63, 64; *Farm Forestry* (*FF*): 21, 22; and one or more Forest Development Corporations (FDC) for each state to support PPI (p. 71). The last one, viz., setting up FDC, was a drastic policy recommendation. Under it 2 per cent of the most potential forest areas is to be clear-felled to make way for fast-growing plantations. NCA is clear that 'the requirement of a specific industry should be the major reason for initiating forestry activity (p. 69), which means a close link between the two, and that 'the forest departments should tend plantations which industries would pay for by giving an economic price for raw material' (p. 74).

Maharashtra was the first to set up a FDC as early as 1974 which coincided with the shrill cry of 'raw material famine' by the PPI elsewhere. In fact, Chandrapur forests were so rich in forest cover with no biotic pressure, that in 1959 the Chief Conservator of Forests, S.S. Buit, conceived the notion of raising economically useful native species in degraded areas in order to increase the financial resources of the Forest Department which is perpetually starved of funds. The Government of India accepted Buit's model and a Forest Development Board was set up to look into this question. It was a forerunner of the notion of FDC. But it was distorted by the FAO expert Von Monroy's pernicious suggestion of cultivating fast-growing species to feed the ever-hungry forest-based industries. Dutifully, *Eucalyptus tereticornis* was planted, but when it mismatched with the site and failed, the foresters resisted the Government of India's imposition, and refused to cultivate eucalyptus in the Forest Development Corporation's (FDCM) area, turning to teak—the native of Chandrapur. In Maharashtra FDCM set up four units, three of them in Chanda. It is important to note that its first activity was to clear fell 2 per cent of the most potential forest area. This was a diaster. The Government of Maharashtra's Forest Department's *Annual Administrative Report* (1981-2: 7) reports that FDCM's working plan was suspended due to sparse growth. The second crop—the plantation crop, was a failure. Technically monoculture plantation cannot be regarded as part of natural forests.

In 1991 FDCM officials were evasive, uncooperative and unwilling to let the research team make even a qualitative assessment of their plantation.

A warning had been given by a sensitive retired forester, Pratap Singh (1971: 87-8). He pointed to the failure in Germany as well as the experiment in Himachal Pradesh to replace old stock by fast-growing species and chided the foresters for taking up this 'foolish' assignment. But in the early 1980s, in Chanda, the foresters fought against eucalyptus and instead, started plantations of native teak on clear felled areas marked for FDCM. The team's observation in Lohana showed that other local species including bamboo have sprung up and are growing well, as second storeys. Teak, after all, is a deciduous species which permits good undergrowth though it is initially suppressed to permit the economically profitable teak to grow. Here nature is given an opportunity to establish its own eco-balance. Social Forestry was introduced in Maharashtra under the tutelage of USAID as late as in 1981. However, in Chanda there was no ready evidence of farm forestry. It is important to note that NCA (1976: 109) recommended that long-term loans for all the above three programmes be made available by the Agricultural Refinance Corporation, now known as National Bank for Agriculture and Rural Development (NABARD). NABARD has honoured this. Thapar, the Managing Director of BILT, is NABARD's client for another of his large paper unit in Haryana where it has a captive eucalyptus plantation. The cost of regenerating bamboo is extremely high (NCA 1972: 861). Its edaphic conditions should be suitable for eucalyptus.

Forest resources of Chanda are primarily managed by the FD, which even supervises BILT's harvesting in lease areas. Bamboo resources of Chanda seem unlimited today and incredibly low-priced. There is thus no reason why BILT would buy expensive hardwood from either social or farm forestry programmes.

Pratap Singh's warning was correct, and the government had to retract in 1986. It called a halt to clear felling by FDCM (see letter no. FDC/1086/CR-75/F dated 16 September 1986). The clear felled trees by FDCM's language as 'crop one', had no takers. It could not be sold even as firewood. This destroys the popular myth that villagers fell trees for firewood. In fact the highest bidder for a selected species of the clear felled wood was BILT. This was a surprise. The senior-most Conservators of Forests debated hard and finally accepted the deal (Maslekar, PCCF, private communication, Feb. 1994). 'Crop one' was eventually purchased by contractors to be converted into coal. As the quality was poor for industrial consumption, no offers were received in the Pranhita, Jimulgutta and Sironcha project divisions where it was auctioned as wood-stock. FDCM desperately took to social forestry and distributed seedlings and bamboo cuts to villagers. It used the Government of Maharashtra's Employment Guarantee Scheme funds to pay wages to forest labour (FDCM 1980-1: 6).

The most potential forest lands had earlier failed to yield the desired industrial timber. Yet, in 1987, the government took a decision that FDCM should undertake afforestation in wasteland development project areas (Govt. Res.1086/303/CR-C/FT dated 3 June 1987). NABARD went ahead and sanctioned resources for its 30,000 ha project. The next year FDCM again asked NABARD for resources for another 10,000 ha plantation which was not sanctioned (FDCM, *16th Annual Report*, 1988-9: 11). What matters is that wasteland development is not reforestation by FDCM in the right sense of the word, but only a plantation programme to provide industrial raw material. The study makes this observation because NCA (1972: 69) is clear that production forestry is business. FDCM was therefore made into a corporation, with all but two shares to be held by CCF and the Director of FDCM, while the rest were held by the Government of Maharashtra (FDCM, *Status Paper*, April 1991). The objectives of the FDCM are (a) to get a high return for its investment and (b) production of wood for industry particularly PPI 'with its mounting demand for manufacture of newsprint, paper and packaging material' (NCA 1972: 68-9). Yet FDCM's available annual reports from 1980/1 to 1989/90 (barring two years 1986/7 and 1988/9) report losses (FDCM, *14th and 16th Annual Reports*, pp. 3 and 7). This was largely due to the extension of FDCM in Maharashtra.

In 1994 the situation appears to have radically changed for FDCM. It has created a market for itself by cultivation of teak, bamboo and *khair*. Young teak that is cut in thinning operations is sold as poles, *khair* is auctioned as firewood to middlemen (Sharma, personal communication, Feb. 1994). Now that BILT pulps every type of wood it perhaps buys firewood from the middlemen too.

It was as late as 1988 that the Government of Maharashtra directed FDCM to take up exploitation in the bamboo felling series in Gadchiroli and Wadsa divisions on an Agency basis (G.R. no. FND/1288/pk/121/88/F-1 dated 30 December 1988.

The *16th Annual Report* of FDCM (1988-9: 16) shows its turnover as follows:

Particulars of goods	Units	Quantity	Amount in Rs.
Bamboo			
(a) In number	Nos.	22,80,779	60,80,138
(b) In bundles	Bundles	25,76,094	2,75,20,149
(c) In metric tonnes	mt	39,25,339	4,51,414

Since each *nistar* and commercial bamboo costs Rs. 7.25, the price in the last column does not tally with the quantity sold, which is unbusinesslike and goes against the first principle of FDCM. The taker of the bamboos could only be BILT. It is also interesting that funds for harvesting were provided by the Government of Maharashtra (FDCM 1988-9: 12). The assumption that all the bamboo goes to BILT on a heavily subsidized royalty rate is based on the statement on the same page (FDCM's *16th Annual Report*: 16), under the head 'List of Contingent Liability' is mentioned 'Interest on Deposit by BILT in view of Supreme Court's interim order in respect of the Sale Tax matter of Rs. 2.85 lakhs'. In February 1994, FDCM had a stock of 6 lakh bundles of bamboo, but BILT was unwilling to bid for it in the auction that year, although invited to do so. This is because it can buy bamboo from the FD at half the price of the *nistar* bamboo made available to the *Bardoods* and villagers (Sharma, personal communication, February 1994. The question of differential price was raised in the Vidhan Sabha, saying that the price BILT pays is only Rs. 500 per tonne since 1991, whereas the market rate is Rs.1,800 per tonne.

That bamboo is sold at a heavily subsidized price is not surprising as it is government's policy to support the industry. FDCM is not self-sufficient. In 1988-9 it received Rs. 290 lakh from the FAO's World Food Programme for its 60,000 fuel wood/bamboo plantations (FDCM 1988-9: 7, 11). Today the revenue from the sale of its products is ploughed back into its plantations, unlike the parent FD whose revenue is clubbed with the State's Revenue Department (Sharma, personal communication, Feb. 1994). In other words, an easy supply of bamboo will grow and keep pace, even though BILT has almost ceased to manufacture its committed quota of levy paper, i.e. cultural paper for writing and printing. The Government of India has listed paper as one of the basic industries of importance in its industrial policy resolutions (1948). It was also one of the items covered by the Essential Commodities Act 1955. BILT's position was further secured in 1980 a ban being imposed on the export of bamboo outside Maharashtra.

BILT is the only forest-based PPI in Maharashtra which is permitted to work its leased areas.

SECTION 3

THE BALLARPUR PAPER INDUSTRY LIMITED (BILT)

World War II had created a scarcity of paper. Prices were soaring. This study has shown in Section 2, that Chanda had not only bamboo but all the necessary natural resources for setting up a profitable paper manufacturing unit. True, there were a series of control orders relating to price distribution and usage of paper during 1944-6, but these pertained to wartime economy measures and to government purchases. 1945 therefore seemed a propitious year for the managing agency of Bedi & Company to float a paper mill. It chose Ballarpur as its site because of plentiful supply of water and coal, besides bamboo of course. Bedi & Company entered into a deal with the Madhya Pradesh Government to finance its unit. At this time, Ballarpur town in Chanda was a part of Madhya Pradesh.

The initial cost was estimated at Rs. 200 lakh for the plant and Rs.15 lakh for working capital. M.S. Ramachandran, Secretary, Government of Madhya Pradesh, was appointed to prepare the project on behalf of the government. Initially the government sanctioned Rs. 90 lakh for the project, against mortgage of assets. The loans were to be repaid in 15 years, beginning 3 years after commencement of production. And, as was the practice the world over, a long 40-year leasing agreement was entered into in 1947 for the supply of bamboo. Meanwhile, the company had appointed a committee headed by A.D. Shroff to study the feasibility of the project. It had talks with the Conservator of Forests for a regular supply of bamboo, and with the Great Indian Peninsular Railway for the construction of a railway siding. In spite of all this, the company failed to take off, and Bedi & Company resigned on 8 November 1949.

Thapar's Ballarpur Industry

In April 1951 Messrs Karamchand Thapar and Bros. Ltd., of Calcutta was appointed the new Managing Agency (MA) for the next 20 years. Collection of bamboo from the leased area began in the same year. The managing agency undertook the responsibility of constructing the plant in Ballarpur. The pulp mill started its production in June 1953 at the rate of only 20 tonnes per day (Ahuja: 22). The company's 'product line' was stationery. This was in keeping with the government's policy. The Industrial Policy Resolution of 6 April 1948, recognized the paper and newsprint industry as an important basic industry for which planning and regulation of production, distribution and price by the central government was regarded as necessary in the national interest. It was the heyday of 'Socialist planning' with a deep concern for rounded development which colonial rule had obstructed. It was for the same reason that the planners considered it necessary to give a boost to the up-and-coming Indian industrialists to contribute to the development of the nation.

It is not clear why Thapar and not Dalmia, Birla, Bagoria, the Mysore State, or J.K. Singhania, who had all struggled against the British bourgeoisie, and set up their paper units as early as in the thirties had not entered Ballarpur. Thapar had a paper unit in Haryana in the mid-1940s. But it is clear that Chanda had strong links with Calcutta through its coal trade and powerful forest contractors, and Thapar's business was Calcutta-based. It was also clear that in 1950, with the withdrawal of the paper and price control order, this nascent industry had to give an assurance that it would not introduce an upward revision of prices without prior consultations with the government. Nationalist sentiments were still high in the early 1950s. But once production began to rise and the nationalist concern weakened, the paper industry revised prices repeatedly, in fact five times by 1958, till the government imposed control over paper prices in January 1960 at the recommendation of the Tariff Commission. This lasted only till May 1968 (Dagli 1979: 269). It would not be out of place to take a brief look at the history of the growth of the Thapar's the third largest business house in India by 1993 and the largest holder of the paper industry.

History of the Thapars

The Thapars are late entrants into India's industrial history. They built their industrial empire through acquisition of sick units (Ahmed 1991: 52). The Thapars are Hindu Khatris from Ludhiana, whose business, however, was concentrated in the east. R.K. Hazari (1967: 48) traces their business history to the early years of this century when they were wholesale merchants and coal agents in Calcutta. Karamchand Thapar consolidated his business in the 1920s and early 1930s. He acquired some collieries, perhaps a small paper mill, and also set up a few sugar mills. It was a decade that marked a leap forward for the

Thapars. After the Second World War, with financial assistance from the Nizam of Hyderabad, he bought over Greaves Cotton, set up the General Insurance Company and subsequently secured the Managing Agency of Ballarpur Paper. During that decade he acquired a few more sugar and coal companies, as well as the Madhusudan Cotton Mill in Bombay, which he later sold to Bangur, whose style of growth and entry into the paper scenario through acquisition of sick units is almost identical.

In 1962, Karamchand Thapar's business instincts outdid his family traditions and he 'bequeathed' to his third son, Lalit Mohan, the Chair of Ballarpur Industries. 32-year old Lalit Mohan had an acute business sense which was further strengthened by training in a US-based business school. Lalit Mohan (LM) followed his father's trail and between 1990-2 bought over four sick units in 1992 reaching the premier position of a paper king. These units consisted of one in nearby Ashti, two in Orissa and one in Andhra. Unlike the other business houses, Thapars have remained undivided.

Rapid Growth of BILT

During the financial year 1951-2, the company entered into an agreement with a German contractor for the supply of pulp machinery and two paper machines.

Production doubled from 6,000-7,000 mt per annum in 1953 to 13,006 mt in 1963. The company's growth was particularly rapid during 1962-8 when demand for paper began to increase. This was also due to the commissioning of two new paper machines in the early 1960s: an MG machine with a capacity of 22.5 mt of industrial paper in 1963, and an MF or cultural paper machine in 1964 with a capacity of 56 mt. During the accounting year 1964-5, production increased by 60 per cent and sales by 65 per cent. The company began making new products, laminated boards, Hi-Fi maplitho, opaque printing papers and acecoated grease proof paper. The fourth machine was commissioned during 1965-6. Sizeable quantities of paper were exported to Egypt, Kenya and Iran (*Bombay Stock Exchange Directory*, 1988).

Merger of Ballarpur and Yamunanagar Units

Bamboo was readily available at Ballarpur, but production of paper meant enormous finance. Shri Gopal Mills at Yamunanagar in Haryana had large resources but its raw material was drying up. This led to the merger of the two units in 1968 (*Financial Express*, 9 Feb.

1981). This study is concerned only with the Ballarpur unit. In 1969-70 new value-added varieties of paper, Three Aces, Sunlit Bond, etc., were introduced in the market. In 1970 the managing agency system was abolished by the government and Lalit Mohan Thapar was appointed the Managing Director.

In 1972 BILT felt an acute shortage of water, despite the fact that Chanda (see Section 1) is rich in water resources with five perennial rivers. A temporary dam had to be built across Wardha river to feed BILT. A radial well had to be dug in 1982. One cannot overlook the fact that the Indian paper industry uses water wastefully, says Neelam Singh (1990: 428). She points out that steam, water, chemicals and raw material consumption ratio are high. Steam consumption per tonne of paper is 11-14 tonne in India compared to 6.5-8.5 tonnes abroad. Digesters, evaporators and machines account for 82 per cent of the steam consumed in a paper mill. Continuous digesters have lower steam power and chemical consumption, and give higher pulp yields compared to batch digesters which are used in India by the private sector units.

Growth of Output of Industrial Papers and the Paper Control Order, 1974

In 1972 another new MG machine for production of industrial paper was commissioned. The company was tending to move in this direction which was highly profitable. This was true of the industry, the large forest-based mills as a whole causing acute shortage in cultural paper. Publishers and the educational sector complained of both 'non-availability' and exorbitant prices. Also the Directorate General of Supplies and Disposal (DGSD), reported the marked reluctance of the paper mills to honour their commitment (Rao 1989: 153-4). Under the Essential Commodities Act, the Production Control Order was introduced in 1974 and an informal price control was agreed upon. Strangely the order was effective only for white printing and not for industrial papers which are high value-added. Precious bamboo, sold at a subsidized price, was not put to a strictly socially useful purpose by BILT. The government's method of coping with the situation was to introduce a whole range of concessions to induce new entrepreneurs to set up small paper units with the use of non-forest based material. The response was tremendous. The small units aggregate production grew from a mere 13.34 per cent in 1977 to 43.02 per cent in 1985 (*Business Standard*, 12 June 1987). The government too, in anticipation of the reluctance and the oligopolic movers of the big paper mills, had set up the Hindustan Paper Corporation (HPC) in 1970 to produce paper as a mass consumption item. These two moves created contradictions between HPC and the small paper mills.

BILT on a Forward March

Meanwhile, in 1973, BILT in its forward march set up a caustic soda and chlorine plant in Ballarpur. It also entered into an agreement with Mysore Industrial Development Corporation for another caustic soda chlorine plant in Karwar, which at the Planning Commission's suggestion was transferred to BILT in 1976. BILT's salt works in Gokarn went into production in 1978 (*Stock Exchange Directory*, 1985). It also harvested salt from Singhash in the Kathiawar region. In other words, it was establishing profitable backward linkages with raw material units needed for paper production in Ballarpur as well as in its other unit in Haryana and rayon production in Andhra Pradesh. But Thapar did not care to conserve the forest and other natural resources. BILT continued to use the obsolete sulphite chemical method in Ballarpur for pulping, although at this point of time in the 1970s, a rework sulphite process which was less polluting, and a raw material saving technology had come into use in the West as a result of pressure by environmentalists. But change in pulping technology also implied fresh investment. Instead, BILT invested in other projects—on the one hand in a joint venture abroad, first with Indonesia to set up Bhakti Coated Mills in 1976-7; much later, in 1983-4, they invested in Thailand to provide machinery and technical assistance to set up a *kenaf*-based Phoenix Pulp and Paper Co. Ltd. (*Stock Exchange Directory*). Incidentally at this point this study will ignore BILT's diversification into various industries unconnected with PPI. At Ballarpur it added yet another MG machine to produce high value-added industrial paper, which however, basically used the same wasteful chemical method. Effluents from high value-added paper contain more chemical pollutants than ordinary writing and printing paper.

BILT's Reaction to the Regulation of Production Order 1979

In 1978 the government revoked the 1974 and 1975 price and production control orders and reintroduced the Regulation of Production Order 1978, but amended it in 1979. This referred to production of

30 per cent cultural paper and was applicable to large units using forest raw material. Cultural paper included cream wove laid paper, coloured printing paper, duplicating paper, off-set or duplicating and typing paper. BILT filed a suit against the price control order and asked for a revision of prices. It held that the prices set by the government were uneconomical and that the government was overestimating the demand for cultural paper. The company complained that it had suffered a loss in the production of cultural paper; its officials mentioned under-utilization of capacity as a result of this order. The government responded to the mill's demand by raising the price of white printing paper. This is evident from Table 4.1.

In spite of the rise in prices the mills did not keep to their commitment (Rao 1989: 163). This seems like a deliberate protest against government's order to supply white paper. The government in 1979 raised the price of cream wove from Rs. 3,000 to Rs. 3,785 per mt, which was being marked at an ex-factory price of Rs. 5,400 per mt (Rao 1989). In 1981, production of paper by BILT reached a record figure, capacity utilization reaching 99 per cent (*Stock Exchange Directory*). In 1983 the company added still another new MG machine and with it the installed capacity was increased still further by 11,500 tonne per annum of lower grades, industrial grades and specialities (*Stock Exchange Directory*). In 1983, in response to the Government of India's demand for environmental control, an effluent treatment plant was set up. The factory continued to use the old highly polluting sulphate process, installing a caustic soda plant the same year for this purpose, and even added a coal fired boiler to generate more electricity. The low-grade local coal with high ash content continued to add to both

TABLE 4.1: CHANGING PATTERN OF PRICE OF WHITE PRINTING PAPER

Date of Revision of Prices of White Printing Paper	Price of White Printing Paper ex-factory rate/Rs. per tonne
30-6-1979	Rs. 3,000
29-11-1980	Rs. 3,500
24-12-1981	Rs. 4,200
11-4-1983	Rs. 5,400
11-5-1984	Rs. 6,400
16-1-1986	Rs. 7,200

Source: Y.A. Rao 1989, *The Paper Industry in India, Status and Prospects.*

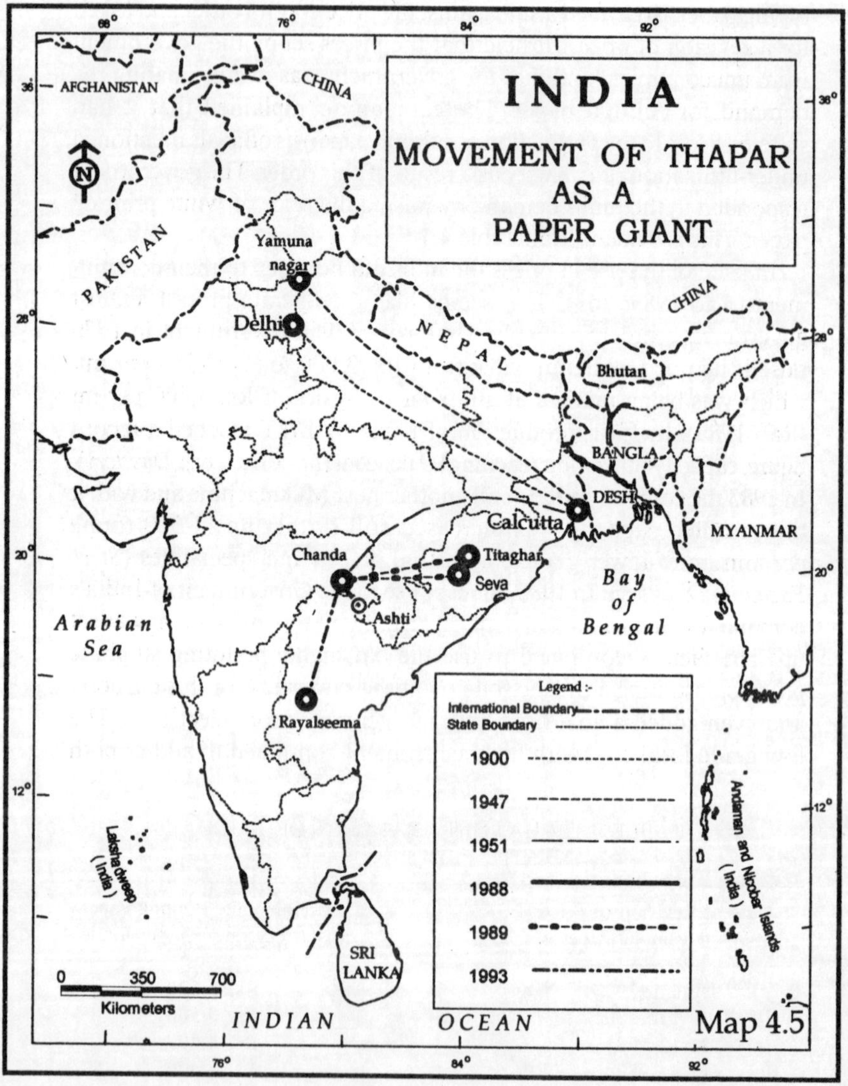
INDIA
MOVEMENT OF THAPAR AS A PAPER GIANT
AFGHANISTAN
CHINA
PAKISTAN
NEPAL
CHINA
Bhutan
BANGLA
DESH
MYANMAR
Yamuna nagar
Delhi
Calcutta
Chanda
Titaghar
Seva
Ashti
Rayalseema
Arabian Sea
Bay of Bengal
INDIAN OCEAN
SRI LANKA
Lakshadweep (India)
Andaman and Nicobar Islands (India)
Legend :-
International Boundary
State Boundary
1900
1947
1951
1988
1989
1993
0
350
700
Kilometers
68°
76°
84°
92°
36°
28°
20°
12°

Map 4.5

water and atmospheric pollution, all of which shows BILT's unconcern. The cement industry intensified the pollution. Incidentally, cancer cases had jumped from 42 in 1962 to 132 in 1988 in Chandrapur district (GOM 1961: 1, 1989: 2).

The industry's concerns were basically different. BILT was out to make quality paper and be in the lead over others in the race to do so.

Resurrection of BILT

The *Bombay Stock Exchange Directory* (1988), notes that there was an under-utilization of BILT's units in 1983-4. The industry was hit by a recession which was of its own making. The recession, however, affected the large wood-based units, and lasted till 1988. A number of small units had risen in response to government's call and in 1986 they produced 60 per cent of the total output. There was a glut in the wood-based paper industry. But the government's budgetary policies in 1987, with the imposition of 25 per cent excise duty on imported paper on which the small units depended, and the extension of the MODVAT scheme to the paper industry, upset the small paper mills and pushed them into the red. A senior FAO official claimed the responsibility of advising GOI to put an end to small scale units (personal communication, October 1994). To add to this, the government repealed the Paper (Control) Order 1979 and the Paper (Regulation of Production) Order 1978, and January 1987, and decided that the supply of white printing paper to the educational sector henceforth be made by the public sector Hindustan Paper Corporation. BILT and some other large wood-based units began to improve.

The End of Social Obligation

Barring expensive value-added papers such as Bond, by 1989-90, Ballarpur had stopped manufacturing stationery—a broad, important component of cultural paper as is evident from Table 4.2.

In March 1980 an agreement was reached that the industry would honour its commitment to supply white printing paper to the educational sector (Rao, 1989: 159). But BILT went on to manufacture products which yielded higher profits,[1] as is evident in Table 4.3.

As already stated, the PPI was protesting against the Control Acts by lowering its production. As a standby, perhaps, BILT ventured into other profitable fields like oil in 1978-9. It diversified into several

TABLE 4.2: BILT'S SALE OF STATIONERY, 1980-1 TO 1989-90

Year	Quantity in mt
1980-1	418.87
1981-2	441.76
1982-3	436.97
1983-4	542.04
1984-5	611.74
1985-6	664.25
1986-7	648.60
1987-8	573.57
1988-9	192.93 (9 months)
1989-90	35.88

Source: SEOD 1990.

TABLE 4.3: CONSUMPTION OF BAMBOO AND RISING VALUE OF BILT'S PAPER PRODUCTS, 1973-4 TO 1989-90

Year	Quantity in mt	Value in Rs.
1973-4	1,19,978	1,74,63,212
1974-5	1,23,867	2,06,50,557
1975-6	1,27,096	2,20,15,998
1976-7	1,27,364	2,14,59,387
1977-8	1,34,049	2,30,10,388
1978-9	1,31,943	2,49,69,651
1979-80	1,41,531	2,90,19,126
1980-1	1,54,739	3,45,40,460
1981-2	1,38,996	3,60,75,121
1982-3	1,48,089	4,01,89,003
1983-4	1,50,227	5,54,95,106
1984-5	1,71,065	6,66,65,379
1985-6	1,79,720	7,33,24,487
1986-7	1,71,932	7,79,93,928
1987-8	1,59,747	8,14,07,696
1988-9	1,63,657	8,36,53,904

Source: Schedule XII – BILT Balance Sheets from 1973-4 to 1989-90.

fields, which varied from shipping to shoes, even entering into ventures abroad—palm oil in Malaysia, glass in Nigeria, paper in Indonesia. These foreign ventures, however, failed (Ahmed 1991: 50-5). BILT justified its diversification in terms of profitability and attracting of investors. But our concern here is different, viz., BILT's relations with forests and the mode of use of this precious raw material on which the whole eco-system and life itself depends. BILT is reluctant to invest in modern pulping technology; it continues the use of obsolete sulphate and does not use even the modified sulphite process that became available in the mid-1970s; it does not use the oxygen delignification process and the eco-friendly thermo-mechanical pulping process available in India since 1981. BILT had the advantage of not having to add hardwood and the process would yield over 95 per cent of cellulosic raw material instead of the obsolete sulphate process which yields only 35-40 per cent. Incidentally, this study has shown elsewhere that only the public sector units have introduced the raw material saving the thermo-chemi-mechanical process of pulping. The chemical component was necessary only because of the eucalyptus component in the raw material.

Modernization of Ballarpur Unit vs. *Shree Gopal Mills*

The large forest-based paper units began to forge ahead after the Paper Control Order was lifted. The Ballarpur unit only looks modern; it is automated, but in the strict sense it is not modernized. Its pulping technology which would save the forest raw material has not been upgraded as a whole, although the government had made provisions for financial institutions to provide assistance for modernization at concessional rates (PTI, *Economic Service*, 15 September 1989). It is reported to have been done elsewhere but not in Ballarpur. In the words of its senior Vice-President, Hemant Luthera, a Rs.120 crore, 2-year modernization effort is on at its Shree Gopal Mills in Haryana. This effort is more comprehensive as it involves replacement of all the four paper machines with an integrated one which will result in considerable raw material conservation (Saikia, 1991). This is understandable. After all, in Ballarpur there is an ample supply of bamboo over which BILT has almost a monopoly right. In 1981 there was a ban on interstate sale of bamboo (*Business India*, 18-31 March 1991). But in Haryana there is an acute shortage of raw material (see Figure 4.3, and Table 4.4 which give a picture of the types of raw material used in the two units taken together, as the balance sheets do not separate the data). Bamboo is believed to be consumed exclusively in the Ballarpur unit. BILT has now set up a farm nursery, 'Tree tech' in Yamunanagar in collaboration with Plantex Australia to get an assured supply of raw material. Its first batch of 3,00,000 eucalyptus

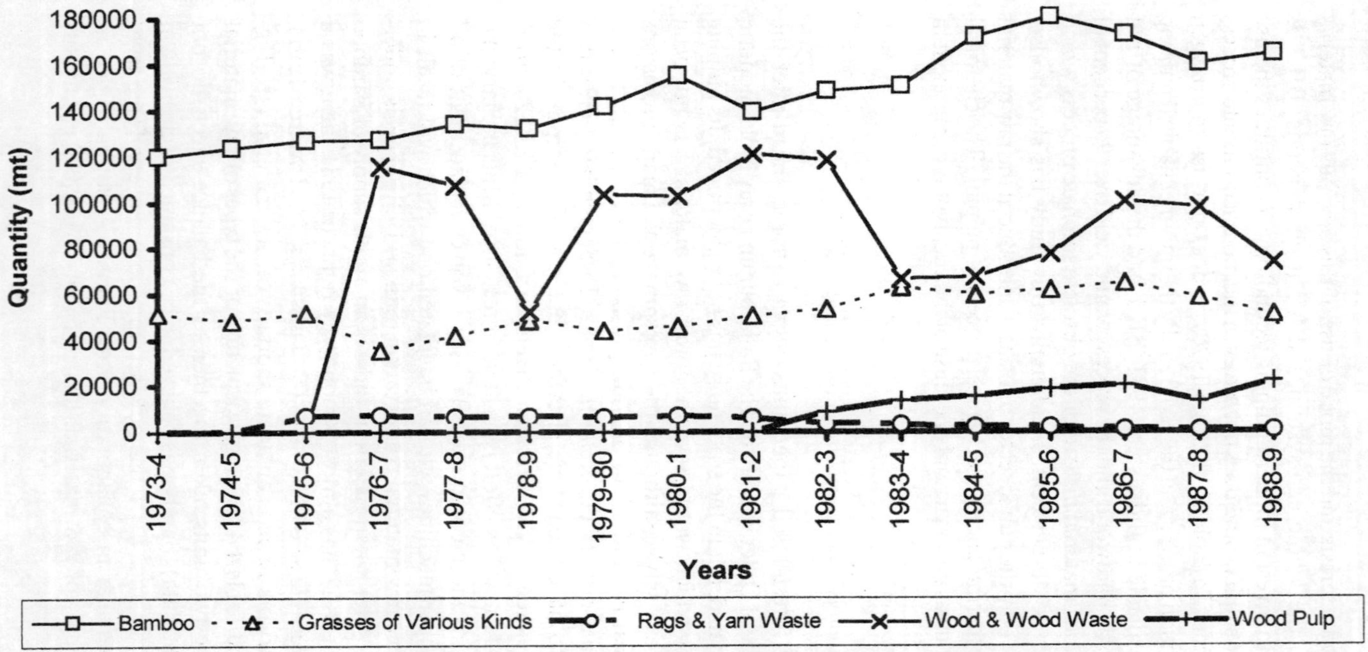

Source: BILT Annual Reports 1973-4 to 1988-9

Figure 4.3: Figure showing the Quantity of Raw Material used by BILT

TABLE 4.4: CHANGES IN UTILIZATION OF VARIOUS RAW MATERIALS BY BILT, 1973-4 TO 1988-9

Year	% Bamboo	% Grasses of various kinds	% Rags and yarn waste	% Wood and wood waste	% Wood pulp	Total	Total qty of raw materials (mt)
1973-4	70.0	30.0	—	—	—	100.0	1,71,511
1974-5	72.0	28.0	—	—	—	100.0	1,72,146
1975-6	68.0	28.0	4.0	—	—	100.0	1,85,711
1976-7	44.6	12.4	2.5	40.5	—	100.0	2,85,825
1977-8	46.2	14.4	2.3	37.1	—	100.0	2,89,887
1978-9	55.1	20.2	2.8	21.9	—	100.0	2,39,464
1979-80	47.9	14.9	2.1	35.1	—	100.0	2,95,387
1980-1	49.8	14.8	2.3	33.1	—	100.0	3,10,627
1981-2	43.9	15.9	2.0	38.2	—	100.0	3,16,512
1982-3	44.6	16.1	1.1	35.6	2.6	100.0	3,32,152
1983-4	50.6	21.3	1.1	22.4	4.6	100.0	2,96,717
1984-5	54.1	18.9	0.7	21.4	4.9	100.0	3,15,998
1985-6	52.9	18.2	0.6	22.8	5.5	100.0	3,39,620
1986-7	48.0	18.0	0.3	28.0	5.7	100.0	3,58,474
1987-8	48.4	17.7	0.2	29.7	4.1	100.0	3,30,125
1988-9	52.5	16.4	0.3	23.7	7.1	100.0	3,11,446

Source: Calculated from BILT, *Annual Reports*, 1973-4 to 1988-9.

Notes: (1) Bamboo is used in BILT's factory at Ballarpur while the others are used at Shree Gopal Mills, Haryana.

(2) Figures for 1951-2 to 1972-3 are not available.

saplings will go to the farmers under its farm forestry (FF) project which is supported by NABARD (BILT, 1991).

The industry as a whole was looking forward to a profitable future. In 1990 in the Ballarpur unit, the paper machines were renovated, and another pulping machine, though obsolete, added to cope with the anticipated rising demand. Also a 'new' chlorine dioxide bleaching process was introduced to produce internationally comparable paper of 88 per cent whiteness. By the end of 1993, the company stopped dreaming about making such high grade speciality papers.

BILT as a Paper Giant

In order to increase its share in the paper market in 1990 BILT acquired assets of the Titaghur Mills no. 3 and Sewa. 'This' said the Chairperson, L.M. Thapar, 'will give an additional capacity at roughly half the cost of setting up an entirely new operation' [but one-third according to others (*Facts and Figures*, April 1993: 15)], and it will also mean an easy access to raw material, something that is impossible with new units and closeness to the eastern market. Both the units are in Orissa. Titaghur would add 22,000 mt and Sewa 45,000 mt capacity and were expected to start production in mid-1993 and 1994 respectively (BILT *AR* 1990-1: 3, 9). In Ashti in Ghadchiroli district that is within Chanda, a 33,000 mt per annum unit was taken over to produce speciality paper. It is in reality a sack kraft mill, and was taken over in anticipation of a large export market (*AR* 1991: 8) tough long sisal fibre can be used for this purpose, as well as for plaster boards and bitumenized paper (Lock 1974, 1962: 284). L.M. Thapar bided his time and acquired the Rayalseema Paper Mills by 1992 although the BIFR had awarded it to G.P. Goenka (Ahmed, 1991: 52). Thapar generally structured deals in such a way that the money that BILT would have paid in taxes was used to rehabilitate the acquired unit. BILT's growth is not through modernization, but through rationalization of marketing and shrewd acquisition of units which total to five paper and one rayon grade pulp unit. Thapar was extremely proud of his premier position as the leader of the paper industry up to mid-1993.

The scene now rapidly changes because big business, which of course included the Thapars (India's third largest industrial house in 1993), was chafing under what it unfortunately should have seen four and half decades later, as the 'Licence Raj', the Raj which in reality was protective and condoning its misdeeds.

Big business in India wanted liberalization and got it. It is no coincidence that when the tentacles of capitalism were penetrating and gripping the world in its stranglehold, Congress policies in India were also changing. Thapar's loyalty, like that of some other big business houses, was moving away from the Congress towards the Bharatiya Janata Party (BJP). But liberalization has its drawbacks; these have now terrified the PPI, particularly since mid-1993 when the recession in the paper industry of Europe and US deepened. Rather belatedly, BILT has discovered that India is not only a convenient dumping ground for its unsold paper products but also a most congenial region for multinationals (MNCs) to set up their paper units. BILT, which was proud of being the largest manufacturer of paper and the sole manufacturer of certain value added speciality papers, confesses that its products are no match for the imported varieties either in terms of quality or price, and that its products are likely to suffer from competition with the superior brands manufactured by MNCs.

The West has stolen a march on India with its continued advancement of technology, and also thanks to the environmentalists' protests against wastage of precious biomass and against pollution.Historically, the more developed and industrialized units of the Western countries and operations through economies of scale make Indian paper units look like expensive old toys, to be handled with care. Its immediate threat comes from an Indonesian-based multinational paper corporation, Sinar Mas. It has set up a Rs. 300 crore plant in Pune with an installed capacity of 200 lakh tonnes of high grade writing and printing paper. It was expected to start production from 1995. The Indian paper makers are upset with the Government of India for its prompt clearance of the Sinar Mas project. And Thapar is upset with Sinar Mas for luring away his ace Managing Director, M.K. Raina. Incidentally, Sinar Mas can undercut the existing units as its capital cost is 60 per cent less than the investment usually required by the local industry. Besides, it will import its own high grade pulp from Indonesia (*Business Standard*, 8 Feb. 1994). The Indonesian group will house 51 per cent stake in the equity and the remaining 49 per cent would be from the public and from financial institutions with no collaboration with an Indian partner (*Business Standard*, 1994).

The only alternative that BILT has, though a lead paper unit, is to revert to the manufacture of the much needed printing and writing paper, but of good quality, in which no multinational is interested. It also plans to make its units 'lean and tall' and the workers are being prepared for this structural change. 'There would be no protest, they have no options open' said a senior executive of BILT in Feb. 1994. Modernization is imperative. It implies investment. The rate of interest is high in comparison to the European interest rate of 5 per cent. The government which has liberalized the economy has asked the industry to find Euro dollars. And it has. But NABARD (1994: 13) points out the fallacy in BILTs argument. It points out that the argument needs to be viewed in the context of real rate of interest, i.e. prime lending rate net of inflation. Inflation in selected industrialized countries is as low as 1.8 per cent in Japan, 3.5 per cent in USA, 3.85 per cent in UK. The Indian rate of interest is comparable with 4.5 per cent with the real interest in UK, Germany and Holland.

According to reports, BILT has forged links with multinationals (*Facts for You*, 1993: 15-19). The most recent one is with Owen, Illinois of USA to produce glass containers at Pondicherry. This is meant to tap the Far East and West Asian market. This linkage helped BILT to salvage two old glass units. Another link is with Ytong International of Germany to manufacture building material with flap ash, a waste product, at Palwal in Haryana. Still another is with Reitzel of Switzerland. The unit has already begun to export gherkins and other vinegar-preserved foods. The most significant one is the Project Engineering Division (PED), an in-house consultancy organization of BILT's tie-up with Pt. Risjadson Holding & Inv. Co. in association with Chemtex International USA, to set up a 380 tpd rayon/paper grade pulp mill in Indonesia. They propose to upgrade the mill to 1,000 tpd and to set up a joint venture to grow 18,000 ha of casuarina and eucalyptus (Gopal Krishnan 1993). Thapars are also planning at a later stage to set up a paper plant somewhere on the east coast at a point close enough for transhipment of pulp from Indonesia (Vishwanath 1993). Thapar has confessed that his earlier 40 tpd paper unit in Indonesia had failed when other giant units of international size went into production. It is not clear why the bamboo-rich Ballarpur unit has asked the forest department for 18 hard and soft wood species including three hardwood Australian species (see Appendices I and II). These are meant for PED's R & D work. PED is housed in Ballarpur. Indonesia has clear felled much of its rainforest to make room for fast-growing eucalyptus species for its pulpwood plantations. Eucalyptus is a native of not only Australia but also of the far eastern islands of Indonesian archipeligo. These islands share fauna and birds with Australia.

Lalit Mohan Thapar has other plans for diversification of BILT. One of them is a Rs.1,000 crore investment in the area of power units, one on the tributary on the river Beas in Himachal Pradesh (HP) and the other on a river near Dandeli in Karnataka. Incidentally, there is already a large hydel-power station on the Kali river near Dandeli, not too far from the West Coast Paper Mills. BILT is reported to have signed a Memorandum of Understanding with both the HP and Karnataka governments for the two hydel projects.The HP hydel project is in collaboration with Voraralberger Illewerke and Ilban (both Austrian). BILT has also sent a proposal to the International Finance Corporation to involve it as an equity partner in the project. 60 per cent of the project cost would be financed from overseas sources, 25 per cent comprising of equity. BILT proposes to mobilize about 15 per cent of the cost from the capital market (Banerjee, 8 May 1994). The fairly detailed account of BILT's activity is mentioned for two reasons—its desperate attempts to survive and grow, and secondly, because large hydel power implies submergence of forest areas, and which will stand at least in theoretical contrast to felling of bamboos, a renewable resource for manufacturing paper.

Summary and Conclusion

Tree-Tech is a case of BILT's limited 'enlightened' self-interest. Its meticulous harvesting of bamboo in Chanda which ceased in 1991 casts doubts on this. BILT is unmindful of the larger issues. It over-uses water and bamboo. It pollutes the atmosphere and water which could all be controlled or reduced considerably with investment in modern pulping technology especially with the oxygen delignification process. It is equally unconcerned with rounded development, and its production of cultural paper has tapered and ceased. BILT's response has been totally negative to the government's multi-dimensional support in the form of highly subsidized raw material, its monetary support in the form of loans at highly subsidized rate from four major financial institutions, and finally, even dramatic policy changes in 1987 when the price of paper was rising. The situation, however, has now dramatically changed. BILT, which was so proud of its achievements and its exclusive control over certain types of speciality paper which includes sack-like packaging, has now decided to reverse its production pattern to ordinary printing and writing paper, not out of any concern for a rounded development but as necessary strategy for sheer survival.

Recommendations

1. From the ecological point of view BILT must use a combination of an oxygen-delignification and thermo-mechanical pulping process. The double advantage of this is saving of scarce raw material and reducing pollution of the rivers to a minimum level. BILT has the advantage of monopoly use of bamboo and does not use eucalyptus.
2. The Forest Department must continue to be vigilant and protect its precious bamboo forests and keep up with the tradition of resisting the central government's imposition of ecologically damaging rules. FD should continue to carefully monitor BILT's felling of bamboo

and learn and apply the technique of conservation of moisture and help in the growth and spread of bamboo rhizome.

3. FD should turn down BILT's demand for supply of local hardwood species. Acceding to such a demand can totally denude the forests.
4. The government must compel BILT and the other extractive industries to pay the social cost for having extracted the natural resources from the area. A good beginning would be to protect labour from economic exploitation, lay-offs, and also from health hazards.
5. Local inhabitants have not shared in the high profits of BILT or other industries in this district. Technical as well as liberal education must be provided, and opportunities made available for them to catch up with the major development processes of the country. This is an alternate for regional development.

NOTE

1. The paper control order was lifted in 1987. After a couple of years, BILT produced expensive Bond Paper which also comprises stationery. By the end of 1989-90 the sales jumped to Rs. 322.61 lakh, in 1990-1 to Rs. 341.06 lakh, and in 1991-2 to Rs. 423.68 lakh.

REFERENCES

Ahmed, Feroze 1991, 'Thapar Group Growing Stronger', *Business India*, April-May.

Ahuja, S.P. (n.d.), *Paper Industry in India*, Institute of Economic Growth, New Delhi.

Anon. 1981, *Financial Express*, 9 Feb.

——— 1993, Ballarpur Paper Industries Ltd., *Facts for You*, April.

——— 1993, 'Ballarpur Industry's Net Profits Soar', *Economic Times*, 31 May.

——— 1993, 'Nuts & Bolts Management', *Business Standard,* 25 May.

——— 1994, 'Sinar Mas Plans Put Paper Units in a Flag', *Business Standard*, 24 June.

Banerjee, Soma 1994, 'BILT To Set Up Two Hydel Power Units', *Economic Times*, 8 March.

Bergbie, L.F.B. 1910, *Central Provinces District Gazetteer Chanda*, vol. IA, Pioneer Press, Allahabad.

BILT, *Annual Reports*, 1973-4 to 1993-4.

Centre for Monitoring Indian Economy, 1985, District Profile, *Chandrapur*, Bombay.

Champion, H. and H. Osmaston 1962, *Stebbings Forests of India*, OUP, London.

Council of Scientific and Industrial Research, 1988, *Wealth of India*, New Delhi.

Dagli, Vadilal 1979, *The Committee on Controls and Subsidies*, GOI, Ministry of Finance, New Delhi.

FDCM 1981, *Eighth Annual Report for the Year 1980-1.*

——— 1989, *Sixteenth Annual Report for the Year 1988-9.*

Ghiara, N.B. 1992, *Freedom to Grow, Business Today*, Feb.-March.

GOI 1976, *The Report of the National Commission of Agriculture*, vol. IX, p. 4, Ministry of Agriculture and Forests, New Delhi.

Gopal Krishnan, T.E. 1993, *Business Standard*, 23 August.

Government of Maharashtra 1961, *Socio-Economic Review Chandrapur District*, Bombay.

——— 1973, *Maharashtra District Gazetteer, Chandrapur District*, Directorate of Printing and Stationery, Bombay.

——— 1989, *Socio-Economic Review, Chandrapur District*, Bombay.

Guha, B.N. and Adarsh Kumar 1977, 'Inter-related Effects of Temperature and Moisture on D. Strictus Seed Germination', *Indian Forester*, vol. CIII, no. 3, Dehradun.

Hazari, R.K 1967, *The Structure of the Corporate Private Sector*, Asia Publishing House, Bombay.

Hewetson, C.E. 1962, in *Stebbings Forests of India,* vol. IV, H. Champion and H. Osmaston (eds.), OUP, London.

Khan, Waheed (ed.), 'On Royalty', *Indian Forester*, vol. XCVI, no. 6, Dehradun.

Kothari, P.S. 1971, 'Royalty on Pulpwood', *Indian Forester*, vol. XCVI, no. 6, Dehradun.

McClure, F.A. 1966, *The Bamboos—A Fresh Perspective*, Harvard University Press, Cambridge, Massachusetts.

Rao, Y.A. 1989, *The Paper Industry in India—Status & Prospects*, OUP, New Delhi.

Rustogi, Krishna 1977, 'Long Term Forest Lease for Land, Its Economic Evaluation', *Indian Forester*, vol. CIII, no. 4, Dehradun.

Sagreiya, K.P. 1971, 'Royalty on Pulpwood', *Indian Forester*, vol. XCVII, no. 9, Dehradun.

Singh, Neelam, A.K. Kochak, Nilam Malhotra 1990, *Paper and Paper Board Industry: Productivity*, vol. XXXI, no. 3, New Delhi.

Singh, Pratap 1963, 'Disastrous Disforestation', *Indian Forester*, vol. XCIX, no. 8, Dehradun.

——— 1971, Development of a New Integrated Forest Policy, *Indian Forester,* vol. XCIX, no. 1, Dehradun.

Stebbing, E.P. 1921, *Forests of India,* vol. 1, John Lane Bodley Head Ltd., London.

Vishwanathan, T.S. 1993, *Economic Times*, 24 September, Bombay.

Yadav, LS.P. 1963, 'Site and Soil Characteristics of Bamboo', *Indian Forester*, vol. XCIX, no. 3, Dehradun.

Watts, H. and T.S. Vishwanath 1992, 'Legal Battle for A.P. Rayon's Control Threatens Rights Issue', *Economic Times*, 11 September.

APPENDIX I

List of pulpable species that BILT asked for from the Chanda Forest Department

1. Bhirra, *Chiroxylan snietenig*
2. Mowar (*Moyan*), *Lannea grandis*
3. Awala, *Embelica officianalis*
4. Chent, *Zizyhpus xylopyrus*
5. Bija, *Pterocarpus marsupium*
6. Chegar, *Garuga pinnata*
7. Salai, *Boswellia serrata*
8. Eucalyptus spp.
9. Subabul, *Leucaena leucocephals*
10. Garai, *Cleistanthus collinus*
11. Heti, *Susbenia grandis*
12. Suru, *Casuarina equisetifolia*
13. Anjan, *Terminalia arjun*
14. Lendiya, *Lagerstroemia parviflora*
15. Surya, *Xylia xylocarpa*
16. Kasul, *Bridelia retusa*
17. Shesham, *Dalbergia latifolia*
18. Rohan, *Soymida febrifuga*

The Forest Department has pointed out that eucalyptus, casuarina and subabul do not grow in the Chandrapur forest.

APPENDIX II

List of unwanted non-pulpable species (submitted to the Forest Department, Chandrapur, by BILT)

1. Ain, *Terminalia tomentosa*
2. Tendu, *Diospyros melanoxylon*
3. Moha, *Bassia latifoli*
4. Dhanda, *Anogeissus latifolia*
5. Khair, *Acacia catechu*
6. Khirni, *Malinkara hexandra*
7. Dhoban, *Dilbergia pennicula*
8. Anjun, *Hardwickia binata*

CHAPTER 5

Andhra Pradesh

A Rock Country and Four Pulp Mills?

It's an ancient country, a sacred country,
an enchanting green country.
It's the home of Krishna and Godavari
where, a huge bare rock stands upon
another huge bare rock precariously.

But, this is no Grimms Fairy Land
Yet nevertheless a strange land,
where the Teddy Bear found its shelter,
but grew into an ugly Grizzly Bear
which ripped the bamboos that gave it shelter,
this was not Baloo our friendly bear.
This is no Alice's Wonderland
Though the Mad March Hare
was invited to dinner,
But the wild Boar gobbled it up.
It cannot be Disney's land
where Mickey and Jerry can play pranks together,
And Ferdinand sniffs the daisy.
It's now a strange rock country
where the Lioness adopted the Monkey
but, the Tiger snatched it away.
And when the Giant doffed his hat
The Lioness invited him in.
It no longer seems like the country,
the ancient sacred country,
where, Krishna and Godavari flow
It's very like Kipling's country
Where, the rule of the jungle is broken
Where, everywhere huge big rocks stand upon
other bare rocks precariously.

M.S.

SECTION 1

Political Economy of Forest-based Paper Industry

There are four large forest-based paper units in Andhra Pradesh (AP). They are more or less evenly distributed in the three mutually exclusive areas, viz., two in the erstwhile feudal Telengana, one in the once politically active delta area and the fourth in the dry backward ceded districts, now known as Rayalaseema. The distribution of these units, particularly through the setting up of the last two was a late political decision in order to maintain a balance among the three regions.

It is also important to remember that the paper industry has to maintain its linkage with the primary sector to acquire steady supply of raw material. The main consideration of a mill when choosing its site, is easy availability of pulpable wood, free-flowing water, lime and coal if possible, and infrastructural facilities like power and communications with, of course, 'political stability'.

It is not surprising that the first paper unit was set up in the British stronghold of Rajahmundry, East Godavari district, as early as 1920 by a British Forest Officer. This was at a time when the Forest Research Institute's (FRI) laboratory as well as its pilot plant experiments were just beginning to prove that bamboo was the raw material which would enable the British to establish their paper industry in India. The district's *agency areas* were rich in bamboo, but the industry did not make much headway, and eventually in 1948, it was taken over by the Government of India. The second Five-Year Plan gave primacy to industrial development. The wealth in Andhra at this point of time was held by the landed gentry. So in 1956 the mill was handed over on a platter, to a Bombay-based managing agency held by Somani with a the general understanding that 30 per cent of its production would be sold to the government. This was the levy paper meant for the educational sector as a part of Government of India's programme of rounded development of the country under its 'socialistic pattern of society'.

The second unit came up in the bamboo-rich Adilabad district in Telengana. This was a part of the Nizam's domain, where there was little scope for emergence of local entrepreneurs as money and grain-lending were extremely lucrative under the *jagirdari* system (Reddy: 47). The Nizam permitted indigenous industries to survive but made no serious attempt to industrialize his domain. One major exception, perhaps, was to set up a small paper unit in 1938, which coincides in time with the beginnings of the Indianization of the paper industry. At about the same time, the ambitious Birla who had struggled against, and risen anew from the British stranglehold, was eager to establish himself in the profitable paper industry. He had set up his first unit in Orissa in 1938 as soon as it came under Congress rule. He was eager to expand and found that the war was a boon. Perhaps the Nizams domain, rather than a directly British-controlled area, seemed more secure to Birla, although the Nizam was a friend of the British. In 1942 Birla took over the Nizam's paper mill and merely increased its production capacity. In 1948 he installed a forest-based rayon unit, Sursilk, immediately adjacent to the paper mill. Till the 1960s there were no other paper mills in AP, and the reason for the simultaneous emergence of other mills is linked with the political economy of the state.

The linguistic state of Andhra Pradesh was formed in 1956 only after the various contradictions generated between its three regions during the British regime had been partially resolved. This matter will be touched upon a little later. However, what is important to remember is that 1956 also happened to be the year when the second Five-Year Plan with its thrust toward industrial development was released. Much of the wealth generated in the state was from agriculture and there was no industrial base. Nothing much happened, apart from setting up three institutional agencies for promoting industrial development, till the fifth Five-Year Plan. In 1974-7 there was a quantum leap with an absolute increase in industrial investment to the extent of Rs.723.95 lakh, with the cumulative capital standing at Rs. 2121.32 lakh (Reddy: 49).

But industrial development was uneven; it first took root in Telengana. The region was rich in natural resources, minerals and timber, combined with a large army of dispossessed tribals, who were a surplus labour force impoverished and oppressed by pre-capitalist agrarian relations and a feudal form of forced free labour. Such features attracted capital investment particularly in Medak and Khammam apart from Hyderabad itself. Besides, investment in notified backward districts made an industry eligible for concessions and subsidies. Close, proximity to the state capital, Hyderabad, was an added attraction for industrial capital to enter—first the public, and then the private sector. While part of Telengana forged ahead, the backward region of Rayalaseema and even coastal AP was left behind.

Some economic and political balance between the regions had to be maintained. A joint sector for a large forest-based paper industry in Rayalaseema was proposed. A local merchant-cum-industrialist, Gupta, who had accumulated wealth after the post-World War I and II booms, was invited by the Andhra Pradesh Industrial Development Corporation (APIDC) to set up a forest-based paper mill in Rayalaseema in the mid-1970s. This suggestion was believed to have been put forth by the Chief Minister, Vengal Rao,who had served a stint as Industries Minister at the Centre.

Vengal Rao is a native of Bhadrachalam in Khammam district of Telengana. It is not clear whether he approached the Indian Tobacco Company (ITC) or ITC first approached him to set up a paper unit. What is however clear by ITC's own admission, is that it had problems with Dalmia's Rohtas Paper Mills in getting a regular supply of a specific type of paper for its cigarette industry. Setting up its own paper unit would prove to be an useful backward linkage, as well as an added boon with numerous subsidies and concessions offered by the government, since Bhadrachalam is situated in a notified backward area. For the same reason it also qualified for financial aid from APIDC and other financial institutions. AP, after all, was familiar ground for ITC ever since the 1920s, and AP was keen on industrialization. ITC walked into this situation and made use of the opportunity. It was about this time that P.N. Haksar, its new Chairperson, announced his plan of 'Indianization' of ITC by 'setting up a chain of five-star hotels' and of encouraging carpet-making for its export promotion; that only the tender fingers of little children could knot the wool for fine carpets is another matter. His ultimate desire was to enter the core sector. But such explanations do not tally with the fact that ITC diversified into paper but only speciality papers, at a crucial time when PPI was passing through a most turbulent period in its war with the government on the question of price and production of levy paper. Haksar was undaunted; he had the political clout, and now the added economic backing of his multinational corporation and was confident that he would overcome and breakdown the resistance of the government. ITCB did not even care to install a machine which would produce levy paper. The company admitted it was on the forefront of the PPI's struggle to get the control orders lifted.

The state's financial aid is said to have begun to flow faster towards ITCB than towards Rayalaseema Paper Mills,[1] (which needed it more), giving ITCB a head start. Again, Rayalaseema is a dry district unilke the densely-forested Khamman with its rich stock of bamboo. True, there are forests in Kurnool district but here bamboo had to be augmented by the eucalyptus. Plantations were set up to provide the necessary pulpwood.

Regarding the political economy of PPI it is necessary to understand, (a) the historical reasons for regional contradictions, which gave birth to two new units and more importantly, (b) backward linkages of the industry first with the forests from which it secured and exhausted its raw material and later with prime agricultural areas to grow its raw material. What are the consequences of the backward linkages on the soil and on the poor peasants who cultivate the tree crop?

Regional Contradictions

The reasons for regional contradictions in Andhra Pradesh have to be sought in history. Very briefly, historical evidence points out that the Sultan of Golconda conquered almost all the Telugu areas between 1580-1622 to give Andhra a political unity. After 1687, many Andhra areas came under the Mughals. In 1724 Asaf Jsaf Jah, the viceroy appointed by the Mughals emperor who won the title of Nizam-ul-Mulk, threw off all allegiance to his suzerain and made himself virtually independent (K.V. Narayan Rao, 1973: 5). In 1765 the Mughal emperor granted the whole of the Northern Circars, viz., the Delta Region or Coastal Andhra, to the British, while in 1800 the Nizam ceded to the British all its southern territories which came to be known as the ceded districts and later as Rayalaseema. The Nizam held Hyderabad and the region north of it under his control till 1948. Telengana falls in this region. The Nizam's rule as well as British rule were both disasters for the agricultural population. In the British-held areas it was a major disaster for the artisans as well. British rule had snuffed out the local cottage industry to make room for its own industrial products to enter India.

A striking feature of the immediate pre-colonial period was the existence of intermediaries between the peasantry and the state. It was an iniquitous social system sanctified by tradition and legitimized by political authority. The intermediary rentier classes and castes in the pre-colonial period were permitted to make claims and share in the agricultural produce (G.N. Rao 1988: 3). The British did not make any change in this basic structure, but built upon it to draw its revenue.

The Delta Region

The East India Company soon acquired a greater portion of the coastal region from the Mughals. Between 1765 and 1802 the Company was busy consolidating its political power and grasping the various facets of its economy, setting up district collectorates, and experimenting with lease tenures without altering the existing agrarian structure. By this time, the Company had also acquired a monopolistic grip over the local handloom industry. In 1802-4 it introduced a permanent *zamindari* settlement in the greater part of coastal Andhra. It altered the old power balance to tilt it heavily against the peasantry until the *zamindari* system was finally abolished in 1948 (G.N. Rao 1988: 5). Mismanagement by the *zamindars*, distortions in the agrarian structure coupled with failure of monsoons, led to severe famines. The most disastrous was the Guntur famine in 1832-3 when 30 per cent of the population died. A fall in agricultural prices (1826-52), combined with a decline in the handloom industry brought about economic stagnation and decay (G.N. Rao, 1978). Alarmed at the fall of revenue as well as by the fall in land prices, the Company undertook construction of dams across the two major rivers, Godavari and Krishna. The Godavari and Krishna districts became practically immune to famines. Also, trade increased. By 1892 in the Godavari Delta much of the peoples' money went to swell the revenue courts and the 'pleader's purse' (*Imperial Gazetteer*, 1908: 19). Meanwhile, on another front, Western education slowly spread from 1850 onwards under the patronage of the Government of Madras. Colleges were set up in several towns in the coastal district but the university itself was firmly located in Madras city. Christian missionaries and philanthropic *zamindars* joined in the efforts to raise education standards.

The spread of Western education, the rise in the well-to-do class of landowners and traders, gave rise to a spurt of social reforms and political consciousness. The Brahmins with their tradition of learning took to western education and began to practise law. They showed a growing interest in the Telugu language, culture and history. Three volumes of an encyclopaedia were published in 1915, 1916 and 1917. The Andhra Sahitya Parishad was established in 1921 and the Andhra Historical Research Society in 1921 (K.V. Narayan Rao: 67). Andhra was a part of Madras Presidency, so investment in education was heavier in the Tamil areas than in the Telugu. To overcome the backwardness in education, a separate university for Telugu districts was considered necessary. There were no Telugu daily newspapers, all English dailies were run by non-Telugus, the prestigious daily, *The Hindu*, was run by Tamils. The Madras government's bureaucratic postings were largely managed by Tamil Brahmins, the High Court being located in Madras. To overcome backwardness, the major demands were the setting up of an university and a High Court in a Telugu district and the starting of a Telugu daily (K.V. Narayan Rao 1973: 36, 37).

The Andhra movement initially drew its strength from the National Congress, the leadership of the Andhra Congress consisting mainly of Brahmins who were professionals and therefore prominent in public life. It gained the cooperation of the Kammas, the rich agricultural class, the Vaisyas and the Christians in the coastal district. Not a single member of the non-Brahmin Justice Party of the region opposed it (K.V. Narayan Rao: 70). Opposition for a separate Andhra Province from other quarters, viz., from Rayalaseema (K.V. Narayan Rao: 315). Rayalaseema, i.e the ceded districts were acquired by the East India Company in 1800. The period immediately preceding 1800 was characterized by lawlessness and insecurity. The harsh ecosystem, uncertain rainfall, infertile soils barring patches of black cotton soil, low density of population, the disused irrigation system and severe tax burden under the ryotwari system, shattered the already precarious agrarian economy. Munro, the architect of the *ryotwari* system, confessed that if the repressive practice of compelling the ryots to cultivate more land than they could afford to do with the given supply of family labour and livestock, 'they would probably throw up one-quarter of all the lands under cultivation' and that 'four-fifths of the farmers paying nearly two-thirds of the land revenue were subsistence cultivators' (quoted in G.N. Rao 1988: 6). Two events that left a deep impression on the local economy were the failure of the Kurnool-Cuddapah canal, a fraud perpetrated by a Madras irrigation company, and a series of famines. Among the famines of 1801, 1821, 1866, 1876-8, 1891, 1895 and 1901, the Dhatta Karauvu of 1876-8 was the most virulent, causing severe mortality of cattle and humans and the consequent curbing of agricultural growth. Lower levels of literacy coupled with a *Poligar* legacy of factionalism and group warfares dissipated the energies of the rural elite. As a result there were lower levels of capital accumulation for re-investment in agriculture and trade. Also, cotton trade in Cuddapah and Kurnool was already saturated by Arya Vaisyas and Marwaris, leaving fewer entry points for the local rich peasants (G.N. Rao 1988: 13-16).

The ceded districts—the infertile, thinly populated and educationally backward Telugu region of the composite Madras Presidency had little enthusiasm for an Andhra Province. The initial hostility came from the non-Telugus and the non-Brahmins of the ceded districts. They looked upon the movement as a Brahmin movement for securing political and administrative offices and power. In the event of the formation of Andhra Province, they feared that the well-educated coastal Brahmins would fill up all the posts. The Congressmen of the region too began to consider the question seriously, but were not averse to the formation of an Andhra Province, provided Rayalaseema was given certain safeguards. The shifting of Andhra University headquarters from Bezwada (Vijaywada) to Waltair in 1929 instead of to Anantpur during the Justice Party's ministry, widened the gulf between the two regions. But after the formation of the new provinces of Sind and Orissa in 1936, and the setting up of Congress ministries in several states in 1937 under the rule of provincial autonomy, Rayalaseema, too, favoured the formation of an Andhra Province and wrested many concessions from the coastal Andhra leaders through the Sri Bagh Pact by the end of 1937 (K.V. Narayan Rao 1973: 315, 316). But only after much haggling, and the death of Sri Potti Sriramalu, was the new province of AP formed in 1956.

The political scenario began to change. It became one of confrontation between the coastal Kammas and the Reddys of Rayalaseema, both powerful land-owning communities. In the bargain, the Brahmins who were in the forefront of the political struggle in the pre-1947 era, began to lose out, their power passing into the hands of the numerically stronger and the educated sections of the Kammas and the Reddys. Both communities had accumulated their wealth through agriculture; however, their surpluses did not initially flow into industrial ventures. But, with political patronage, state subsidies and assistance from intermediaries, they began branching into urban investment (G.N. Rao 1988: 16).Thus a new peasent class emerged in the country-side, while the under-privileged groups like the Harijans and the tribals in the hilly forested areas and sections of the backward communities in agricultural areas all over continue their grim battle for survival.

Telengana

The Nizams ruled Telengana between 1724 and 1948. There was a continuation of the Mughal politico-economic set up till modern times. The economy was medieval in character with *jagirdars* holding sway over a third of the territory, *jagirdari* being the most dominant form of land-ownership. The *jagirdars* set the trend for the rest of the rich who alone had the capacity for industrial investment. But there was no effort on their part, their vast estates were coupled with labour bonded to land, called *vetti,* meaning service without payment. This yielded a surplus. Their principal 'economic' activity was conspicuous consumption and maintaining a large retinue of servants. Both capital and enterprise were not forthcoming for industrialization although the area is rich in mineral resources especially coal (J.S. Narayan Rao 1987: 210). Sir Salar Jung, the Prime Minister of Hyderabad (1855-78) attempted merely administrative and financial reforms at the top and did little for villages, or for improvement of cultivation or agrarian relations (B.K. Narayan Rao 1960: 27). Added to this, the state followed a closed policy of insulating itself from outside influences till the beginning of the present century (B.K. Narayan Rao 1987: 210). In the first decade of this century cultivable land increased by 23.88 lakh acres, in the second decade by 11.60 lakh acres. This increase was due to reclamation of cultivable land, diversion from forest area and from grass-land by the plains people (B.K. Narayan Rao 1960: 28). The forests were the homes of the Gonds and other tribals who cultivated and lived in harmony with their forest ecosystem (Haimendorf 1982 and 1987: 21-6).

The agrarian scenario has been dealt with in the course of presentation of the history of regional contradictions for two reasons. The Forest Department (FD) and the AP Forest Development Corporation (APFDC) are unable to fulfil the growing and insatiable demand for bamboo and hardwoods to the PPI. Through careless harvesting the industrial units have decimated the bamboo forests, their R&D having concentrated on finding a pulpable species of wood. The industries need land to cultivate the species they have found suitable, but the government is unwilling to lease any forest land to the industry (see Report of the National Commission of Agriculture, 1976, vol. IX). The government's faith has been shattered by the industries' activities in the lease areas of the forest. The industry therefore turned to 'farm forestry'. The broad strategy is to induce farmers to cultivate specific species for the needs of the industry. The small and marginal farmers, the truly down-and-out are those who still hold their titles, are eligible for bank loans, the company subsidizing a project if the proposal is approved by the premier financing agency, NABARD. The question is: what would be the effects of these projects on the small and marginal farmers, if the species recommended by the industry degrades their land?[2] This has happened before, and the organization responsible for it is no other than ITC, which has now a paper unit in Andhra Pradesh.

SECTION 2

The Forests of Andhra Pradesh

The forest catchment areas of Sirpur Paper Mills (SPM), Andhra Pradesh Rayons, ITC Bhadrachalam (ITCB) and the Andhra Pradesh Paper Mills (APPM) are contiguous. They occur on the hill ranges in the northern end of the state stretching through the districts of Adilabad, Warangal, Khammam, East Godavari and Visakhapatnam lying between 78°–84° and 16°35" N and 18°50". This study has omitted forests in the southern part as the Rayalaseema Paper Mill has stopped production. AP Rayon in Warangal district has stopped drawing raw material from the forests.

Rainfall and Temperature

Adilabad has a hot dry summer, the mean daily temperature reaching 42 °C. Thunder-showers useful for the flora bring down the temperature. The coldest month is December when the mean temperature is 29 °C, and the lowest mean is 15 °C. The average rainfall is 700 to 1,200 mm, received during the south-west monsoon. On an average there are 56 rainy days. Rainfall and temperature pattern in Warangal and Khammam are not any different than that of Adilabad. But any depression in the Bay brings sudden thunder-storms in Khammam, the climate begins to

TABLE 5.1: FOREST AREA BY TOPOGRAPHY—CATCHMENT-WISE

Catchment area	Percentage of forest in plains	Percentage of forests in hills	Total area of forest in sq. km	Per cent
Adilabad	49.9	50.1	6340	100
Warangal	43.3	56.7	5940	100
Kothagudam	76.8	23.2	4638	100
E. Godavari	16.6	83.4	5642	100

Source: *Facts and Figures*, 1993: 21, PCCF's Office, Andhra Pradesh.

change in East Godavari and Visakhapatnam. Humidity is high and summer becomes oppressive, the average rainfall is higher, in these districts it is 1,137 mm to 1,300 mm. There is precipitation from both the south-west monsoon and the retreating monsoon popularly referred to as, the north-east monsoon. Temperatures are much lower in the hills.

Drainage

The Godavari which originates near Nasik runs through all the four districts except Visakhapatnam. It gets enriched on the way and becomes a mighty river. In Adilabad it is fed by its major tributaries, the Penganga, Wardha, Pranahita, Kadam and the Peddavagu on which the SPM stands. Godavari then runs south-east into Warangal. The other smaller rivulets—the Satnala, Swarnvaghu and Siddhavagu—also drain into Adilabad district. In Warangal, the other local rivers apart from the Godavari and the numerous hill streams, are the Muner, Pales and Aker, all running further south into Khammam. The Godavari is further enriched by two other large tributaries, the Sabari and the Kinnerasami. Minor streams in the district are Katleru, Talipary and Wyra. It is in the next district a little south of Rajahmundry, that the Godavari widens to a majestic 8 km in width. The other noteworthy river is Yellura formed by three streams rising in the hills of Rampa, Golgonda and Jaddangi.

The most important rivers which drain Visakhapatnam are the Muchkund, Sarada, Varaha, Gosthani, Champavati, Thandava, Saragudugedda and Meghagrigedda.

Relief

The Sahyadri Parvat or Satmala range traverses the Adilabad district from the north-west to south-east for about 281.5 km. Its minor hills and hillocks lie in the eastern portion and stretch into neighbouring Warangal where AP Rayons' erstwhile leased forest areas lie. The elevation of Warangal district varies, the average being around 265.48 m. A hill range known as Kandikalgutta extends from the south-west to Chinnur in Adilabad district. The Chandragiri hills and the famous Hasanparthi iron hills lie 16 km to the north-west of Warangal. A range of hills stretches from Parakal and Singaneri to

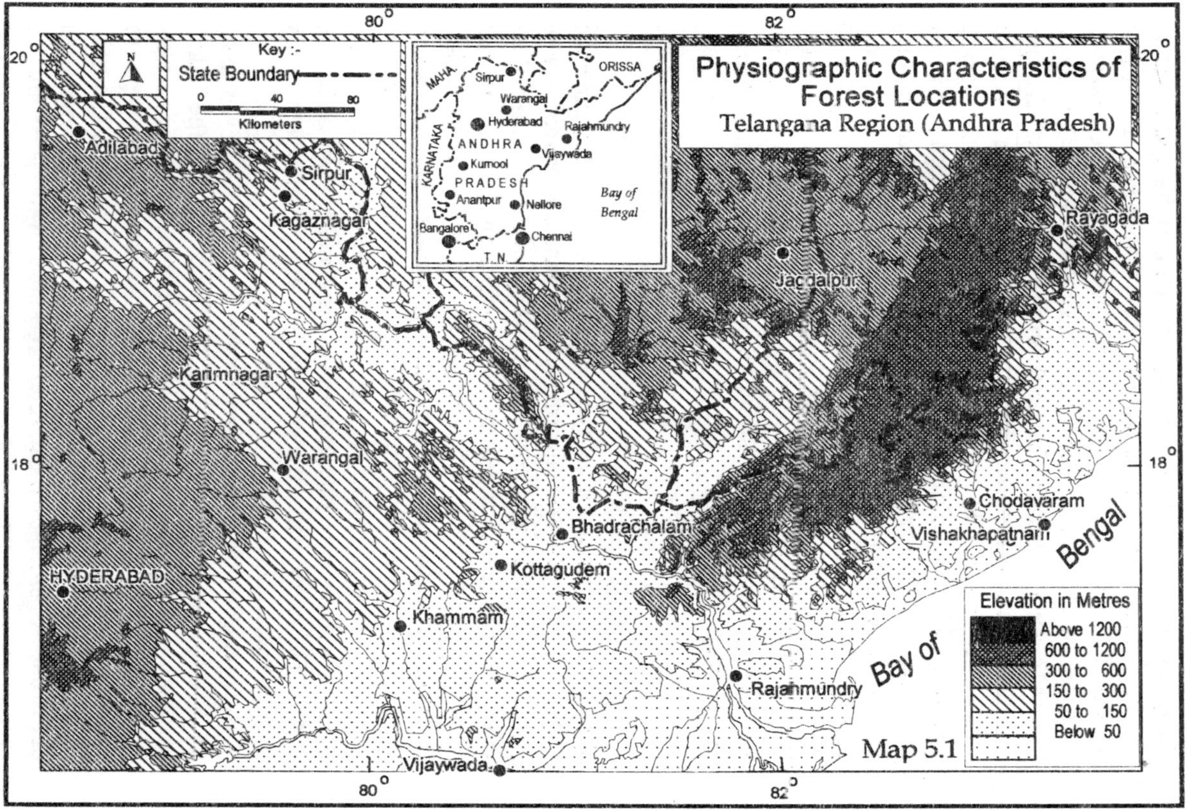

Physiographic Characteristics of Forest Locations
Telangana Region (Andhra Pradesh)
Key :-
State Boundary
0 40 80
Kilometers
N
MAHA.
ORISSA
KARNATAKA
ANDHRA PRADESH
T.N.
Bay of Bengal
Sirpur
Warangal
Hyderabad
Rajahmundry
Vijaywada
Kurnool
Anantpur
Nellore
Bangalore
Chennai
Adilabad
Kagaznagar
Karimnagar
HYDERABAD
Khammam
Kottagudem
Bhadrachalam
Jagdalpur
Rayagada
Chodavaram
Vishakhapatnam
Bay of
Bengal
Elevation in Metres
Above 1200
600 to 1200
300 to 600
150 to 300
50 to 150
Below 50

Map 5.1

Ashwerapeta into the next district Khammam, on its east, bordering lower Godavari valley. The most important hills are Marrigutta, Rajagutta and Yerragutta. ITC Bhadrachalam is in this district and so too the Kothagudam forest division where we made its assessment of bamboo felling by ITCB. Nearly half the district of East Godavari is occupied by the Eastern Ghats. The main hill tracts which are in the mill's Agency Area in the north in the sub-talukas of Rampa-chodavaram and Yellapuram rise up to 1,000 m and gradually descend to sea-level. In fact, Bhadrachalam was earlier a part of the Agency Area of East Godavari, later transferred to Khammam district. Viewed from East Godavari district, the eastern ghats run eastwards and then northwards, parallel to the sea to form the Agency Areas in Visakha-patnam. This tract comprises Chintapalli, Padedu taluka and portions of Narsipatnam, Chodavaram and Sringavaruka taluka. The average height of the hills is over 914 m. Several peaks are above 1,210 m. Foothills and secluded valleys intervene between the countryside and the low hills.

Soils

Various factors influence the type of flora, soil being an important consideration. Soils, after all, are rocks or minerals reduced to fine particles by weathering. Composition and pH of the soils vary. With changes in the geological structure the composition of the flora also changes. A small diversion into the geological history of the region that is under investigation is necessary, to show the extent to which forests are dynamic. Resilient for long stretches of time, they have within a short span of two decades suddenly begun to collapse. The collapse appears to be due to anthropogenic rather than geological factors. The interferences started in the last century and were speeded up in the second half of this century.

According to information given by the Geological Survey of India in 1976 (quoted in Rajgopal 1977). AP forms a part of the Indian (Archaean) Peninsular Shield which has remained stable ever since the formation of the earth's crust about 2,000 million years ago. At that point conditions for existence of life were unfavourable. The rocks during the period were essentially igneous or highly metamorphosed due to intense earth movement. After exposure to agents of weathering for a period of 500 million years, the Pakhal basin covering parts of

Adilabad, Warangal and Khammam districts of AP and Nanded in Maharashtra was formed, and primitive life began. Rocks such as conglomerates, sandstones, shales and limestones were deposited. After about 800 million years the region was lifted, folded and faulted. There was then a lull in the geological history for about 400 million years. At this time peninsular India, together with most of Africa, Australia, Antarctica and South America, formed a supercontinent—the Gondwana Land to the geologists (and the land of the Gonds to the anthropologists). In the earlier period, Gondwana Land was covered by glaciers; the Pakhal basin received glacial deposits. Simultaneously there were changes in the land flora and their growth was accentuated by retreating glaciers. This flora was essentially Glossopteria while the land plants included pteridosperms. Luxuriant vegetation which drifted in and settled got buried by deposits of sandstone and shales; they were transformed into coal in the course of time. There were further deposits of shales, clays, limestones and conglomerates. A gradual change in the weather conditions was accompanied by a change in the flora and fauna. *Ptilophyllum flora* which were cycadean in character replaced glossopteria. These conditions existed about 100 million years ago when there was a gradual breakup of Gondwana land. Flows of molten rock and lava, poured out of fissures from the west coast of India to give rise to flat topped hills known as the Deccan Traps. In between the successive lava flows, sedimentary rocks were deposited in the lakes formed in the depression. There was a revolutionary change in the flora and fauna. Agiospermous plants made their appearance. The area was exposed to atmospheric weathering but there was no significant event in geological history except for the deposition of alluvium in the Godavari river basin. The more recent tree fossils east of Sirpur indicate the presence of deciduous trees, 'the climate during the last 190 million years has not changed, that gave rise to a very stable climatic climax type of vegetation in the districts of Adilabad, Warangal and Khammam' (Rajgopal 1977). Tropical dry deciduous is the climax in regions which enjoy rainfall of 155 cm. A few edaphic climax types are discernible occasionally. The clay calcarious soils produce a gregarious crop of *Acacia leucophloea*; eroded sandy soils, produce *Acacia sundra* or *Somida febrifuga*. On shallow soils and on hill tops in Adilabad *Boswellia serrata* occur almost as a pure crop, while on the stream bank *Terminalia arjuna* grow tall and handsome.

Forest regions of East Godavari and Visakhapatnam districts are an extension of the area that has so far considered. Naturally, therefore, they also bear coal, iron, limestone and clay. Apart from commonality there are interesting variations; manganese is found in the Adilabad district and in Visakhapatnam, dolomite in Warangal and neighbouring Khammam, graphite in East Godavari and Khammam. Khammam is rich in other minerals too—barytes, calcites, chromite, copper, corundum, galena, garnet, kyanite, mica, quartz, slate, woframite and zircon. Visakhapatnam is also mineral-rich with apatite, vermiculite, philogopite and bauxite.

Forests and Forest Management

Large parts of all the six districts were covered with thick forests in the last century. They are broadly divided into dry and moist deciduous with patches of evergreen as one goes eastward and uphill.

The forests of Adilabad lie between 18°40″ and 19°56″ N latitude and 77°46″ and 80° 00″ E longitude. Forty per cent of the total area of the district is said to be forested (Rajgopal 1976), this entire area being classified as southern tropical dry deciduous according to Champion's classification. This type stretches into Warangal and most of Khammam circle. The forests are two storeyed, generally teak and its usual associates. *Terminalia tometosa, T. arjuna, Pterocarpus marsupium*, *Acacia latifolia*, etc., form the upper storey. Climbers such as *Butea superba*, *Acacia instia* and *Cryptopes buchanavs* which damage tree growth are prevalent. There are several species including *D. strictus* as the understorey.

Until 1867 the forests of Adilabad circle were under the revenue department, and those outside the Reserve Forest continued to be managed by it till 1958 when they were transferred to the Forest Department. The constitution of the Reserve Forests was sanctioned in 1894; only much later in 1952 was a forest survey undertaken. 4,600 sq. km are marked as reserved, 2417.73 sq. km as protected and 56.2 sq. km as unclassified forests by *Facts and Figurers* [*FF* (1993: 4)]. Forest management consists of selection-cum-improvement, coppice with reserve and clear-felling of natural forest for teak, some patches of eucalyptus and bamboo plantations.

Since bamboo is of special interest to this study it is noted that the *Pre-investment Survey Report* (*PISF*) show 2,080.81 sq. km being under bamboo, mainly *D. strictus* (p. 24). It further classified bamboo on the basis of the density of growth as well as in terms of quality.

BAMBOO AREA BY OCCURRENCE AND QUALITY—ADILABAD*

(in sq. km)

Scattered 1174.65	Dense 682.42	Pure 223.74
(56.4 %)	(32.8 %)	(10.8 %)
Quality I 374.77	290.87	83.90
Quality II 799.88	391.55	39.90

Source: *PISF Reports* quoted in *Facts and Figures*, 1993: 23-6.
Notes: 21.1 per cent of the total bamboo area of AP lies in Adilabad.

Quality I—Average height of culms should be more than 7 m and average girth more than 4 cm for *D. strictus* and 5 cm for *B. arundinacea.*
Quality II—Average height of culms should be more than 2 m and average girth same as in Quality I.

Forests of Warangal circle are located between 17°5" to 8°36" N and 78°36" to 80°25" E. The forest type is dry deciduous in the eastern half as in Adilabad and scrub type in the western half. Warangal circle covers an area of 24,669 sq. km of which 4,561.41 sq. km is marked as reserved, 1,633.94 sq. km as protected, and 67.67 as unclassified (*Facts and Figures* 1993: 4).

Earlier much forest was lost through thieving under the selective felling system. 'Now there are larger professional thieves' reports *District Gazetteer, Warangal* (1976: 19), possibly implying the forest based industry. The main type of bamboo is *D. strictus* found in some of the dry deciduous divisions like Chityala, Bondala and Chintugudem. *B. arundinacea*, is sometimes seen along the banks of streams in moist valleys. *D. strictus* had flowered in 1896, 1899, 1906, 1921, 1939 and 1952 gregariously and died, but given way to profuse natural regeneration.

Khammam Forest Circle has five divisions. Khammam, Kothagudam, Polvoncha, Bhadrachalam North and Bhadrachalam South. According to the Forest Department, 6493.49 sq. km are reserved, 1,041.02 sq. km protected and 789.65 sq. km unclassified. The forests are classified into two types: (a) dry deciduous which harbours *D. strictus*—497.33 sq. km and in Kothagudam area 68.07 sq. km is under *B. arundinacea* (*FF*, 1977: 23); and (b) tropical thorn which includes some mixed deciduous species of low density, poor girth and height (*AP Gazetteer Khammam*, 1977).

The silvicultural system followed was (a) simple coppice, (b) coppice with reserve, and (c) a selection system used only for timber areas of inaccessible forest. 'Hundreds of acres of mature, over-mature and irregular forests are felled year after year and replaced with teak and fast growing species like eucalyptus and bamboo. To augment and replenish forests, large scale plantations have been undertaken in the district from 1959—further plantations of bamboo and eucalyptus are taken up with a view to supplement the requirement of pulpwoods to paper mills at Kagaznagar and Rajahmundry' (*AP Gazetteer*, Khammam, 1977). Felling of timber trees ceased in 1987-8 with the new Forest Policy. But bamboos are still being harvested by and for the paper industry. It is the present catchment area of the ITCB, ostensibly established to give a boost to industrialize a backward district in 1978. The Pre-investment Survey found Kothagudam a rich source of bamboo. According to it, bamboo in that division was classified as follows:

BAMBOO AREA BY OCCURRENCE AND BY QUALITY—KHAMMAM

(*in sq. km*)

Scattered	319.00 (56.5 %)	Dense 209.40 (30 %)	Pure 39.65 (6.5 %)
Quality I	78.53	62.32	10.47
Quality II	240.82	146.58	413.58

Source: *PISF Report*, quoted in *Facts and Figures*, 1993: 25.

Note: Quality I & II classifications are explained under the head Adilabad Forest Division.

Forests of East Godavari: According to the latest figures, the forests of Rajahmundry are classified into 3,900.25 sq. km reserved and 807.26 sq. km protected which cover 17.26 per cent of the total geographical area (*Facts and Figures*, 1993: 5). Apart from some beach and mangrove forests, the forests in this district are largely confined to the hills which are outliers of the Eastern Ghats flanking the main ranges. The *District Gazetteer* (1970) reports that good miscellaneous dry deciduous and semi-evergreen forests are found in the Rampa Reserve. It also states that they are a good type of forest, with a height ranging from 10 to 35 m. The evergreens and sub-evergreens are confined to moist and more sheltered spots in the southern tropical dry deciduous forests in the uplands and in the Agencies. The greater part is situated on hill-sides with shallow stony soil, the growth being more or less of one type. The remaining forests are in the shallow valley or gently sloping ground and vary with configuration, rock and soil. Patches of good thick forests (*lankas* or islands) are found in the midst of ordinary forests close to the foothills where good soil accumulates and is deep and fertile. The top canopy in these forests are: *Tectona grandis, Pterocarpus marsupium, Acacia latifolia, Terminalia tomentosa,* etc., with varieties of shrubs and *D. strictus* as understorey.

Southern tropical secondary moist deciduous forests are closed forests of medium to good height. The secondary storey includes some evergreens. This type occurs on hill slopes having an elevation ranging from 30 to 900 m with gneiss as underlying rock in Rampachodavaram and Yellavur talukas. The species falling under this category are *Terminalia tomentosa T chebula T bellerica, T ariuna, Pterocarpus marsupium, Lagerstomia parviflora, Xylocarpus* spp., *Albizza* spp. and *T. grandis.* According to the 1979 *District Gazetteer,* the composition of the forest has remained unchanged for over a century except for its share of bamboo which has gone down. The *Gazetteer* also states that the Forest Department has taken up plantations of teak and bamboo. The team's observation of the unchanged nature of the forest in the Agency Areas of Rajahmundry division does not tally with this. There were pure bamboo forests in very good condition but it is not clear whether they are plantations or natural formations—primary, secondary or climax. In the Southern Tropical Thorn Forest, both thorn and hardwoods predominate. They are found in Kogida, Murari, Rameshwaram, Tirumalayapalem, Sudikonda, Mallavaram, Purushotapatanam, Rowthulapudi, Rajeepeta and Thotapalle blocks. Prevalence of *Pterolobium indicum* and *Acacia pinnata* are an indication of human and animal interference, as these forests are conditioned by edapic and biotic factors. The forests in this district came under British rule in 1877. The Madras Forest Act came into force in 1883 and reserves were constituted between 1891 and 1900.

The first working plan was prepared in 1899 for the Mallavaram Reserve. Forest produce was allowed to be exploited on the basis of permits issued by the Revenue Department (*District Gazetteer, East Godavari*, 1979: 12). In 1904-5 the 'total revenue of the district from forests amounted to Rs. 2,00,000, the highest being derived from the sale of bamboo which was Rs. 56,000 while timber was Rs. 35,000. The working plans for the district were rewritten between 1934 and 1944, the final plan revised in 1991-2 by K.V. Prakash Rao.

Some of the significant objectives laid down in forest management were (a) to obtain a sustained yield of timber, fuel, bamboo and minor forest produce to the fullest extent, (b) to meet the agricultural and pastoral needs of the people living in the vicinity of the forests (c) to protect and conserve, as efficiently as possible, large areas of Reserve Forest in the inaccessible Rampa Agencies which form the watershed for the tributaries of the Godavari. But today, several thousand hectares of plantations of economic and fast-growing species such as teak, bamboo and eucalpytus are raised to meet the ever-increasing need for timber and pulpwood for paper (*Gazetteer, East Godavari*, 1973: 13).

The Pre-investment Survey in 1968-9 (Prakash Rao, 1991-2: 2005/6, part I: 152) noted and classified the bamboo forests according to its usual qualitative and quantitative categories. It states:

BAMBOO AREA BY OCCURRENCE AND BY QUALITY—EAST GODAVARI

(*in sq. kms*)

Scattered	1,790.00 (42 %)	Dense	1,272.00 (30 %)	Pure	1,170.00 (27.6 %)
Quality I	498.00		304.00		773.00
Quality II	1,292.00		406.00		397.00

Source: *PISF Report*, quoted in *Facts and Figures*, 1993: 25.

It is not clear which of these forests are primary, secondary or just plantations. The yield of plantations is said to depend on soil, rainfall, survival and protection afforded to them from biotic and anthropogenic factors, and the successful yield, 10 tonne per ha in each felling circle (Prakash Rao, *East Godavari* 1991-2 to 2005/6, Part I: 152). The same report shows that a great deal of bamboo was exported in 1991, 2.5 lakh tonne and 2.7 lakh tonne of short bamboo (p. 28). Till 1987 a large amount of wood was supplied to plywood and match-wood industries, while firewood was supplied to several sugar and sago factories in the district. The 123 saw mills drew their raw material from neighbouring Maharashtra, Orissa and even MP and from private *patta* lands of AP (Prakash Rao: 31).

Visakhapatnam Forest Circle: This has been classified by the Forest Department—Reserve Forest 3,374.40 sq. km; protected 3,081.17 and 90.17 as unclassified. They together constitute 36.38 per cent of the geographical area (*Facts and Figures*, 1993: 5). *District Gazetteers* of 1908 report that the forests on the hills, 'are most luxuriant in Anant-giri, Chintapalli, Minimulura and Siluru'. The forests are reported to have a graded diversity in composition as well as in quality. The team's observations in August 1992 were very different. The valleys in the forests are green, lush, washed clean by the monsoon; some of them are dark green, almost black, and others of different shades of green. They are by and large, well-tended plantations of coffee, cashew, etc. The *District Gazetteers* report on forests, based on Champion's classification are (a) the Southern Tropical Evergreen forest localized in small pockets in valleys and near banks of perennial streams where climatic conditions are favourable with plenty of humus and moisture. Here trees range in height from 21 to 30 m, girth above 1.2 m. The forests are reported to be post-climax and represent the highest floristic evolution. A part of the top storey is deciduous and the second storey evergreen. Bamboos are generally absent, and where they are present they constitute the middle storey to the exclusion of other species. (b) Southern Tropical Moist Deciduous Forests occur in regions with rainfall above 1,016 mm and altitude above 610 m. The girth of the trees is between 12.2 and 21.3 m, the trees are tall with straight boles. The soils are deep and rich. The main species are *Agnoeissus latifolia, Terminalia tomentosa, T. chebula* and *T. belerica, P. marsupium, P. albezzia, P. lebbeck, A. amura, A. odoratissima*, *Lannea grandis*, *Cedrla toon*, etc. Bamboos are present as the middle storey.

Southern Tropical Dry Deciduous Forests occur in uplands where soil conditions are poor, rainfall varying from 635-1016 mm; the average height of the tree varies from 6-12 m and the girth from 305 to 904 cm. Some of them *A. latifolia, A. odoratissima, T. tomentosa, Strychnos nux vomica, D. strictus*, occur widely but vary in density and growth. They are found in all exterior hills, valleys and between altitudes of 244 and 610 m in the less moist and more exposed slopes of Vijayanagaram, Visakhapatnam, Madugulu and Narsipatnam ranges.

Southern Tropical Dry Evergreen and Thorn Forests occur in the plains and coastal areas with a rainfall of 635 mm and where the altitude is less than 244 m. The two types are mixed. There are no bamboos. Grassy Savannahs are found on the upper portions of several ranges above 1219 m.

Status of Catchment Areas of PPI: Assessment of Catchment Area of Sirpur Paper Mills (SPM)

According to a very senior executive of SPM, the mill is located in a place that is backward and locally referred to as 'Kalapani'. It was densely wooded, and tigers are believed to have roamed here till just ten years ago (Mishra, personal communication, Oct. 1992). The huge trunks of a few trees that stand on either side of the circular road in the elite quarters and the neem tree with a girth of about 450 cm behind the guest house are the vestiges of the old forests. The mill area is now referred to as Kothapeta (new town).

SPM (1938) and Sirsilk Rayon (1946) are located in Sirpur with their earlier catchment areas in Sirpur, Warangal and Khammam circles.

SPM now uses both hardwoods and bamboo. The quantum of hardwoods supplied by the Forest Department and AP Forest Development Corporation (APFDC) is given below. Although the 1988 Forest Policy put an end to the supply of hardwoods, the Forest Department continues to supply bamboo. Despite the stoppage of the supply of hardwoods, SPM continues to use it as is evident from the mill data—the source according to the mill management is the open market.

TABLE 5.2: SUPPLIES OF MIXED HARDWOOD BY FOREST DEPARTMENT AND AP FOREST DEVELOPMENT CORPORATION AND ROYALTY, CIRCLE-WISE TO SPM 1983-4 TO 1991-2

Year	Department/Corporation	Quantity in mt	Royalty Collected (Rs. in lakh)
1983-4	Adilabad Circle	12,332.00 (24,644 cu.m)	7.393
	APFDC	21,335.50 (42,671 cu.m)	12.801
		33,657.50 67,315 cu.m	20.194
1984-5	Adilabad Circle	11,368.00 (22,736 cu.m)	11.529
	APFDC	3,016.00 (6,032 cu.m)	
		14,384.00	
1985-6	Adilabad Circle	14,342.00 (28,642 cu.m)	33.176
	APFDC	NIL	
		14,342.00 (28,684 cu.m)	33.176
1986-7	Adilabad Circle	1,656.50 (3,313.50)	4.227
	APFDC	NIL	
		16,56.50 (3,313 cu.m)	4.227
1987-8 to 1988-91		No supplies of hardwood made	
1990-1 to 1991-2		NIL	

Source: *Facts and Figures*, 1991-2, pp. 101-2.

A senior FD official at Sirpur blamed the mounting pilferage of hardwoods from the local forests on unemployed ex-employees' of Birla's Sirsilk—a rayon unit lying adjacent to the paper mill. Sirsilk has been under a lockout since 1983. Many of the employees continue to live in their mill quarters. 'As part of goodwill' SPM supplies electricity and water free of charge to these employees. There is no prospect of the dilapidated mill reopening. The locked out employees have to survive and the only readily available sources for survival are hardwood trees which can be felled and sold (personal communication, Oct. 1992).

The Sirpur Paper Mills draws its formal supply from five circles which includes Kurnool, the former lease area of the Rayalaseema Paper Mills. Sirpur Paper Mill's production capacity is 71,100 tpa. Its assured bamboo supply is 75,000 mt per annum (*Facts and Figures* 1993: 97), see Tables 5.3, 5.4 and 5.5 and Maps 5.2, 5.3 (Sources of Raw Material for SPM).

SPM's immediate lease area is in its vicinity in Kagaznagar division in Adilabad circle. In October 1992, a 18 km drive south to assess the condition of the forests revealed that there were no forests, not even scrub forests. The earth was bare and dry, the only relieving features on the calcarious soil were *Pavetta indica* and *Cleistanthus collinus* and occasional farmlands. The drive towards Vampalli forests in the opposite direction also revealed totally degraded land.

The bamboo yield in Adilabad circle dropped from 71,616 mt in

TABLE 5.3: SUPPLY OF BAMBOO TO SIRPUR PAPER MILLS—KAGAZNAGAR, CIRCLE-WISE AND LEASE YEAR-WISE 1986-7 TO 1991-2

Name of circle	1986-7	1987-8	1988-9	1989-90	1990-1	1991-2
Adilabad	71,616.00	46,032.00	43,896.130	52,533.680	29,823.384	39,582.110
Warangal	—	11,617.00	1,141,660	741.820	3,774.740	5,639.970
Guntur	—	—	—	—	2,445.320	2,027.580
Hyderabad	—	2,720.00	406.920	4,463.260	6,893.900	2,186.880
Kurnool	—	—	—	—	18,713.340	23,097.540

Source: Progress Reports of the Conservator of Forests, Concerned Divisional Forest Officers and the FRI Section of PCCF.

Note: Working of Lease Year = Period starts from 1 October of each year to 30 September of the successive year (*Facts and Figures* 1993: 101).

Sources of Raw Materials for S P M : 1990-91

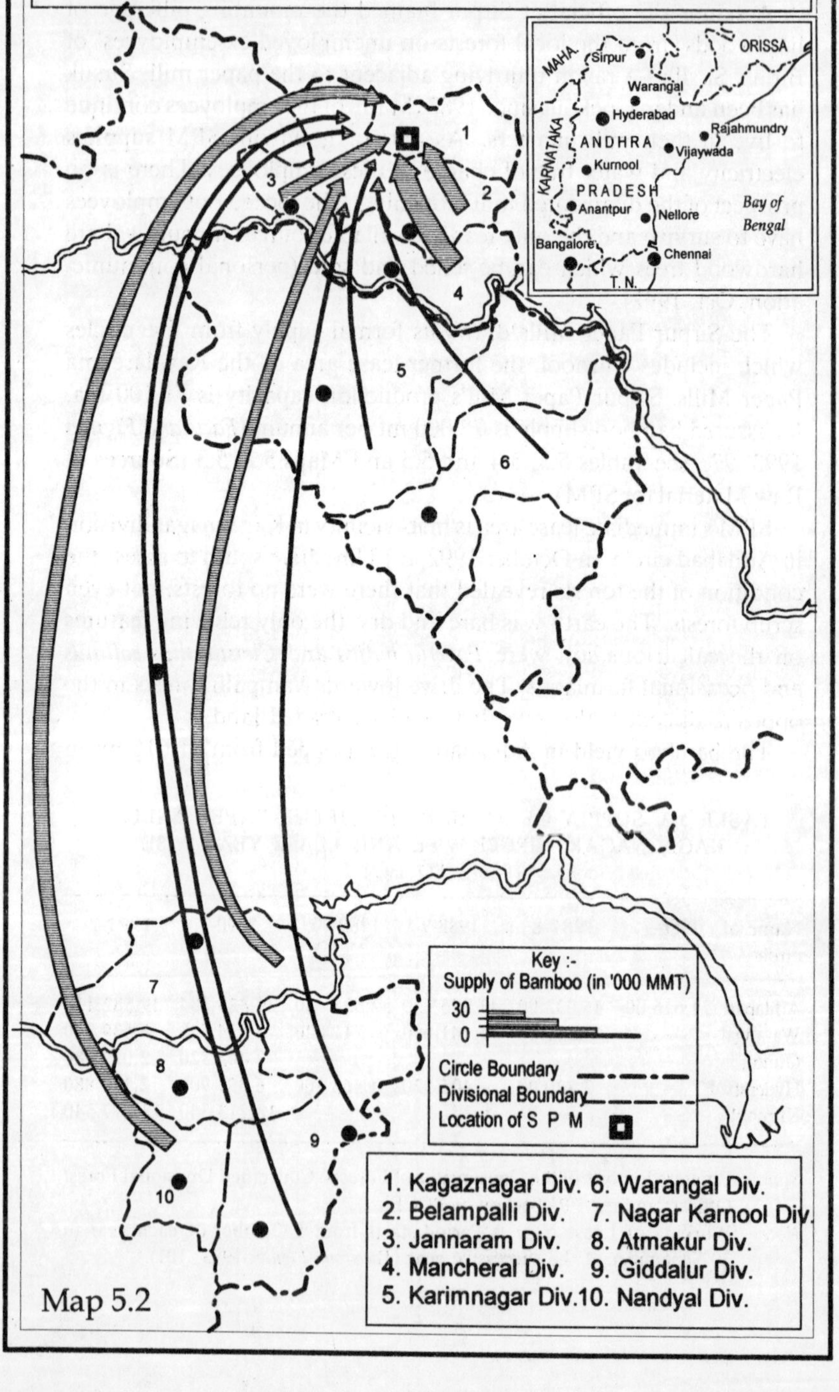

Map 5.2

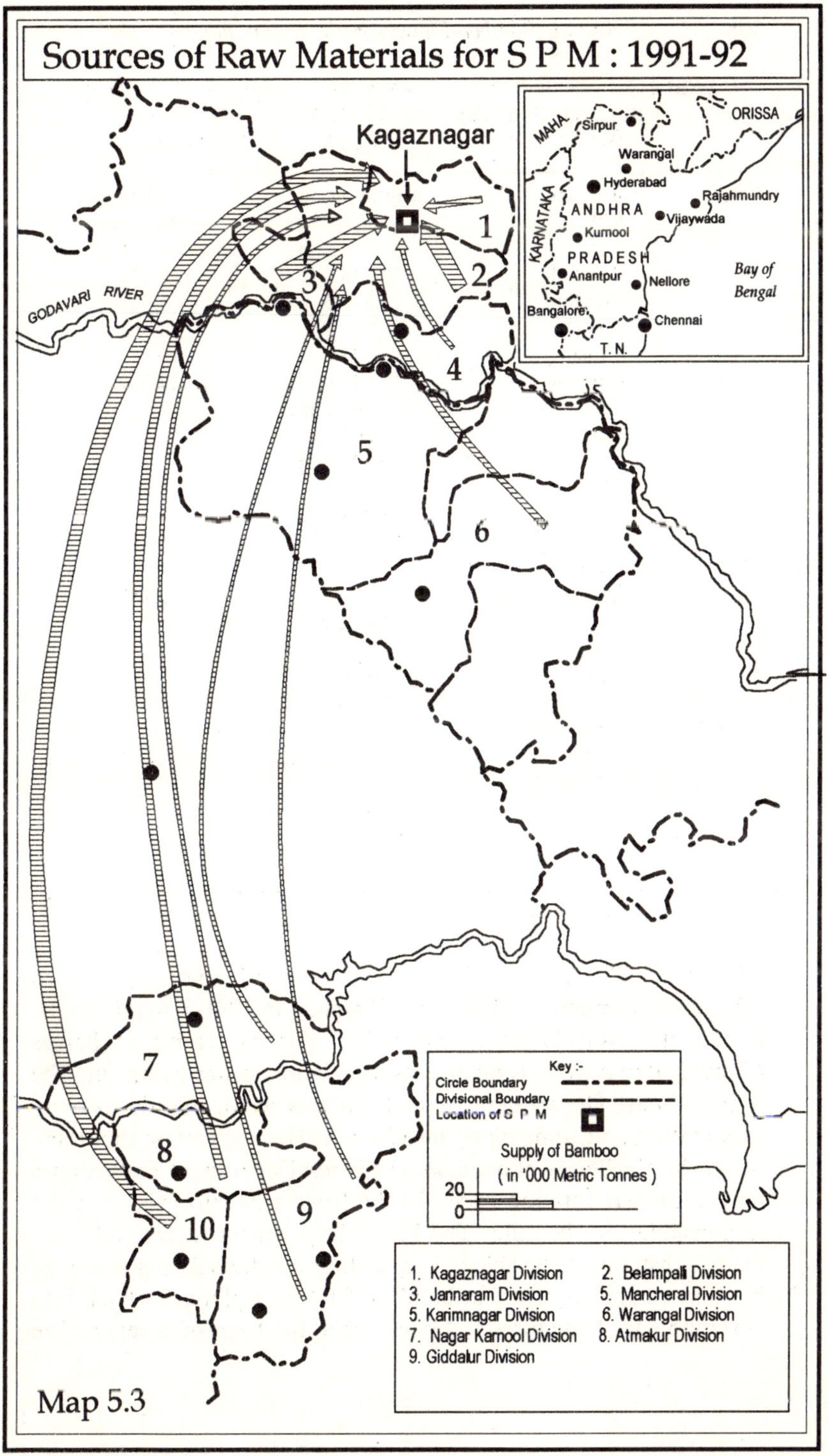

Sources of Raw Materials for S P M : 1991-92

Map 5.3

TABLE 5.4: SREE RAYALASEEMA PAPER MILLS, KURNOOL, SUPPLY OF HARDWOOD BY THE FD AND APFDC 1983-4 TO 1991-2 DIVERTED TO OTHER MILLS

Year	Circle and APFDC	Quantity in mt	Royalty paid (Rs. in lakh)
1983-4	Kurnool	19,948.3	5.125
	Guntur	17,845.5	5.022
	APFDC	Nil	Nil
	Total	37,793.8	10.147
1984-5	Kurnool	19,643.0 (39,285.0 cu.m)	20.337
	Guntur	10,303.5 (20,607.0 cu.m)	—
	APFDC	Nil	Nil
	Total	29,946.0 (59,892.0 cu.m)	— —
1985-6	Kurnool	—	—
	Guntur	200.0 (400.0 cu.m)	0.437
	APFDC	Nil	Nil
	Total	200.0 (400.0 cu.m)	0.437
1986-7	Kurnool	Nil	Nil
	Guntur	Nil	Nil
APFDC	Nil	Nil	
	Total	Nil	Nil
1987-8 to 1991-2		No hardwood supplied	

Source: FRI Section of PCCF Office.

1986-7 to 39,582 mt by 1991-2. The furthest area is Kurnool from which SPM drew as much as 23,097 mt of bamboo in 1991-2, thanks to the closure of the Rayalaseema Paper Mills (Table 5.3). Transport charges of raw material were rising even as local bamboo was dying out. The senior Forest Department official put this down mainly to over-exploitation and some biotic interference. He was certain that there was no local pilferage of bamboo. The Forest Department allots coupes on a four-year rotation cycle. In the lease area supervision is the responsibility of the mills. The Forest Department works under too many constraints to be able to keep a close watch on felling practices. No part of the royalty that bamboo fetches is ploughed back into development of the forests; it goes out in the form of revenue. The

TABLE 5.5: SIRPUR PAPER MILLS, KAGAZNAGAR SUPPLY OF BAMBOO CIRCLE-WISE AND FINANCIAL YEAR-WISE 1987-8 TO 1991-2

(quantity in mt)

Lease Years	Adilabad	Warangal	Hyderabad	Kurnool	Guntur	Total
1987-8	—	—	—	—	—	—
1988-9	43,896.13	741.66	406.920			4,544.710
1989-90	52,533.68	741.66	4,463.260	—	—	57,736.780
1990-1	29,823.384	3,774.740	6,893.900	18,173.340	2,445.320	61,650.684
1991-2	39,582.110	5,639.970	2,186.880	23,097.540	2,027.580	72,534.080
Total	69,405.494	9,414.710	9,080.780	41,810.880	4,472.900	1,34,184.76

link between the Forest Department and the revenue department has not been broken.

It is true that in AP the Forest Department has evolved a technique to salvage bamboo forests, a simple one which consists of piling earth at the base of the clump to conserve moisture, while simultaneously, a shallow trough is dug around to trap rain water. Rs. 2.5 lakh were said to have been allotted for this purpose, but the technique has been tried out only in some areas of Adilabad.

The enormous quantum it draws from each circle and division is worth noting (see Tables 5.5 and 5.6).

Observation of Felling Practice by SPM

Observation of the Mancherial Forest Division to assess felling practices revealed contempt for basic silvicultural rules. High straight cuts instead of low slanted ones were often seen opening out the clump for fungal attack. Again, not all were sharp clean cuts but careless cuts leading to 'splintering'; injury to the bamboo is quite common. No water conservation technique for multiplication of the culm had been introduced in any of the coupes the team visited in Adilabad.

One of the sensitive foresters (DFO, Madhusudan Rao) gave us the other side of the picture. According to him, the maximum bamboo went to the SPM. It was becoming one of the extinct species due to repeated exploitation. Bamboo's natural regeneration was affected by careless harvesting and by biotic interference. Since it was difficult for forest department workers to go deep into the forests due to 'extremists

TABLE 5.6: SUPPLIES OF BAMBOO TO SIRPUR PAPER MILLS, KAGAZNAGAR DIVISION-WISE AND FINANCIAL YEAR 1990-1

Supplies made from lease year coupes	Divisions	Quantity in tonnes	
1987-8 to 1989-90		Information not available	
1990-1	Bellampalli	12,891.964	(221.504 Burnt)
	Mancherial	22,700.430	(45.150 Burnt)
	Jannaram	14,660.990	(18.010 Burnt)
	Adilabad Circle Total	29,823.384	
	Warangal (S)	2,865.840	
	L.D. Karimnagar	909.200	
	Warangal Circle Total	3,775.040	
	Achampet	6,893.900	
	Hyderabad Circle Total	6,893.900	
	Nandyal	8,516.540	
	Atmakur	854.070	
	L.D. Nandyal	8,342.730	
	Kurnool Circle Total	18,713.340	
	Markapur	143.300	
	L.D. Giddalur	2,302.020	
	Guntur Circle Total	2,445.320	
	Bellampalli	17,631.620	
	Jannaram	12,835.480	
	L.D. Kagaznagar	5,347.570	
	Mancherial	3,767.440	
	Adilabad Circle Total	39,582.110	
	L.D. Karimnagar	442.540	
	Warangal (S)	4,127.470	
	Warangal (N)	1,069.960	
	Warangal Circle Total	5,639.970	
	Achampet	2,027.580	
	Hyderabad Circle Total	2,027.580	
	L.D. Nandyal	3,233.660	
	Atmakur	2,819.150	
	Nandyal	11,466.630	
	L.D. Atmakur	5,578.100	
	Kurnool Circle Total	23,097.540	
	Markapur	166.560	
	L.D. Guddalur	2,020.320	
	Guntur Circle Total	2,186,880	

Source: *Facts and Figures*, 1993: 100.

problem', the supervision of bamboo harvesting was done by the mills. The mills in such circumstances tried their best to extract as much as possible. Also they did not take up the responsibility of following silvicultural practices for which they were compensated, by paying 7.5 per cent more royalty (for bunding of soil). This extra 7.5 per cent went to the government in the form of surplus revenue but was never reinvested for bamboo regeneration. Consequently, now there is no bamboo anywhere in the Forest Department's areas. Mills have now turned to eucalyptus as a new substitute for bamboo. Madhusudan Rao was of the firm opinion that eucalyptus is not a desirable species as it affects the soil adversely; only when land is unfit for any type of cultivation should eucalyptus plantation be attempted.

Both the study team's qualitative assessment, as well as the quantitative data from the Forest Department indicate depletion in the bamboo stock.

Augmenting Forest Resources for SPM

Raw material had to be replenished. SPM has the clout and the know-how to carry out its farm forest projects in degraded forest lands. The mills were in the heart of Adilabad forests now totally devoid of trees. The SPM Farm Forestry booklet refers to it as vast areas of wastelands. SPM holds *Eucalyptus camaldulensis* and *E. tereticornis* as the species, vegetative propagation as the technique, and Brazil as a model for fulfilling its targeted needs. As early as 1955, Sukum Thirawat, a Thai representative with FAO team to Australia had noted that all eucalyptus forests of economic significance lie more or less within 100 miles distance following the coastline. They grow on undulating grounds, gentle slopes and plateaus, in other words in well-drained localities with light soils and only some species especially *E. camaldulensis* thrive in water logged locations (Thirawat 1955: 232). Sirpur is arid. The mill also seemed to be unaware that later FAO rejected both species *E. camaldulensis* and *E. tereticornis* as unsuitable for cultivation in India and their pulpable qualities for sulphate pulp as low. (FAO 1980, vol. I, 73.211). Secondly, it is not clear why the growth of eucalyptus in Brazil surpasses its growth in its native Australia. Land to cultivate the new species was not an insurmountable problem. According to a SPM official the mill managed to get hold of different categories of local farmers to cultivate eucalyptus for it as a farm forestry programme. Its activity began in 1982, and within 10 years, 700 acres of land had been brought under eucalyptus (SPM has a forestry wing to manage production of this raw material). The land belonging to local farmers whose holdings varying from 7 to 30 acres, are taken on lease by the mills. The local labour is condemned as 'lazy' by SPM and therefore coastal labour is brought in. This study has shown in another section that the poor peasantry of the coast is impoverished and is brought in as migrant labour. It was not very clear who the original owners or controllers of the local lands were, but it was often report locally, as well as in Hyderabad, that many of the old *jagirdars* and landlords had fled Telengana and settled in Hyderabad due to the fear created by CPI(ML) groups. CPI(ML) threatens the oppressive landlords who have either taken over the lands of the poor peasantry or reduced them to tenants through fraudulent moneylending systems.

SPM's official also reported that in 1964, 4,000 ha of high forest area was cut down to make room for East Bengal refugees. The Principal Chief Conservator of Forests (PCCF) is said to have protested in vain. The area has now been turned into 5-acre farm-lands for each family. These farmers have taken to farm forestry to produce eucalyptus for SPM, and are reported to have earned Rs. 14,000 each. However, what the team noticed was that SPM has a large number of Bengali workers in its factory. But SPM's farm forestry programme itself is far from impressive. An interesting observation is that the trees growing on the bunds are healthier than those in the fields, the latter being thin and straggly. SPM also reported that coppice growth is healthier than the first growth. But the FAO/Indian team, found the cellulosic content of better growth of the coppice is even poorer than the parent plant of *E. tereticornis*. SPM's explanation was 'with passage of time the roots were able to dig deeper for its moisture and nutrients' Whatever the validity of this hypothesis, it is clear that the once rich and densely wooded tropical forests with their rich bio-diversity have given way to small unhealthy patches of monoculture on degraded land. The causative factor as well as possible consequences of growing eucalyptus on degraded forest has been discussed in the section on Karnataka, and therefore is not repeated here. The difficulty of eucalyptus yielding a good biomass on degraded land by the Tamil Nadu Forest Development Corporation (TNFDC) has also been presented in the relevant section in Tamil Nadu. To use *E. tereticornis* as a Tangya system is unproductive, says the Punjab University agricultural scientist, R.K. Koli. This has been discussed in detail in the plantation section in Karnataka.

The team's observation was that the soil in the Adilabad region was calcareous and stony with no plant cover except for some straggly weeds. But under the eucalyptus canopy even weeds did not grow. If the leaves are swept up as they regularly are in the Nilgiris in order to extract essential oils, then perhaps the weeds would grow. The leaves of *E. tereticornis* are allelopathic to many species. In and around the area of Sirpur only occasional fields of *jowar* are seen. *Jowar* is a coarse grain grown on depleted soils. Fruit trees such as mango, papaya and guava are occasionally to be found on farms. The forestry wing of SRM encourages farmers to use NPK fertilizers. It does not however recommend matching organic fertilizers to the local needs of the soil.

A *patta*-holder, or a farmer with clear title, is eligible for loans through banks by the NABARD scheme, so that the mill is absolved of even its financial responsibility to grow the raw material. This has been discussed in the report on NEPA.

Catchment Area of ITC Bhadrachalam Papers Ltd. (ITCB)

According to a very senior executive of ITCB, 90 per cent of the present factory site was covered with forests and only 10 per cent with agricultural land. The peasants were persuaded to sell it to the company in return for employment in the factory. The surrounding forests, he said, were 'abandoned' by the Forest Department due to the fear of 'Naxalites' and were given on lease to the company for 'exploration' of bamboos. The foresters did not express any such fears. The production capacity of the mills is 81,944 tpa and the assured yield of bamboo 60,000 mt per annum (*Facts and Figures*, 1993: 103). See Figure 5.1—the Forest Department and the APFDC supplied hardwood, the figure till 1986-7 just a year before the ban for felling of hardwoods was imposed is indicated in Table 5.7.

The supply of raw material supplied from Kurnool forest circle which was initially meant for Rayalaseema, is indicated in Figure 5.2. This material has been diverted to other mills. But it is not clear why ITC paid only half the royalty due for hardwood. The total supplies are indicated in Table 5.7 and the supply of bamboos is indicated in Table 5.8.

A senior company executive expressed his appreciation and gratitude to the government which gave them not only good forests

TABLE 5.7: SUPPLY OF MIXED HARDWOOD TO BHADRACHALAM PAPER BOARDS, FROM FOREST DEPARTMENT (DIVISION-WISE) AND APFDC 1986-7 TO 1991-2

Year	Department/APFDC	Quantity payable (mt/cu.m)	Royalty payable (Rs. in lakh)	Royalty collected (Rs. in lakh)
1986-7	Forest Department			
	1. Paloncha (L)	12,516.00 (mt)		
		25,032.00 (cu. m)	30.351	15.176
	2. Bhadrachalam	496.00 (mt)		
		992.00 (cu. m)	1.203	0.601
	Total	13.012.00 (mt)		
		26,024.00 (cu. m)	31.554	15.777
	3. APFDC	Nil	Nil	Nil
	Grand Total	13,012.00 (mt)		
		26,024.00 (cu. m)	31.554	15.777
1987-8 to 1991-2			No hardwood supply	

Source: Report of Conservator of Forests, Khammam. See Report no. 9254/86/ M4 dated 21 Dec. 1987 and FRI Sections of PCCF Office (*Facts and Figures*, 1992/3: 104).

but also the necessary electricity to run the mills. But he was annoyed with the former Forest Secretary T. Seshan, who had refused to give permanent land sites in the forest for the mill's captive plantations. Seshan had even turned down the request for an allotment of area in the Chambal valley on 'a 50:50 basis'. Evidently, what the official meant by the considerate behaviour of the government was the 20-year lease granted in 1977-8 to collect as much as 60,000 mt of bamboo per annum. Interestingly, the agreement with the industry was not signed by the representative of the state, generally the Principal Chief Conservator of Forests, nor by the State Secretary of the forests, but by the Governor. This was not the case with the other two mills owned by Birla.

The official in charge of the mills' raw material also clarified that the deal for raw material was not in terms of space but in the quantum of bamboo. The mills had to work the coupes annually from the lease area allotted to it by the Forest Department. During our visit in October

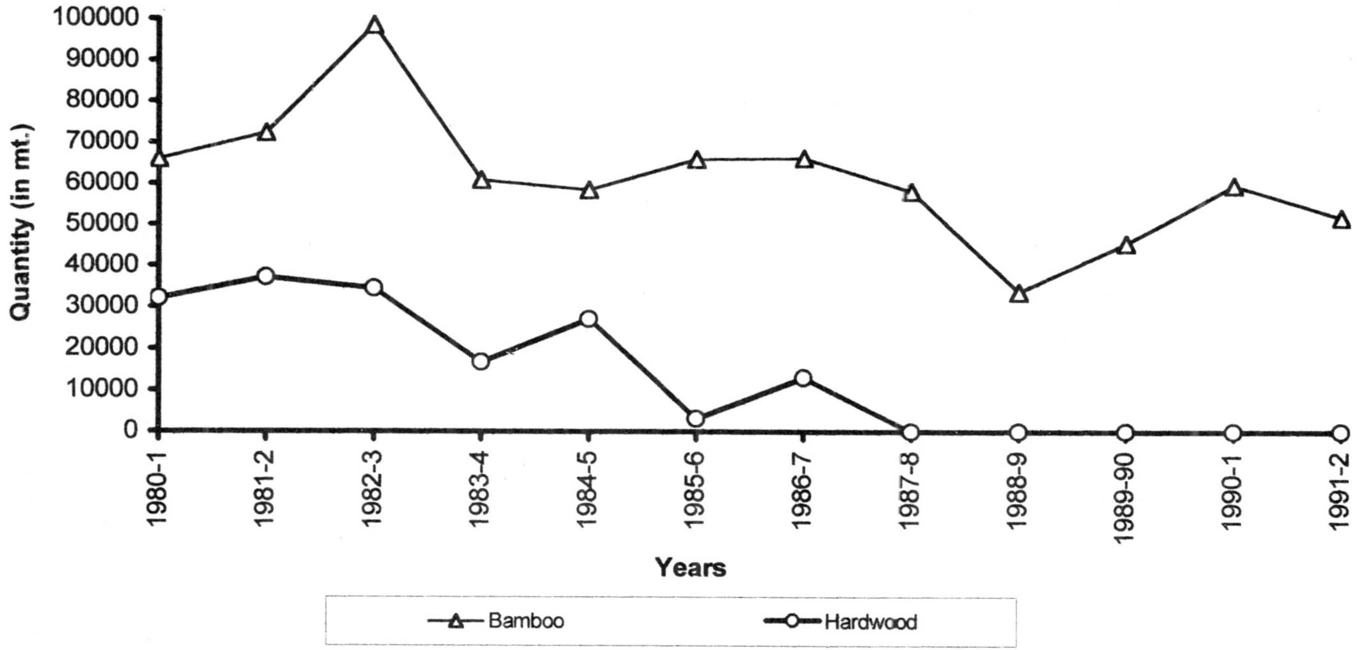

Source: PCCF Office - Andhra Pradesh Forest Department, 6 September 1992.

Figure 5.1: Quantities of Bamboos and Hardwoods Supplied to Bhadrachalam Paper Boards Limited 1980-1 to 1991-2

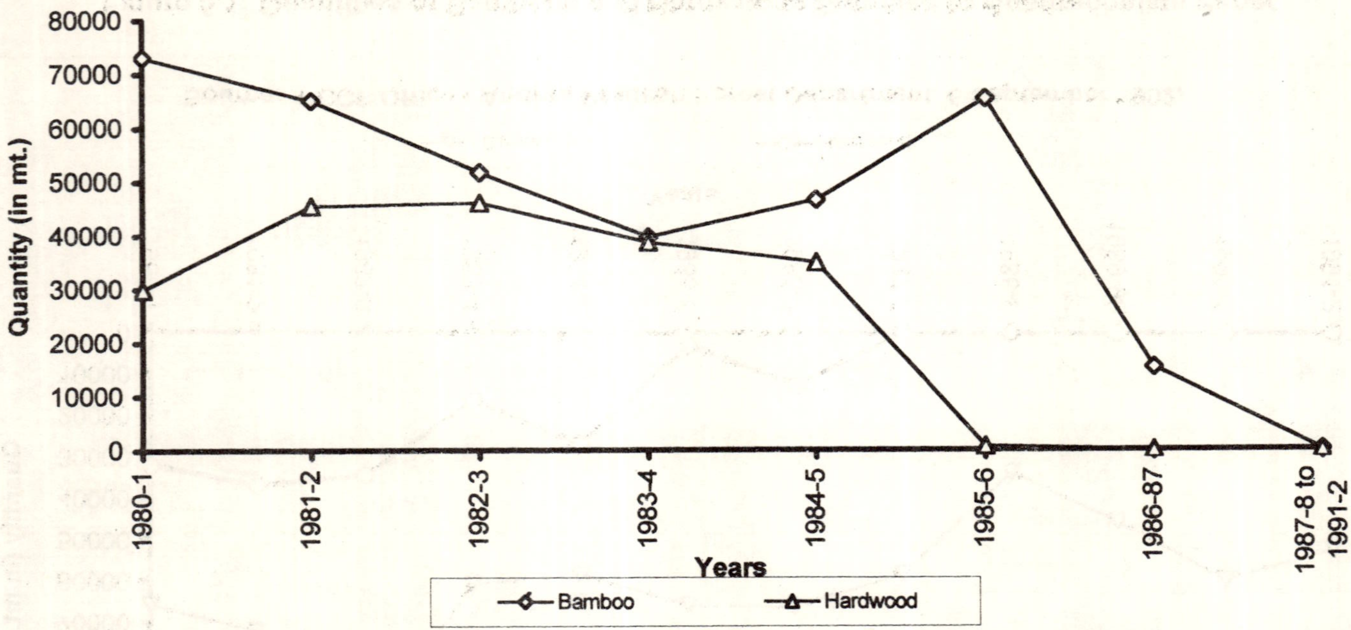

Source: PCCF Office of AP Forest Department. September 6, 1992.

Figure 5.2: Quantities of Raw Material Supplied to Sree Rayalseema Paper Mills

TABLE 5.8: SUPPLY OF BAMBOO TO BHADRACHALAM PAPER BOARDS, AND ROYALTY COLLECTED YEAR-WISE 1987-8 TO 1991-2 FOR KHAMMAM CIRCLE

Working year of the leased coupes	Circle	Quantity (in tonnes)	Royalty collected (Rs. in lakh)
1987-8	Khammam	58,198.428	253.09
1988-9	Khammam	33,651.894	162.44
1989-90	Khammam	45,569.858	244.61
1990-1	Khammam	59,728.126	382.67
1991-2	Khammam	36,832.588	262.12

Sources: Report of Conservator of Forests, Khammam and FRI Section of PCCF's Office Working Year, 1 October to 30 September of Successive Years (*Facts and Figures*, 1992-3: 104).

1992, he complained that the 1990-1 allotments were made so late, that the company was compelled to buy bamboo from Maharashtra. A retified Principal Chief Conservator of Forests was very sceptical about such a statement. Besides, this statement not only does not tally with the figures given in the tables, but goes against the information that was generated from the Forest Department in Chandrapur, the bamboo-growing area in Maharashtra. All its industrial bamboo is allotted only to BILT.

One of the mill's officials was annoyed that the best of the coupes were allotted to the Buroods, the artisans who work on bamboo. The bamboo artisans need mature upper portions of tall bamboo to work on. The sturdy lower portions are allotted to PPI. PPI needs cellulose. The age of the bamboo is of no consideration, but it is the lower ends so rich in cellulose which are the preferred bits. Incidentally, the quantum of bamboos allotted to bamboo artisans is small in relation to what is consumed by the paper mills. For instance, Forest Department figures show that the number of bamboos allotted to Burood Co-operative Societies in the Khammam circle in contrast to the metric tonnes of bamboos allotted to ITCB for the years 1987-8 to1991-2 are indicated in Table 5.9. Such discrepancy in information has been found in every one of the cases investigated.

The industry's interest is to get hold of raw material. The team's concern is mainly to understand whether the industrial unit harvests raw material from its allotted forest area in a manner which would give it a sustained yield, and if it has not, how it expects to cope with it. Will it follow the beaten path of introducing exotic hardwoods in the farms and in the forests?

TABLE 5.9. FOREST DEPARTMENTS ALLOTMENT OF BAMBOO FROM KHAMMAM CIRCLE 1987-8 TO 1991-2

Allotted	1987-8	1988-9	1989-90	1990-1	1991-2
Burood Cooperative Societies in terms of number of bamboos	6,22,577	6,01,213	10,35,120	10,94,936	n.a.
ITCB in terms of mt	58,198.428	33,651.894	45,569.858	59,728.126	36,832.585

Source: *Facts and Figures*, 1993: 103 and 107.

Observation of Felling Practice of ITCB

Bamboos with their dual system of propagation are indestructible if harvested with care, and protected when they seed. Poloncha, Kothagudam, Khammam, and Bhadrachalam South Division of Khammam circle, are the mills' allotted areas for felling. The team's observational visit in October 1992 was to Barlamadgu forest of Poloncha Division in Khammam circle. This was situated in the fertile foothills of the Eastern Ghats. Most of the forests on the 50 km south-east tract from the mill premises to the bamboo forests of Barlamadgu were either partially or totally degraded. The prevalence of *Cleistanthus collinus* and *Pavetta indica* were obvious signs of degradation of the soils which had once harboured thick forests. There were also vast patches of *Babul*, *Acacia nilotica*, close to the mill premises *en route* to Barlamadgu forests. It is again a species which thrives and colonizes in degraded areas. As we entered the foothills, the condition of the forests began to improve, but only deep inside did the moist deciduous turn three storeyed with *D. strictus* as the second storey. As we entered deeper, it turned into almost pure bamboo, a climatic climax according to the forest manager, but more likely a bamboo break due to edapic conditions. The terrain was uneven, the soil deep and moist with small rivulets criss-crossing the narrow pathway cut for one-way truck transport through the uneven terrain. There was no evidence of cattle or human interference in the deep forest, except from the mills.

The bamboo in the area had been felled the previous season and stacked in little inlets and slopes in order not to block the narrow pathway. The bamboos were being loaded on to trucks to be transported to the mills. The area had been cleared of all debris. The

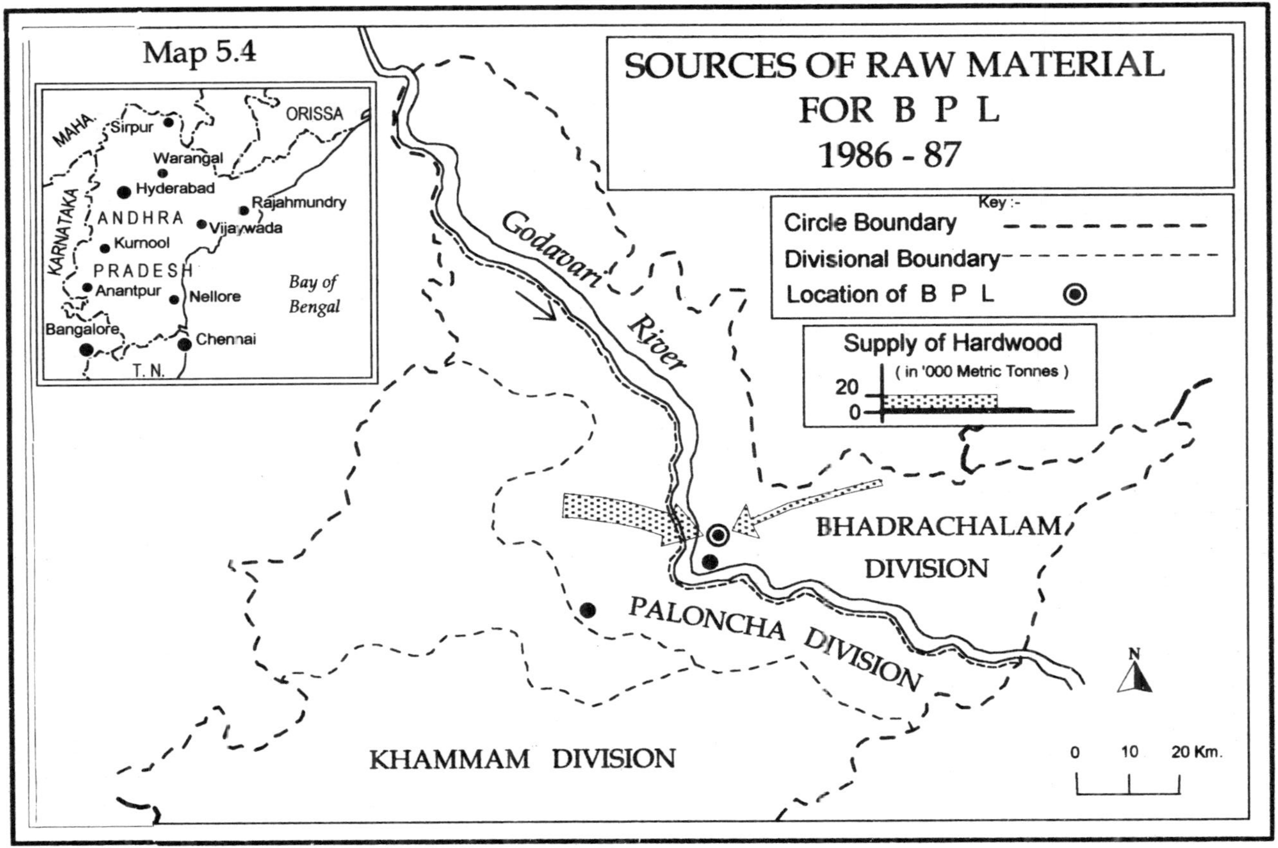
Map 5.4
SOURCES OF RAW MATERIAL
FOR B P L
1986 - 87
Key :-
Circle Boundary
Divisional Boundary
Location of B P L
Supply of Hardwood
(in '000 Metric Tonnes)
20
0
Godavari River
BHADRACHALAM DIVISION
PALONCHA DIVISION
KHAMMAM DIVISION
N
0
10
20 Km.
MAHA.
Sirpur
ORISSA
Warangal
Hyderabad
Rajahmundry
ANDHRA
Vijaywada
KARNATAKA
Kurnool
PRADESH
Bay of Bengal
Anantpur
Nellore
Bangalore
Chennai
T. N.

bamboos did not appear healthy. They were on an average about 10 cm in girth no more than 10 ft tall, particularly those close to the pathway. Their girth and height increased progressively on the steeper slopes. The clumps close to the pathway were small, five to seven culms on an average, but the size of the clump too, increased on the steeper slopes.

The bamboos were harvested by daily labour from the mills. The mill's representative claimed that it pays higher wages than the Forest Department and the productivity is, therefore, higher. Such claims by other mills often meant clear felling. But this was not the team's observation here. The last season's bamboo appeared to have been felled with little knowledge of basic silvicultural rules, since neither the number of culms to be retained nor the specific height and angle at which they ought to have been cut were being followed. Nor was there any evidence of protecting the bamboos by piling up of soil, or digging a trough around the clump. Although one could not see any sporadic flowering, there were quite a few healthy new seedlings in damp spots. A narrow rivulet that meanders through this area is said to keep the valley bed moist till February, by which time the new seedlings would perhaps establish themselves. The climatic conditions at that point of time appeared suitable for *D. strictus* to flourish.

Like all other PPI units, ITCB had lost faith in bamboo, and turned to other species like eucalyptus and casuarina. Both were expensive and purchased in the free market. Its Research and Development wing (R and D) had established that while casuarina has better pulpable qualities than eucalyptus, it is also more expensive. 'So like beggars we buy only its lops and tops'. While a senior executive says 'we pay a couple of rupees more than other mills and buy it off, another mill to its south which has a 'putting-out' system to cultivate casuarina, complained of poaching on its territory. In Bhadrachalam the terrain and climate are not suitable for the growth of casuarina.

Farm Forestry

ITCB has turned to eucalyptus cultivation. The relative price of eucalyptus when purchased is high. It is in the ratio of 1:9 as compared to the cost of bamboo from the leased area. Experiments in cultivation began in the late 1970s. The walled area of the ITCB is dense with trees. Some are vestiges of the forests, while most others are various species of pulpable hardwoods, while there are eighteen species of eucalyptus. The company finally opted for *E. tereticornis*, as its experiment with *E. camaldulensis* had failed, or rather, given up after a vicious fungal attack. A senior executive thought that eucalyptus was no longer a controversial species. That the mills' *E. camaldulensis* plantation was attacked by fungus and destroyed did not seem to him to be a point of controversy.

Like all other PPI units, this unit too has concentrated its R and D on production of raw material. The forestry section of ITCB made a proud display of its cloning technique. The technique consists of simple crown cuttings of the best trees of *E. tereticornis* induced to root with root hormones and protected in a mist chamber. The rooted seedlings are gradually exposed to sunlight, given a bit of inorganic fertilizer and planted in polythene bags. In about 35 days the seedlings stabilize and are ready to be distributed to farmers.

True, Forest Secretary T.N. Seshan had turned down ITCB's request for prime forest land to be transformed into pulpwood plantation. But there were other opportunities that the industry exploited for instance, the oppressive landlords of Telengana, fearing the Naxalites who had organized the poor indebted peasants, fled to Hyderabad, leaving the lands under their control lying unfarmed. This is significantly connected with agrarian relations in backward areas which have certain features: (a) leasing-in and leasing-out of land by both the landlords and the poor peasants is common; (b) the poor peasants virtually lose control over their leased-out land through indebtedness; (c) however, tribal lands cannot be alienated. Bhadrachalam lies in the tribal belt, as a result the poor peasants hold their title but have no control over their land. The landlord, operating from Hyderabad finds a tree crop with little labour input to turn to farm forestry on the leased-in land of the poor tribal peasantry. The industry provides the seedlings, the title deeds of the poor peasants which he has now in his possession, entitles him to bank loans through a NABARD scheme put forth by National Commission of Agriculture in 1976. This scheme is beneficial to both the industry and the landlord who can reap a reward at the end of the eighth or the ninth year, when the eucalyptus crop is felled and permitted to coppice for a second round. The gains for the poor peasants on whose land the eucalyptus is cultivated are far from clear. True, NABARD gives the poor peasant a right to sell his crop to the highest bidder. But this right is probably only theoretical. He has neither control over his land nor the crop on the land. Through his indebtedness he has forfeited his rights. In the NABARD tripartite agreement, the poor peasants on whose land the trees are cultivated get some sort of a survival or living wage till the crop is harvested. But what would be the status of the land where eucalyptus is cultivated? Empirical experience in other areas, such as well-cared for plantation areas of the Mysore Paper Mills under the immediate concern of a Chief Conservator of Forests, or the Pudukottai plantations in Tamil Nadu under the guidance of a Chief Conservator of the Forest Development Corporation, shows that the new cycle of even the eucalyptus crop is dismally poor, evidently due to further degradation of the soil by *E. tereticornis* itself. FAO when confronted by GOI confessed the futility of cultivating this particular species. Will the hopelessly-degraded land yield food or any other crop for the poor peasants to survive after a couple of rotations of eucalyptus? Does the industry care, does NABARD know? In August 1995, the officer-in-charge of forestry, a FRI trained forester, confessed he did not know while the IDBI Chief of Project Finance was sceptical (personal communication, August 1992).

Catchment Area of Andhra Pradesh Paper Mills (APPM)

East Godavari Forests are rich in bamboos. 42.8 per cent of the total bamboo area in this circle, 4,57.00 sq. km under *Bambusa arundinacea* and 3,775.00 sq. km under *D. strictus* (*Facts and Figure*, 1993: 23). The pre-investment survey in 1969-70 classified these bamboo forest as:

	Scattered	Dense	Pure	Bring the total to
	1,790 sq. km (42.3%)	1,272 sq. km (30.1%)	1,170 sq. km (6.5%)	4,232 sq. km 100%
Quality I	498 sq. km	804 sq. km	773 sq. km	2,075 sq. km 49%
Quality II	1,292 sq. km	468 sq. km	397 sq. km	2,157 sq. km 51%

Source: *Facts and Figures*, 1993: 25.
Note: The meaning of Quality I & II have been explained under the head Adilabad Forest Division in the earlier section.

APPM is located in Rajahmundry. The mills production capacity is 1,01,447 tpa and the raw material assured by the Forest Department is 75,000 mt per annum (*Facts and Figures*, 1993: 97). The mills' supply from the leased areas in the Rajahmundry and Visakhapatnam circles and Kakinada and Narsipatnam Forest Division from the year 1979 to 1992, and the royalty collected, are given in three consecutive Tables 5.10, 5.11 and 5.12. There are some slight discrepancies in

TABLE 5.10: SUPPLIES OF BAMBOO TO ANDHRA PRADESH PAPER MILLS LEASE YEAR-WISE 1979-80 TO 1991-2

From the leased coupes during the working year	Circle	Quantity (in tonnes)	Royalty paid (Rs. in lakh)	Remarks
1979-80	Visakhapatnam/			
	Rajahmundry	69,599.00	—	—
1980-1	Visakhapatnam/	66,201.69	39.721	
	Rajahmundry	772.51	0.463	Burnt Material
	Total	66,974.20	40.184	
1981-2	Visakhapatnam/	95,631.06	94.597	
	Rajahmundry	2,318.80	2.783	Burnt Material
	Total	97,949.86	97.380	
1982-3	Visakhapatnam/	70,356.95	98.500	
	Rajahmundry	9,285.67	13.000	Burnt Material
	Total	79,642.62	111.500	
1983-4	Visakhapatnam/	15,138.33	104.784	
	Rajahmundry	49,353.63		Burnt Material
		64,491.96		
		7,472.31		Burnt Material
		71,964.27	104.784	
1984-5	Visakhapatnam/	28,027.96		
	Rajahmundry	45,649.07		
		73,677.03		
		2,385.86		
	Total	76,062.89	110.498	
1985-6	Rajahmundry	61,524.00		
1986-7	L.P.C. Rajahmundry			
	S.C. Visakhapatnam			
1987-8	S.C. Visakhapatnam	49,244.00	214.16	
1988-9	S.C. Visakhapatnam	41,255.00	199.15	
1989-90	S.C. Visakhapatnam	65,000.00	344.88	
1990-1	Rajahmundry			
	Visakhapatnam	51,415.00	329.42	
1991-2	Visakhapatnam	24,461.00	166.90	

Sources: Progress Reports of Conservator of Forests, Rajahmundry and D3 Section of Office of the CCF Office, in *Facts and Figures*, 1993: 98.

Note: Working year of lease of the successive year; period year starts from 1 October to 30 September of the succedding year.

TABLE 5.11: SUPPLIES OF BAMBOO TO ANDHRA PRADESH PAPER MILLS FINANCIAL YEAR-WISE 1987-8 TO 1991-2

From leased coupes during the working years	Circle	Financial year	
		Quantity (in tonnes)	Royalty paid (Rs. in lakh)
1987-8	1. Rajahmundry 2. Visakhapatnam	49,244.00	
1988-9	Visakhapatnam	41,255.00	199.15
1989-90	Visakhapatnam	69,000.00	348.86
1990-1	Visakhapatnam	n.a.	
1991-2	Visakhapatnam	n.a.	

Source: Progress Reports of the Conservator of Forests and Chief Conservator of Forests, from *Facts and Figures*, 1993: 99.
Note: Working Year: 1 October to 30 September of the successive year.

the published data given by the Forest Department. So no attempt has been made to collate it, the sources from which it has been compiled being different, also there were no means to check, counter check or correct it.

We do not know whether the hardwood data includes 452 mt supplied from Visakhapatnam circle. The company paid a royalty of Rs. 0.657 lakh (*Facts and Figures*, 1993: 99).

Assessment of APPM's Felling Practice

This study holds that bamboo, not hardwood is the easily renewable resource for manufacture of paper. Our assessment of the felling practices of bamboo is based on the assumption that if the rules of the game are followed, bamboo would continue to yield and even increase and the, rhizome spread. This is not a difficult task in the case of *D. strictus*. The team's observation was made in early November 1992 on the way to Maradmelli from Rajahmundry. Now and again adjoining the road, sometimes on the right and sometimes on the left, the study team saw dense bamboo, *D. strictus* brakes, neat and in excellent condition. Are they natural or secondary formations on almost flat surfaces through which a motorable road has been cut, or are they plantations? Good, well-maintained plantations of coffee, cashew, etc., abound in Andhra, and secondary bamboo formations elsewhere have been observed to be very dense but not 'neat'. Evidently these were FD's bamboo plantations. There was also much confusion in the account the study team was given, of who had felled the bamboos in this area and stacked them on the roadside. Whoever did it was totally oblivious of the basic silvicultural rules laid for felling. The bamboos were indiscriminately hacked at an average height of three feet above the ground level. The age of the bamboo was also ignored. These faults were gleefully pointed out by the representatives of the mills as a folly committed by the Forest Department. This information does not quite tally with that of APPM, which has a right to fell within its catchment areas. And this was its catchment area. There was no evidence at any point of heaping of soil and building of low troughs around the clump to conserve water and increase the vegetative growth of the rhizomes and the culms about which Forest Department officials had spoken so proudly at the head office. A third puzzling fact was the assertion made by a representative of the industry that the mill had not cared to collect the bamboos that were felled in June. Truckloads of bamboo did pass us, going towards Rajahmundry. A stationary one, heavily loaded with bamboo seems to have created a great deal of anxiety for the officials who had accompanied us on the field trip.

Tables 5.12 and 5.13 show that the quantum of available bamboo is reducing. The Forest Department has ceased to provide hardwoods therefore APPM sought an alternative.

From the East Godawari circle alone the quantum of bamboo and hardwoods was as shown in Table 5.14.

In seeking an alternative it chose casuarina which is not new to Andhra. It was introduced by the Forest Department of the old

TABLE 5.12: SUPPLIES OF BAMBOO TO ANDHRA PRADESH PAPER MILLS FINANCIAL YEAR-WISE 1987-8 TO 1990-1

Supplies made from the lease year coupes	Division	Quantity (in tonnes)
1991-2	1. Kakinada	17,244.045
	Rajahmundry circle	17,244.045
	2. Visakhapatnam	1,160.495
	3. Narsipatnam	19,937.047
	Visakhapatnam circle	21,103.902

Source: Facts and Figures, 1993: 99.

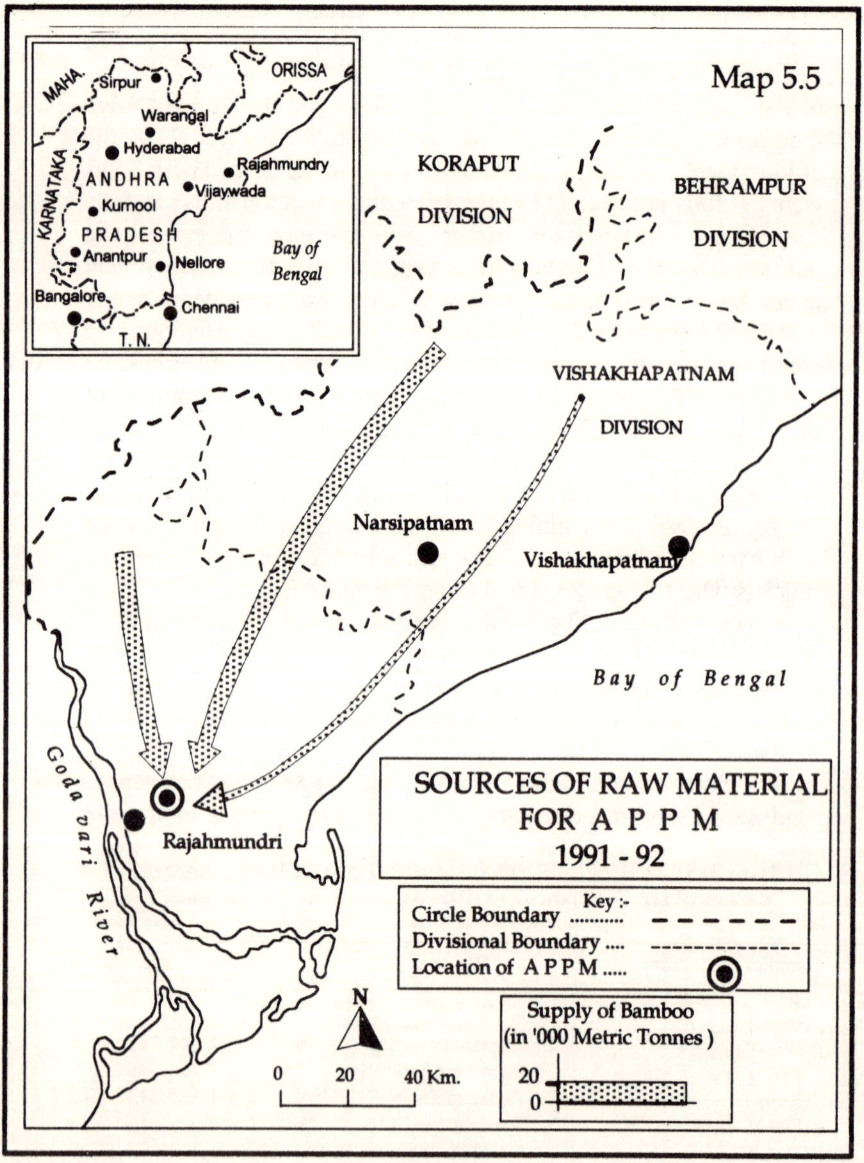

Map 5.5
MAHA.
Sirpur
ORISSA
Warangal
Hyderabad
Rajahmundry
KARNATAKA
ANDHRA
Vijaywada
Kurnool
PRADESH
Anantpur
Nellore
Bay of Bengal
Bangalore
Chennai
T. N.
KORAPUT DIVISION
BEHRAMPUR DIVISION
VISHAKHAPATNAM DIVISION
Narsipatnam
Vishakhapatnam
Bay of Bengal
Godavari River
Rajahmundri
SOURCES OF RAW MATERIAL FOR A P P M 1991 - 92
Key :-
Circle Boundary ....
Divisional Boundary ....
Location of A P P M .....
N
0 20 40 Km.
Supply of Bamboo (in '000 Metric Tonnes)
20
0

TABLE 5.13: RAW MATERIAL SUPPLIES TO ANDHRA PRADESH PAPER MILLS FROM 1980-1 TO 1992-3

(in tonnes)

Year	Bamboo	Hardwood
1980-1	2,12,583	58,422
1981-2	96,001	38,612
1982-3	64,451	14,558
1983-4	66,074	49,696
1984-5	75,892	53,777
1985-6	50,007	59,845
1986-7	70,829	45,375
1987-8	60,891	12,230
1988-9	41,043	2,702
1989-90	72,017	—
1990-1	45,425	—
1992-3 (up to Dec. 1992)	56,682	

Source: PCCF Office, AP Forest Department, 6 September 1992.

TABLE 5.14: QUANTUM OF BAMBOO FROM EAST GODAVARI CIRCLE TO ANDHRA PRADESH PAPER MILLS 1978-9 TO 1989-90

Year	Bamboo supplied	Hardwood supplied (in cu.mt) for its timber and veneer
1978-9	37,310	1,07,824
1979-80	31,824	—
1980-1	56,536	1,03,339
1981-2	37,339	72,314
1982-3	21,514	28,375
1983-4	42,405	74,558
1984-5	35,635	79,804
1985-6	34,261	1,16,279
1986-7	42,647	78,503
1987-8	33,915	—
1988-9	13,424	—
1989-90 (up to the end of Feb. 1991)	26,085	—

Source: DFO Bamboo Weighment Division Rajahmundry, quoted in *Working Plan for Forests of* East *Godavari*, 1991-2 to 2005-6, Part I, K. Prakash Rao.

Note: Only Brushwood from pulpable species. The trunk is for veneer or timber.

composite Madras Presidency as part of its programme to create fuel-wood, and planted all along the coast which included Andhra and part of Orissa in the east coast, Malabar (Kerala) and Karnataka on the west coast. The coastal plantations served an additional purpose as a shelter-belt for crops. Much later it was taken up by Maharashtra and Gujarat.

All the regions in India where casuarina is grown have a tropical climate with a mean annual temperature of 28° C and are frost free. The topography ranges from coastal flats to undulating terrain; the soils vary from coastal sand, shifting sterile sand, river alluvium, red gravelly loam with a high water table, red loam, hard laterite, etc. The species of choice in India is *Casuarina equistefolia* which is not very drought resistant.

Farm Forestry and Effects of Casuarina on Farmers and Farmlands

In 1986 APPM had started using *C. equistefolia* as its raw material. Its consumption rose from 19,605 mt to 68,390 mt by 1989. Casuarina is an expensive hardwood much in demand by the construction industry and for sturdy mine props. To the paper industry, as the price of raw material is of critical concern, it had to find some means to overcome the high price. This it found in farm forestry. Social forestry programmes for a paper industry are supported by NABARD as tripartite agreements. But APPM could not avail this, said a senior executive, since the farmers had some problems with their land titles. Without a clear title a bank will not loan money for farming. But a middle-ranking official of the corporation explained that small and marginal farmers cannot afford to take on a long plantation tree crop. Yet many small and marginal farmers lease out their land on a 50:50 basis to richer farmers; they provide labour and get 50 per cent returns. It is the landlord who invests in the small and marginal farmers' land which he leases. NABARD cannot step into these deals and make farm forestry more widespread as the landlord does not have the title to the leased-in land on which casuarina is grown. The poor peasant who is often indebted to a bank can demand no credit. The question therefore is—what is the long-term effect of growing casuarina on the land; can poor peasants afford to permit this deterioration of their land?

The company's estimated requirement is 30,000 acres for a sustained yield in order to achieve self-sufficiency in raw material.

This, of course, the mill's official conceded, would reduce the total output of rice in the region. The delta region is considered the rice bowl of south India. With rising wage-rates the absentee landlords have taken to casuarina.

An attempt was also being made by the APPM in 1992 to take on 1,000 ha of land on a 5-year lease to start its plantation. A junior official explained that the rent for land ranges from Rs. 1,000 to Rs. 2,000 per ha depending on the type of soil. A deal for a 400 ha plot was said to have been almost finalized at the end of 1992. In the leased area a well would be bored, for if saplings are watered a couple of times a month in summer, their growth and yield are hastened. The cost of the bore-well, around Rs. 1 lakh, is deducted from the rent.

There are other classes of farmers who have contributed to the spread of casuarina-farm-forestry programme for APPM. According to an official, landless scheduled caste farmers formed a society and asked the district collector to grant them land, as casuarina can grow on barren degraded land. It thrives on porous soil. The members of the society who have titles are entitled to bank loans. It is not very clear how the groups of scheduled castes got hold of this ingenious idea. Different sources give different figures and different explanations for the expanding APPM farm forestry programme. According to *Business India* (March 1992), the company has distributed about 10 million saplings to marginal farmers in the delta region of East and West Godavari districts and in Visakhapatnam district to cover an area of over 3,333 ha which would ultimately yield 10,000 tonnes of wood. The reason for the shift in the raw material, according to this source, is two-fold. The 30-year lease with the Forest Department of Andhra Pradesh came to an end in 1975. The fresh lease for 1 lakh tonnes of bamboo was at an increased rate of Rs. 60 per tonne which was further hiked to Rs. 210 per tonne from October 1980, and Rs. 30 per tonne for hardwood. From 1988 onwards the Forest Department stopped its supply of hardwoods. Bamboo is expensive. The cost of transport of bamboo is also expensive. Therefore the company began to progressively change the raw material mix of 70 per cent bamboo and 30 per cent hardwood to the present 70 per cent casuarina and 30 per cent bamboo.

Casuarina is said to have several advantages over eucalyptus. According to APPM officials the wood can be pulped without debarking, thereby saving the cost of a single operation. Secondly, the pulp is brighter than the eucalyptus pulp. The cost of bleaching and effluent treatment is reduced. The yield of the fibre is seven per cent more and the strength higher. There is of course a catch, two in fact. The wood can be stored only for three or four months as it hardens under prolonged storage. Secondly the disadvantage to the farmer is that its coppicing ability is poor and the young sapling is grazed by animals (Subramaniam 1992: 104)

The work put in by the Central Government Research Station—The Institute of Forest Genetics and Tree Breeding (IFGTB) Coimbatore shows that high-yielding varieties could be produced by vegetative propogation. The yield per acre of such a 4-5 years old plantation would be 8 to 10 tonnes, and the returns as high as Rs. 7,400 per acre (Subramaniam 1992: 104) as demand is rising. Thapar's AP Rayon buys casuarina, so does Birla's Harihar Polyfibre—as much as 1 lakh tonnes at the rate of Rs. 740 per tonne from Tamil Nadu. APPM is however sore that other PPI units poach in its area. ITCB confesses it has learnt the trick; 'pay a couple of rupees more and get hold of the raw material' as transport of bamboo from other states is expensive. Competition is the game the corporations play.

C. equisetifolia is an angiosperm belonging to the family of casuarinacea. Morphologically, it is characterized by long needle-like branchlets divided into regularly-spaced nodes. At each of the nodes is a ring of teeth-like structures which are reduced leaves. They arise in a whorl at each node and fuse along the inter-nodes forming longitudinal ridges on the branchlets. Between the ridges are grooves in which the stomata are located.

The flowers of casuarina are unisexual and greatly reduced. Male flowers are produced in catkins. Each flower consists of a single stamen enclosed in a pair of deciduous tepals in the bud stage and two bracteoles which are frequently deciduous. The pollen is light and plants are wind-pollinated. The female flowers have no perianth and are produced in small cones. Each flower consists of a single unicellular gynocecum formed of two carpels subtended by two lateral bracteoles and a bract. *C. equisetifolia*, is both dioecious as well as monoecious (Subramaniam, 1992: 1).

The scientists from the central government-sponsored Institute of Forest Genetics and Tree Breeding, Coimbatore, had set in motion the practice of intensive breeding of *C. equisetifolia*. APPM which was interested in this species as its pulpable raw material concentrated its R and D on vegetative propagation of the species, to enable transfer of superior phenotypes from the selection sites to the clonal bank.

'But', says Forester A.L. Rao, 'the benefits accruing from good but unproven genotypes selected is minimal' (A.L. Rao, undated). The clones are now used for its seed production farm at Palacharla. It is a 7-acre plot said to be taken on 'lease cum sale basis' in April 1992, whatever that might mean. The whole area is divided into 25 plots, each plot holding 450 saplings. Its own 30-day old clones, both male and female, are planted in a certain pattern in the ratio of 75 female and 25 male which is meant to cut off pollination from other genetically inferior ones from outside. The farm is intensively managed to produce a frequent, abundant, easily harvested seed crop. The clones are from plus trees selected on the basis of morphological characters such as predominant height and diameter, strong apical dominance and a straight bole. Seeds are planned to be collected clone-wise, and information on phenological characters collected to establish a one-parent progeny trial. Once the growth phenology, heritability and combining ability are obtained, the inferior clones are to be weeded out. This is meant to eventually establish a second generation seed orchard (A.L. Rao undated: 18).

But evidently the mill cannot wait for such long drawn-out programmes. The seed referred to as 'good hybrid seeds that first experimental batch yields is proposed to be collected and distributed to the farmers in place of saplings' said a mill officer. This project looks not just at the growth of the wood-based paper industry but also at the impact of the industry on the forest ecosystem as well as the ecosystem where pulpwood trees are attempted to be cultivated. The forest ecosystem has been disturbed through over-felling and therefore the wood-based raw material that is sought to be generated outside the forests and its consequences on the ecosystem is also part of the concern of this project. Humans are a part of the ecosystem, and the lands of the small and marginal farmers on which the trees are grown are also a part of the same concern. Under the prevalent pattern of agrarian relations, the small and marginal farmers have legal title to the land, thanks to the land ceiling acts, yet through indebtedness have lost their control over their own lands. Earlier this study has stated that many of them leased out their land to richer farmers, and it is in such lands, that, by and large, casuarina, eucalyptus and some doubtful species such as tobacco are grown, exposing the land to degradation.

Neither the richer farmer nor the mill which has been distributing saplings (and in another five to eight years will distribute seeds to farmers) care to match the species with the soil. And if we go by the evidence of 50 years, single species plantations face a dual threat—a fall in the second and subsequent yield due to alteration in the nutrient status and the physiological properties of the soil (FAO 1991, *Report* no. 103: 13). Most species of eucalyptus are extremely attractive and majestic trees, but the straggly plantations that abound in this country today are a mute testimony to the mismatch of the species to the soil.

Except for a small beginning, not much work has been done on *C. equisetifolia.* K.K. Suresh and Vinaya Rai report that farmers have voiced concern about the harmful effects of these trees on cultivated lands and on standing crops. Their comparative study of this species along with two other pulpable species, *E. tereticornis* and *L. leucocephala*, on Sorghum (Co_{24}), Cow pea (Co_4) and Sunflower (modern elite), show that the maximum inhibition of germination and seeding growth is by eucalyptus. It is of comparatively smaller magnitude in the cases of casuarina and leucaena. No difference is manifested between the two species (Suresh and Rai 1987: 109-15).

A.L. Rao (p. 21) reports that in the APPM catchment area inter-cultivation of groundnut, til and pulses is also carried out in the year of raising of plantation, which yields a revenue of about Rs. 1,000 per ha in a normal year. But inter-cultivation is not possible or economical in the second year of plantation which, according to him, is due to 'such close spacing'. How spacing changes in the space of two years is not clear. Is Rao rationalizing the failure of crop to yield with the growth of casuarina over time'? As early as in 1949, FAO had warned against the danger of high casualties in the Taunga crop when inter-planted with *Casuarina equisetifolia.* In its words, 'An intensive programme of cooperation between the silviculturists and mycologists and soil scientists alone can help to meet the danger that threatens the very foundation of forests'.

That casuarina changes the status of the soil cannot be denied. Dr. Kadambi suggests that in areas which are fairly fertile, crop mixture rather than crop rotation, be practised. After growing casuarina for one or two rotations, the soil may be put under broad-leaved species so that it regains whatever fertility it has lost (Puri 1958: 90). Casuarina's nitrogen-producing nodules alone that the APPM boasts about, do not necessarily improve the fertility of the soil. Casuarina like any other species has both a positive and a negative impact on the soil. A.Y. Venkatachalam et al. (undated), reports that casuarina brings down the pH of the soil. It has therefore very positive action in both alkaline and saline soils which do not yield paddy. This was found in E. Kandasamy Pandian's holding no.194, in Cherankoilpathu village in Ambasamudram taluka in Tirunevelli district. E.K. Pandian was able to grow paddy after harvesting a casuarina crop. The pH of the soil which had been 9.4 went down to 8.3, which the researchers explain could have been due to absorption of calcium and magnesium by the trees. Paddy can grow only in soil with pH below 8.3 and conductivity 3.1 milli hom cm^{-1}. Since the species also has a well-developed lateral root system with nearly 60 per cent of root weight relegated to the top 30 cm, the increased nitrogen content observed is thought to be due to dinitrogen (N_2) fixation and less through return of litter fall. But the researchers found considerable depletion of potassium especially from 15 to 30 cm depth in plots that were under the tree. There is an export of potassium through harvested produce.

The second threat of monospecific plantation is increase in the risk of a serious attack by insects, pests or diseases. Unless other habitat requirements of the attacking organisms can be managed (for details see Subramaniam et al. 1992: 62-6), would APPM or the rich farmer who control the poor farmer's crop come to the latter's rescue should need arise? APPM's consultant A.L Rao lists not only fungal diseases that attack nursery seedlings that are vulnerable, but also those that attack young and mature trees in plantations, the most common being *Phomopsis casuarina* which causes stem canker and eventual death of the tree; *Corticium salimouicolour*, the pink disease; *Dolhloretta* species, *Phoma* species, which cause root infection and death; and *Trochosporium nesiculosom* which creates a black blister bark. He identified the causes which range from edapic and climatic conditions to overuse of fertilizers in the nursery stage and suggests methods to limit and cope with the spread of fungal diseases. Then, there are the insect pests (a) *Cryptotheria crameri* (bag worm) is the major one, and (b) the difficult to cope with bark-eating caterpillar *Inderbella quadrintaca,* (c) two species of termites—*Odontotermia obsus* and *O. wallonesis* which damage the root system in young plantations, (d) meal bugs which debilitate and kill and cause considerable damage to young plantations, (e) The beetle *Celosterma scabrator* which is still not a very great threat in AP (Rao: 12-13). Under ideal conditions a casuarina plantation in the APPM catchment area should in a five-year period, yield an income of Rs. 40,000 per ha, twice as much as a goad cereal crop. Apart from the soil which is sandy loam and red loam ideal for the growth of a beautiful casuarina tree, everything else is far from ideal. The very concept of monoculture, except perhaps in the sand dunes, according to the team's present knowledge, is dangerous. Besides, there are contradictions in the current pattern of agrarian relations which would adversely affect only the small farmers on whose land the casuarina is cultivated as farm forestry, for it would degrade their land and impoverish them further. In the larger context too, it is important to ask whether prime delta land, the rice bowl of the country, can be allowed to degrade.

Role of Forest Development Corporation

It is not just the industries which are augmenting their raw material resources but also the Forest Department. The first major effort was through setting up of the Forest Development Corporation in 1976. The Forest Development Corporation has an advantage over the State Forest Department, in so much that it can borrow money from the major financial institutions like NABARD and also draw from numerous agricultural programmes to pay for its labour force. The greatest advantage is the almost free land, the very best of forested land all over the country. In AP the lease rent for 47,288 ha of land transferred by the government was Re. 1 per ha per annum (CCF. Ref. no. 3407/87 D2 dated 11 February 1987) and the lease for 9,926 ha of land [GO Ms. no. 229 dated 24 June 1987 of FA & F (For V)] adopted at Re. 0.50 per annum. It was only in 1989 that the rate was raised to Rs. 100 per ha per annum with retrospective effect from 1987 (AP Govt. see GO Ms. no. 380 dated 12 October 1989) (APFDC Ltd, *Annual Report* for 1987-8: 34 presented on 5 September 1991).

FDC is mainly meant to generate raw material for forest-based industry. It has quietly gone beyond to introduce various types of plantations such as coffee and cashew, besides of course, bamboo. This is the only state which has bamboo plantations set up by its FD. These are well-tended as was evident all along the road from Visakhapatnam to Araku valley and from Rájahmundry to Maradumilli. Eucalyptus is certainly not its favourite but an FD obligation to the GOI. They are prominent near every railway station but not in the forests, except sometimes in Adilabad.

Only trees do not make a forest and a forest is a complex ecosystem. This study is interested in understanding how APFDC tries to augment PPI's raw material. According to the latest available information from APFDC *Annual Report* for 1989-90, 2,465.59 ha were planted during the year. Certain premature plantations had to be felled due to attack of an unknown disease. During 1988-9 Rs. 277.34 lakh compared to earlier Rs. 51.97 lakh worth of eucalyptus and Rs. 62.06 lakh worth of bamboo were sold. Due to the introduction of the Forest (Conservancy) Act in 1980, the state government has transferred only degraded land to APFDC. This has brought down APFDC's revenue by Rs. 0.57 lakh (APFDC, *Annual Report,* 1987-8: 9). The *Working Plan for the Forests of East Godavari District* 1991-2 to 2005-6, Part II, GOAPFD for instance, reports that a pulpwood plantation working circle has been created. The object of this working circle is to improve the value of degraded forests by planting fast-growmg pulpwood species like eucalyptus. The APFDC was set up during 1975 to create forests of industrially utilizable fast-growing species, mainly to meet the pulpwood requirements of the industry in the state. The APFDC Ltd, has raised 7,345.77 ha of eucalyptus plantations in both dry deciduous and moist deciduous forest areas. Plantations are raised in Valamirulu, Bodaluveru, Kakuvery, Kundata and others. The technique prescribed is clear felling the existing growth in degraded forest areas, uprooting stumps, tractor-ploughing and planting with six to seven-month old seedlings (APFDC, *AR*, 1987: 3).

The objective of the bamboo overlapping working circle is to meet the raw material requirements of APPM in respect of bamboo as well as that of local people. Thus the Working Circle (WC) includes all productive bamboo-bearing areas in almost all the blocks of the district. The WC includes 2,86,167 ha under forests. It is estimated to obtain an average annual yield of 34,600 tonnes of bamboo. It is proposed to supply annually about 30,000 tonnes of bamboo to APPM, Rajahmundry and the balance to meet domestic needs. Thus WC includes 907.6 ha of bamboo plantations also (APFDC, *AR*, 1987: 4).

In its GO no. 605, Forests and Rural Development Department dated 19 Sept. 1975, the Government of AP agreed to make available 1 lakh tonnes of bamboo annually to Andhra Pradesh Paper Mills for a period of 20 years from 1 October 1975 to 30 September 1995. It was also clarified that the mills have to use hardwoods and softwoods to meet their additional requirements of raw material. The mills have to pay a royalty of Rs. 60 per tonne of bamboo supplied to them from 1 October 1975, the rate being reviewed from time to time. The rate in force in the year 1989 was Rs. 536.71 per mt, (APFDC, *AR*, 1987: 124). In Maradumilli alone six plantations were raised from 1976 to 1979 which yielded both pulpwood and faggot wood, and also bamboos

TABLE 5.15: YIELD OF PULPWOOD FROM PLANTATIONS HARVESTED BY APFDC 1976 TO 1981 IN EAST GODAVARI

Year	Zone	Yield obtained as pulpwood (in tonnes)
1976	Sudikonda, Rajarommangi, Ultimodalu, V.S. Wada, Kuduru Bit I & II Elliwada	23939.50
1977	Lakkonda, Molleru, Mallavaram, Labbarthi, Ultimodalu, Rajarommangi Bit I & II, V.S. Wada, Kunduru, Bandapalli	31688.00
1978	Sudikonda, Rajarommangi, Maradumilli	35001.00
1979	Sudikonda, Rajarommangi, Maradumilli	37642.50
1980	Sudikonda, Rajarommangi, Maradumilli	21937.80
1981	Sudikonda, Rajarommangi	11840.00

Source: *Working Plan for the forests of East Godavari District* 1991-2 to 2005-6, Forest Department of AP, Part II, K. Prakash Rao, Rajahmundry.

Note: Details of yield obtained contain yields of faggot wood and number of props which are not taken into account in this table.

that could be used as props. The yield obtained as pulpwood is listed in Table 5.15.

AP's Forest Department's Research in Resources for PPI

Despite the controversy on monoculture particularly regarding exotic species which do not match with the sites, much research and use of biotechnology continues to be concentrated on producing, economically rather than socially useful or ecologically meaningful species, by the parent institution—AP Forest Department. This is highlighted in its colour brochure *Forestry Research in Andhra Pradesh 1992, Marching Towards 21st Century.*

The Forest Department also carries out a great deal of research spread over its various districts in the state. It either uses its own resources or turns to funding agencies and collaborators for its tree improvement programmes. Its interest is primarily in local and to some extent in exotic species as well. Its experiments go beyond the tried and tested processes to modern biotechnology. Experiments in afforestation techniques include agave plantations in the difficult dry regions of Cuddapah and Chittor districts. The study mentions this because agave is productive and useful for both paper and pulp making. Like bamboo it is self-perpetuating. APFD also carries out silvipastural experiments (AP 1992: 5-15).

TABLE 5.16: INDUSTRIAL PLANTATIONS IN ANDHRA PRADESH

Name of the industry	Total Planting targets (ha)	Annual Planting (ha)	Total Project Cost (Rs.)	20% interest free loan amount as equity capital	80% loans (Rs.)	Species				Period of repay-ment	Project revenue before loan pay-ment	Surplus the end of the year (Rs.)
						Euclaptus (ha)	Bamboo (ha)	Others (ha)	Total (ha)			
Bhadrachalam Paper Boards Ltd.	12,250	1,750	1,549.69	309.92	1,239.69	1,750	—	—	1,750	13 yrs.	4,853.41	1,880.75
Andhra Pradesh Paper Mills	2,970	—	427.11	85.42	341.69	—	—	—	2,970	11 yrs.	845.89	26.45
Sirpur Paper Mills	12,390	1,428 to 1,429 Eu. 265 Bam.	1,459.63	291.92	1,167.71	1,428 to 1,429 Eu.	265	—	1,649	15 yrs.	3,452.51	165.96

Source: Sirpur Paper Mills, November 1992.

SECTION 3 (A)

Sirpur Paper Mills, Andhra Pradesh

History

Sirpur Paper Mills (SPM), a Birla concern, established its mills at Sirpur Kagaznagar in the forest-rich Adilabad district. The mill is the second oldest in Andhra Pradesh. The date of the takeover of the mill by Birlas from the Nizam of Hyderabad is far from clear. According to one account it was in the early forties, and according to the mills own handout a decade later. Information for the initial years indicate that year of incorporation was 1938—7,500 shares issued to H.E. Nizam's government, 5,000 to Hyderabad Construction Co. Ltd., 15,000 to directors and associates of Hyderabad Construction Co. Ltd., 1,900 shares issued to Court of Wards Department while 15,600 shares were offered to the public. The position as on 30 September 1992 is as follows:

	Category	No. of shares held	% of total
I	Foreign holdings	24,245	0.41
II	Govt./Govt. sponsored financial institutions	36,88,333	61.91
III	Bodies corporate (not covered under I & II)	8,09,973	13.00
IV	Directors & their relatives	4,794	0.08
V	Top 50 shareholders (not covered under I, II, III, IV)	2,85,653	4.79
VI	Others	11,44,369	19.21
	Total	59,57,369	100.00

Source: *Stock Exchange Official Directory* (*SEOD*), Bombay, 1992: 25 (iv) 84740.

According to official documents SPM was established in 1938 and went into production in 1942-3 under the management of the Hyderabad Construction Company. In 1949 the management was taken over by the Industrial Trust Fund of the erstwhile Hyderabad government. In 1953 it was entrusted by the Hyderabad state to Birla Brothers (Pvt Ltd) as managing agents for the company. At that point of time it was still a small single machine (14 tpd) mill capable of producing cultural paper. Slowly it began its expansion and modernization programme and claimed that it was ready to take off in 1970 when the managing agency system was finally abolished. However, it is necessary to go back in time to understand the paper scenario and why Birla was attracted to Sirpur and how he coped.

During the Second World War for quite some time the supply of paper had fallen short of demand, particularly in respect of cultural varieties of paper, and the government had to control prices sometimes formally, and occasionally on an informal basis. From the point of view of reasonableness of prices and availability, the focus of attention of the government had been on the supply of white printing paper required for the production of text books and exercise books used by the student community (Dagli 1979: 295).

It is therefore important to see that though the paper industry was being recognized as one of the 'core' industries of Indian economy, and despite receiving protective tariffs as well as concessions from the Government of India for its smooth growth, it has used these concessions only to its own advantage.

Statutory price control by way of the Paper Price Control Order 1945, was withdrawn in 1950. Between 1950 and 1958 the industry revised prices five times. As a result, the government imposed control over paper prices with effect from January 1960 on the basis of recommendations made by the Tariff Commission.

Growth of PPI in the 1960s

The paper industry as a whole was unhappy about the steep rise in the cost of production, and retaliated by reducing production; SPM rationalized the problem and stated that production of paper declined in the year 1965-6 on account of restrictions imposed on the consumption of power by the State Electricity Board.

In the following year, 1966-7, SPM again stated that the company's water pumping station was completely submerged due to floods in Peddavagu river that September, resulting in a loss of production for about a fortnight. Peddavagu is a perennial stream from which SPM draws its supply of water. In addition, the shortage of essential imported spares and intermittent power cuts was said to have resulted in frequent breakdowns and stoppages in the plant which further adversely affected production. Every PPI unit behaved in a similar manner. In other words, the PPI created a paper famine. Its argument was that while there was a steep all-round rise in the cost of production, the prices of paper and board remained static due to price control.

According to the company it had to absorb this rise in costs, which again adversely affected its profit margin.

Until May 1968 there was a relatively small increase in paper production, barely adequate to meet the growing demand. Hence control over prices was removed that year. The company immediately started its programme of expansion for introduction of new products. This was an unsettling process and production output became uneven. It was also a period of world recession. However, due to decontrol of paper prices, the sales realization of SPM immediately improved but the company was unhappy that high sales were offset by the continued increase in the cost of production.

The company also stated that production declined due to labour problems. According to the *Stock Exchange Directory* (1978), even before the agreement of the company with the labour union expired on 30 June 1968, the union prematurely terminated the said agreement and adopted a non-cooperative attitude which affected working efficiency. The workers adopted 'casual' strikes whatever that may mean and go-slow tactics for higher wages. After that the company entered into a three-year agreement with the labour union.

Another factor to which the lowered production was attributed, was frequent tripping of the Ramagundam power supply. Production registered a slight increase in 1969-70 and 1970-1. Incidentally, there was an accident in boiler no.10 due to tube failure on 25 September 1970, but no casualty was reported.

It is interesting to see that once there was de-control on paper prices, the production of the company registered an increase. The company also claims that production could have been still higher but for the scarcity of imported spares and the frequent tripping of the Ramagundam power supply. Production also fell because the machines were being renovated to produce new value-added paper in the late 1960s and early 1970s. But the annual reports always have various excuses for declining production.

The company reported, that it experienced an acute shortage of water for about three months as a result of the failure of monsoon in the year 1971-2, and that a 25 per cent power cut was also in force during the summer of that year. Also workers went on strike for seven days. Production declined to 34,400 tonnes in year 1971-2 as compared to 35,300 tonnes in 1970-1. It is important to remember that from 1968 onwards certain renovations were being made to change the product line in order to produce value-added paper.

Production increased considerably to 37,923 tonnes in 1973-4. The company exported paper valued at Rs. 31 lakh. This despite the adverse conditions such as shortage of wagons needed for transport of essential materials. *SEOD* (1983) writes that material had to be transported by road at high costs. In the same year (1973-4) the company developed a new process and set up a consultancy unit for rendering technical services. But the overall production of cultural paper was still low. The Government of India had no alternative but to impose two control orders.

Control Orders and their Consequences

In 1974-5, two control orders were issued by the central government under the Essential Commodities Act. The first, entitled 'Paper (Control of Production) Order' stipulated that paper factories with an installed capacity of 25 tonnes or more per day had to produce at least 30 per cent of their output in the form of white printing paper. The order also stipulated the production of another 27 per cent of paper by way of other varieties of cultural paper. Interestingly, this was not by a DIR order but by an informal agreement. The price of white printing paper was fixed at Rs. 2,750 per tonne because there was a shortage of printing and notebook paper which the paper industry was under an obligation to produce.

Paper was also being diverted by the mills from fulfilling essential needs because it fetched a higher price as either speciality or industrial paper. So the second control order entitled 'Paper (Conservation and Regulation of Use Order) 1975', placed certain restrictions on paper being used for making calendars, diaries, greeting cards, advertisement posters, etc.

Because of the imposition of the paper control orders, that same year, SPM was immediately compelled to increase its production of cultural paper. There was a great deal of political instability, people were restive and the threat of Emergency loomed large. Anarchic behaviour would not be tolerated. When the entire industry responded to the control orders, the supply increased and exceeded the needs of the country, creating a glut. Production exceeded demand but there were no takers. The company, however, complained that it was the excise duty on certain varieties of printing and writing paper which had adversely affected the offtake. Strangely, the demand for paper board also declined in 1975-6. The reason for this is not very clear. With a view to stimulate demand and clear up accumulated stocks, the company had not only to effect significant reduction in the prices of its products, but allow extensive credit facilities to its customers. In the same year the company started a R and D division for developing new processes and equipment in order to achieve economy (*Financial Express*, 5 May 1976).

Expansion Programme

The mills also received an industrial licence for the manufacture of a complete pulp and paper making plant. The company was able to successfully fabricate some of its components as import substitutes. With the improved performance of the newly acquired board machine which produced value-added paper, the total production of paper and board increased in the year 1976-7. Sales went up to Rs. 16.48 crore from Rs.13.34 crore in 1975-6.

Production and sales further rose to 50,362 tonne and Rs. 18.69 crore respectively, in 1977-8. The company began to default on production of cultural or levy paper so the government was compelled to introduce new controls. Under the Paper (Regulation of Production) Control Order which came into force from 1 April 1978, 63 per cent of the company's production was to be of specified varieties. Out of this, 30 per cent was to be of white printing paper to be supplied against government allocations at what the company considered as an uneconomic price of Rs. 2,750 per tonne. The rates of excise duty were also increased. In the circumstances, production achieved for the year was considered satisfactory.

Production increased further to 53,749 tonnes in 1978-9. Also, sales recorded an increase of Rs. 22.58 crore. At the same time, the company lamented that the selling prices of white printing paper, supplied at concessional rates, continued to be unremunerative.

In the next year 1979-80, as a reaction to the price control order, production was maintained at the previous year's level, but due to frequent tripping and higher power cuts in the summer months, sales for the year amounted to Rs. 26.59 crore. With the installation of one more roaster, there was an increase in chemical recovery.

In 1980-1, production was reported to be slightly lower than the previous year due to frequent and erratic power supply, exposition in the soda recovery furnace and an accident in paper machine no. 3. The company complained that though the total sales were slightly higher at Rs. 28.20 crore, profits were actually lower, because of the substantial rise in the cost of almost all the inputs. Also, the State Electricity Board (SEB) once again increased the power tariff from 1 November 1980.

The company resented its obligation to supply white printing paper to government allottees at the concessional rate, and at what it considered as an uneconomic price of Rs. 3,000 per tonne (*Annual Report*, 1980).

In 1981-2 in spite of the continuing power supply problem, the company achieved slightly higher production. Sales turnover for the year amounted to Rs. 43.63 crore as against Rs. 35.88 crore in 1980-1.

The company tended to club cultural paper with industrial paper and boards, in its reports on quantum of production. The production of paper and board decreased in 1982-3. However, sales were maintained at Rs. 43.76 crore in 1982-3 after achieving a huge jump from Rs. 28.20 crore in 1980-1 to Rs. 43.63 crore in 1981-2. Production decline does not mean a decline in profit as is clear from the SPM's own annual report (AR). But for frequent tripping, low voltage and fire/explosions in the roaster in the soda recovery plant, production would have been much higher than in the previous year. Increases in the costs of raw materials, chemicals, stores and spares, power and fuel, transportation charges, and wages and salaries pushed up the cost of production. There was also a scarcity of bamboo and hardwood raw materials during the year.

In 1983-4 production of paper and board increased. The company would like the team to believe that production would have been still higher but for frequent tripping and low voltage. Sales declined to Rs. 34.89 crore. The margins were under pressure due to rise in the cost of various inputs including coal, power tariff and freight.

In 1984-5, in spite of power problems, production of paper and board increased. Sales also rose to Rs. 47.25 crore.

In 1985-6 production was maintained at the previous year's level despite poor quality of coal, and frequent tripping of power and low voltage supply from the Andhra Pradesh State Electricity Board (APSEB). The company continued to incur losses in the supply of white printing paper. Total sales amounted to Rs. 48.06 crore. Floods in August 1986 damaged plant and machinery, buildings and roads there being no production for about a week. Production also suffered heavily for another two weeks for the same reason. A 5 mw turbo alternator set was commissioned in this year. Plans were on hand to modernize the

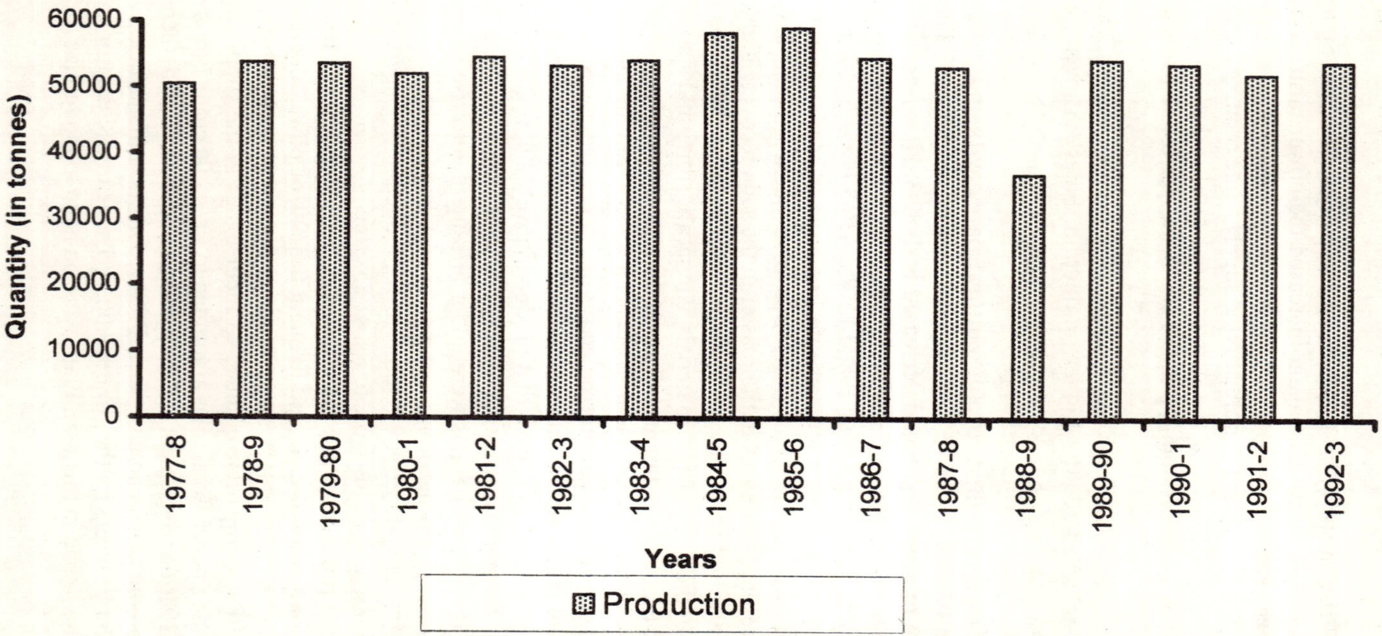

Source: Stock Exchange Official Directory

Figure 5.3: Production Figures of SPM from 1977-8 to 1992-3

existing units including installation of a higher pressure boiler for improving all-round efficiency.

Production was adversely affected in 1986-7 by the floods and fire that had damaged the stocks of bamboo in the previous year. The cost of production rose due to hike in raw material costs, railway freight and power tariff. The overall working suffered, sales standing lower at Rs. 44.57 crore. With the repeal of the Paper (Regulation of Production) Order 1974 and the Paper Control Order 1975, since 22 January 1987, the company commenced manufacture of superior quality paper.

Production of paper and board was unsatisfactory in 1987-8 and 1988-9 due to the severe power cut by APSEB and the now familiar problems of frequent tripping of power supply coupled with low voltage and interruption in supply of coal. The supply of raw materials from the forest leased out to the company by the state government began to dwindle and was inadequate; it had to be augmented from other sources. Alternate sources of raw materials were being explored. For nine months, the performance continued to be unsatisfactory due to power constraints and an increase in the cost of production.

The production of paper and board increased in 1989-90. Once again, production would have been still higher but for the usual shortage of raw materials, frequent power trippings and severe power cuts. Despite increase in the cost of inputs, the company achieved better financial results due to the improved paper market, higher production and better product mix.

Expansion, Modernization or Renovation

Renovation: For sheer survival the mill had to be renovated. It had been set up in 1938, and when taken over, was one of the oldest mills in India. It continues to look like a grim medieval forge even in 1992. All the machines were old and hence consumed more energy. Similarly, the technology (i.e. bleaching process by chlorine, caustic soda) was most outdated this being the crux of its problem. Mill officials felt that it was highly impossible to modernize as the cost incurred would be equal to setting up an entirely new mill, amounting to Rs. 25 crore. All other mills were installing MG machines. Sirpur renovated its old MF machine in order to produce glazed paper which could be sold as value-added industrial paper. Another problem faced by SPM was that the machines were labour intensive. Three labourers had to be employed where one should have done. It was not possible to retrench labour due to the existing labour laws (Mishra, private communication, November 1992).

No financial institution came forward to help SPM because of its poor performance. Birla, the major shareholder holding 10 per cent of the shares did not show any interest in effectively modernizing the mill. As a result there was no radical modernization, only specific renovations to yield value-added paper. Contrary to our observation and information offered by the embarrassed officials in the mill premises, are reports which give a totally different picture.

Apart from the actual process machinery, manufacture of paper requires utilization of a wide range of plant and equipment from raw material storage/transportation to packing and handling of the finished product. In addition, various auxiliary items such as steam boilers, power generation equipment, etc., are also involved.

The information regarding SPM's expansion and modernization programme was available from the *Stock Exchange Directory Papers* from 1965-6 onwards. The company claimed that in 1965-6 the expansion and modernization (E and M) programme for increasing the production capacity to 65,000 tonnes of paper and board per annum was progressing satisfactorily. It also expected that with the commissioning of the rag plant, one board machine and a new paper machine for manufacturing high grade paper, the installed capacity of paper and board would be increased from 36,000 to 44,000 tonnes by June 1967.

First Phase of Modernization: SPM informed us that the expansion programme in 1967-8 had to be slowed down due to recession, and owing to delay in some cases, in the supply of indigenous machinery and equipment.

The price control on paper was lifted in 1968. In 1968-9, the company decided to accelerate the expansion programme. The board machine with the capacity of 60 tonnes per day was expected to be ready for commissioning in the first half of 1972.

The Second Phase: In 1985-6 the company announced plans to modernize the existing units including installation of high pressure boilers for improving all-round efficiency. In 1986-7, the balancing and modernization programme undertaken by the company was almost complete with the commissioning of a drum chipper, double disc refiners, white liquor clarifiers and various other plants and machinery. The company expected that after completion of its modernization programme, the quality of products would improve considerably (SPM, *Annual Report,* 1986).

SPM was sure that the control orders would be lifted. It was the end of a period of shadow boxing. In 1987, further modernization programmes and energy conservation measures, including installation of high pressure boilers, was proposed to be undertaken by the company for improving efficiency to achieve better financial results (SPM, *Annual Report*, 1987).

The 1991 *Annual Report* reported that the company was implementing a modernization and balancing scheme with a view to conserving energy, increasing production, improving quality and product mix and effecting cost reduction. The schemes were meant to include among other things, renovation of paper machines, installation of a FBC boiler and/or retro-fitting of existing boilers, installation of drum chipper, double disc refiners and other equipment for energy conservation and process improvement. Financial institutions sanctioned assistance of Rs. 20 lakh for implementation of this entire scheme.

Expansion needed resources, but on 31 March 1990, the following amounts were outstanding against the term loans from financial institutions (a) Rs. 110.49 lakh from IDBI (b) Rs. 58.50 lakh from IFCI (c) Rs. 58.50 lakh from ICICI (d) Rs. 77.60 lakh from LIC and (e) Rs. 22.50 lakh from GIC and its subsidiaries.

As on the same date, the following amounts were outstanding against the term loans from various banks (a) Rs. 51.43 lakh from Central Bank of India (b) Rs. 20.83 lakh from UCO Bank (c) Rs. 35 lakh from State Bank of Hyderabad and (d) Rs.13.15 lakh from the Andhra Bank.

Raw Material Consumption by SPM and Royalty

The major raw material for integrated forest-based mills is bamboo and now hardwoods including eucalyptus as well. In Section 2 we had shown how severe depletion of bamboo had forced SMP to shift to hardwoods extracted from natural forests and start eucalyptus plantations.

As the Deputy I.G. of Forests rightly pointed out, 'The work culture of Birla is very bad as regards bamboo' (personal communication, March 1994, New Delhi). This has been the team's observation, too, in Orissa, Madhya Pradesh, Karnataka and Kerala, and of course in Andhra. This also implies increased transport costs.

This scenario raises some important queries regarding Sirpur Paper Mill's (SPM) raw material consumption pattern. How does a big mill like SPM satisfy its raw material needs? If the need is satisfied by an agreement between the company and the government, what is the rate of royalty paid by the company? Is the company actually benefiting from the arrangement? What are its other sources of raw material? Lastly, is the company aware of its responsibilities towards the depleting forests?

It may sound reasonable on the part of the company to claim that increase in the rate of royalty is one of the reasons which affected its profitability adversely. However, the other side of the coin seems to give an entirely different picture. The arbitrary fixing of royalty rates was always a matter of dispute between the mills and the Forest Department due to the long-term agreements. Only a few agreements provided for periodic upward revision of royalty rates. Where such revisions were possible, the extent of stipulated increases were nominal and *ad hoc* (ICICI undated: 55).

SPM, which is run by Birlas, used to get bamboo supplies from the nearby forests on a royalty payment ranging from Rs. 3 to 6 per tonne. This agreement was to expire in March 1971, but even three years after the government alert in 1969, the Chief Conservator could not revise the agreement in time. The reasons for this are far from clear. What is even more intriguing is that the government declared that bamboo might be supplied at Rs. 35 per tonne on an *ad hoc* basis 'after obtaining an undertaking from the mills to abide by the terms and conditions including the rate of royalty to be prescribed by the Government'. A new agreement was formally entered into with the mills only on 19 July 1976, when it was agreed that the increased royalty of Rs. 60 should be collected from 1 October 1975 onwards, instead of making it effective from 1 April 1975 when the earlier agreement would have expired.

What was the basis for fixing up the rate at Rs. 60? Even according to the company's auditors, it should have been Rs.100 per tonne. It could have been at least Rs. 80 as recommended by the NCA. But the government not only decided to fix it at Rs. 60, but this revised royalty rate effective nearly five years after the earlier agreement expired. The loss to the government is computed at approximately Rs. 2.25 crore. The revenue ought to have been ploughed back to revive the forests.

Again, the bamboo felled by the mills in the forests is weighed by the mills operated by their own men, the government accepting this weighment. This is despite the stipulation in the agreement that all bamboos shall be weighed on the weighing points prescribed by the department in consultation with the mills. There is no clue as to what could be the actual tonnage of bamboo lost (*Economic Times*, 19 October 1979).

Thus it becomes clear that the company was benefiting from the revised royalty rate as well as the FD's failure to weigh the bamboo, a benefit which is at the expense of both the government as well as of the forests.

According to the company Directors' Report for the year 1980, the company could procure 47,000 tonnes of bamboo from the leased areas in Andhra Pradesh as against the assured supply of 70,000 tonnes by the government. In addition, the bamboo procurement from *ad hoc* areas inside AP state was 14,000 tonnes and purchases from outside the state, viz., from Maharashtra, Madhya Pradesh and Himachal Pradesh, was 24,000 tonnes. The mill was unable to produce any evidence when confronted by the Forest Department which refused to believe that SPM had imported raw material from other states (A Retired PCCF, personal communication, August 1992). And due to short supply from areas within Andhra Pradesh, the company said that it was obliged to use other raw materials like paper cuttings (not imported but available in India itself), wood, pulp, etc., resulting in higher cost of raw materials. But the data available for the raw materials used by SPM for the year 1980 reveals that more than 60 per cent of the raw material used was in the form of bamboo which comes to 85,000 tonnes (AP 47,000 tonnes + *ad hoc* AP 14,000 tonnes + purchased outside AP 24,000 tonnes = total 85,000 tonnes).

The *Annual Report* 1979, carries the information through its balance sheets showing the consumption of bamboo alone to be 73,765 tonnes (i.e. 58 per cent) and other materials constitute only 41 per cent. There are two different quantities given for the same raw material, i.e. bamboo. The only answer to such variations can be that the company wants to show that its costs of raw materials are very high, which in reality does not seem to be so, because bamboo is a major component in the list of highly subsidized raw materials, while the other materials account for 41 per cent. The report also says that even in these circumstances, the financial results were considered satisfactory.

In 1980, with a view to ensuring a steady supply of raw material, the company approached the Andhra Pradesh government and the

Forest Development Corporation to formulate a long-term plantation scheme. Work was underway in 1981 in the Kagaznagar catchment area. SPM had leased forests but had not developed them.

According to the report the company also actively pursued distribution of pulpable plants from the company's Raw Materials Research Centre under the Farm Forestry Scheme (*Directors' Report,* 1980, discussed in Section 2).

The Director of SPM routinely reported in the annual report from 1981 onwards that due to short supply of bamboos and hardwoods from the areas within the state of Andhra Pradesh, the company had to continue to purchase bamboos and hardwoods from far-off places paying higher prices and increased transport costs, so that the shortfall in these basic raw material had once again to be made up by using paper cuttings, wood pulp, etc., resulting in higher cost of raw materials.

The report also mentioned that the revision of the rates of royalty on woods and bamboos was still pending finalization with the Government of Andhra Pradesh since 1 April 1980 and 1 October 1980, respectively.

As compared to 1980 the 1981 figures in the balance sheet showed a slight decline in the use of other raw materials such as paper cuttings, hardwood, etc.

In a notification dated 4 November 1981, the state government increased the rates of royalty. In the case of hardwoods, the rate was raised from Rs. 30 to 100 per mt and that of bamboos from Rs. 60 to 210 per mt with retrospective effect from 1 April 1980 and 1 October 1980, respectively. The revised rates were fixed for a period of five years with provision for a further 11 per cent increase every year.

Accordingly, the company stated that the royalty payable for the year 1 April 1982 and 1 October 1981, was Rs.123 per mt in the case of woods and Rs. 233 per mt in the case of bamboos, with effect from the respective dates mentioned. Being aggrieved by this arbitrary steep increase in the rates of royalty, the company filed writ petitions in the High Court of Andhra Pradesh. The court issued interim orders dated 24 February 1982, staying the operation of the relevant notification, and directing the company to pay royalty at the rate of Rs. 120 per mt for bamboo and Rs. 60 per mt for hardwoods, with effect from February 1982, pending further orders on these petitions. These are still pending before the High Court.

In 1983, the company was unhappy that the market conditions

remained sluggish and that its products could be marketed only by giving heavy discounts and other concessions.

Again in 1984 the company brought up the issue of royalty rates in its report. According to the notification dated 4 November 1981, of the Government of Andhra Pradesh, the royalty for the year under review was payable at Rs. 258 per tonne for the period 1 July 1983 to 30 September 1983 and at Rs. 287 per tonne for the period 1 October 1983 to 30 June 1984. Silmilarly, royalty for hardwood was payable at Rs.137 per tonne for the period 1 July 1983 to 31 March 1984 and at Rs. 152 per tonne for the period 1 April 1983 to 30 June 1984. However, the company was paying royalty on bamboo and hardwood in keeping with the orders passed by the High Court of Andhra Pradesh. The matter was still pending.

It was further said that the relief allowed by the government in excise duty on writing and kraft paper in the budget for 1984 was passed on to the consumers (*DR*, 1984).

The cost of production was further pushed up in the year 1985. The reason given for this by the company was that the state government had revised upwards the rate of royalty per tonne of bamboo and hardwood to Rs. 352.98 and Rs. 218.67 with effect from 1 October 1985 and 1 April 1985, respectively. The revised rates had been fixed for five years with provision for a further increase of 11 per cent every year (*AR*, 1985). In the years 1985, 1986 and 1987 the company stated that it also used raw materials from other sources which increased the cost of production.

In 1988 control orders were lifted. Though there is no logical connection between decontrol and raw material consumption, in 1989 for the first time the company admitted that there was adequate supply of bamboo from the state government lease agreement and that the problem was only one of inordinate delay in getting this bamboo. This caused a severe setback in obtaining adequate supplies of raw materials, thus adversely affecting the production.

Bamboo: Figure 5.4 gives the consumption of raw materials of SPM for the years 1979 to 1991 in real terms and Table 5.17 in money terms. This study deals mainly with the amount of bamboo and hardwood that SPM uses annually. In the case of bamboo, SPM used 87,994 tonnes (68.61 per cent) in 1979. There was a decline by 10 per cent in the very next year when the utilization of bamboo was shown at 73,765 tonnes (58.26 per cent). After 1980, the trend in exploitation of

TABLE 5.17: CONSUMPTION OF RAW MATERIALS BY SIRPUR PAPER MILLS

(*in Rupees*)

Year	Pulp purchased	Bamboo	Wood
1979	86,18,251	2,75,49,604	77, 55,722
1980	87,30,418	2,90,35,155	1,25,74,206
1981	66,77,700	3,83,56,262	1,17,38,852
1982	85,573	4,99,49,105	1,17,99,926
1983	—	5,69,43,483	1,11,57,280
1984	—	6,03,85,923	84,58,385
1985	—	7,03,95,097	57,93,366
1986	2,38,06,572	8,13,00,195	47,65,713
1987	1,54,66,807	8,51,26,719	1,55,06,460
1988	n.a.	n.a.	n.a.
1989	11,83,821	6,95,98,049	2,23,97,357
1990	3,78,390	13,40,24,179	3,29,75,428
1991	58,31,786	16,12,98,027	2,04,03,511

Source: SPM, *Annual Reports*.

bamboo shows a distinct phase of increase up to 1986 when it was 87.32 per cent.

In 1987, the balance sheet showed a more than 10 per cent decline in use of bamboo (from 87.32 per cent to 74.83 per cent). Information for the year 1988 was not available. The very next year there was an increase from 63.57 per cent to 68.55 per cent. The trend continued till 1991 (79.16 per cent). The highest percentage of bamboo used was in 1986 (87.32 per cent).

Hardwoods: Since SPM had exhausted its enormous bamboo resources, it started making use of hardwood in paper production.

The information from the mills available for wood, is from the year 1979 when its percentage use was 22.83 per cent. The year 1980 showed an increase by 6 per cent. But in the later years there was a sharp decline from 30.90 per cent in 1980 to 4.53 per cent in 1986.

For the next two years, from 1987 to 1989, the wood supplemented for paper production increased while the last two years 1990 and 1991, showed a decline in consumption of wood.

If the trends in the use of bamboo and wood are compared, a clear picture emerges; whenever the use of bamboo increased, the consumption of wood declined and vice-versa. 1986 was the year when the highest amount of bamboo was used in paper production (87.32 per

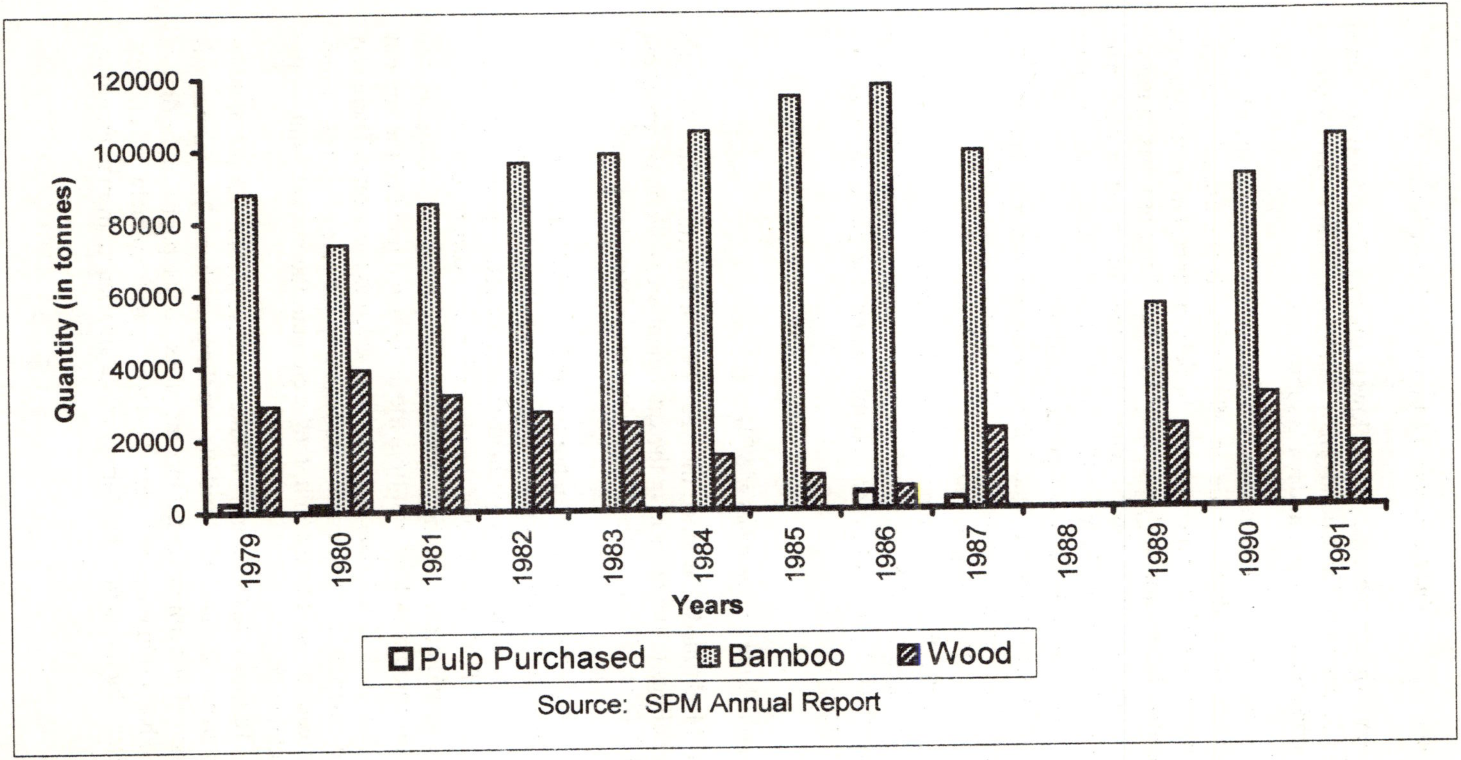

Figure 5.4: Consumption of Raw Materials by Sirpur Paper Mills

cent) while in the same year the lowest amount of wood was consumed (4.53 per cent). These trends prove that bamboo and woods are used as supplementary inputs and not as substitutes.

Other Inputs: In addition to bamboo and wood, the company's balance sheets show that it made an effort towards the use of other raw materials such as paper cuttings, hosiery cuttings, tailor cuttings, raw cotton linters and kraft corrugated sheets. Their use was marginal, starting from 1979. On the other hand, raw cotton linters were made use of only from the year 1984, and showed a steady decline from 1.2 per cent in 1984 to 0.04 per cent in 1991.

Similarly, kraft corrugated sheets were introduced as late as 1989, the 3-year period pattern of use showing an increase from 1.54 per cent to 2.35 per cent in 1991.

Lastly, the annual reports give information of the pulp actually purchased by SPM to supplement raw material needs. The use of purchased pulp was very minimal when compared to other raw materials used.

Farm Forestry Programmes of SPM

SPM claimed that since the year 1978 it has distributed 70 lakh seedlings to the farmers in the catchment areas which covers about 3,000 ha of land. This then is degraded forest land dealt with in Section II.

The farmers, the mill authorities and NABARD had an agreement so that the farmer mortgaged the land to the bank in return for a loan. The mill claims to have given necessary assistance to the banks such as periodical inspection of the plantation, assessment, distribution of loans and other local help. After the harvest, the mills collected the loan amount with interest from the farmers and remitted the amount to the bank. The balance amount was given to the farmer. There was no compulsion for farmers with regard to the sale of output. If they agreed to sell it to the mill, they were paid the market value of the pulpwood.

To motivate the farmers to take up these plantations the mills on their part decided to distribute seedlings to them. The farmer was assured of a minimum net income of Rs. 15,000 per ha at the time of the harvest in the eighth year. This was apart from fuelwood and leaf litter which was free. The farmer could also benefit from intercrop for

TABLE 5.18: NUMBER OF SEEDLINGS DISTRIBUTED AND AREA PLANTED UNDER SOCIAL FORESTRY BY SPM 1979-91

Year	No. of plants distributed	Area planted in acres			Total
		SPM area	Farmers	Government agencies	
1979	1,50,000	20	17	—	37
1980	3,00,000	20	—	—	20
1981	3,86,000	24	2	280	306
1982	2,40,000	87	24	—	111
1983	2,50,000	—	91	10	101
1984	8,00,000	7.5	272	5	248.5
1985	3,30,000	25	183	—	208
1986	6,12,000	46.5	348	—	394.5
1987	3,51,000	12	291	—	8
1988	9,000	4	4	—	8
1989	11,00,000	—	1,026	—	1,026
1990	15,00,000	—	1,500	—	1,500
1991	18,49,500	—	1,781	—	1,781

Source: SPM (SF Activities of SPM).

three years. The second and third cuts (Coppice shoots) that would be harvested with the five-year interval period would also be an additional income (SPM *Farm Forestry Project* [Booklet]).

All these programmes were earlier tried on a small scale by R and D using the Brazilian pattern of vegetative propagation of eucalyptus cuttings.

Table 5.18 shows the number of seedlings distributed and area planted year-wise from 1979 to 1991 by SPM.

A mill employee informed the team that SPM was also carrying on SF and FF programmes on 700 acres of *patta*-land which has been taken on lease by mills from farmers.

It was also emphasized that farm forestry programmes were undertaken to help small farmers in augmenting their income while what we actually found was that the farm forestry plantations were on big plots ranging from 7 acres to 30 acres (i.e. all big farmers) and that most of these farmers were absentee landlords who were engaged in business or trade in far-off cities. Also, most of these plots were leased out to small farmers who were supposed to take care of the land, tend the plantations, etc.

Since the mills paid the rent or money for the use of these plots they presumed that these plots belonged to small farmers and that they were, in a sense, helping them economically.

In reality, however, none of the small farmers could afford to take up the plantations as it takes a long period of eight years for harvesting which is too long for a poor farmer to survive without any alternate source of income. It was quite clear that the rich absentee farmers were gaining from the mills farm-forestry programmes.

Secondly, the mills took an area of 45 ha on long lease from Vampally Reserve forests for carrying out species trial. This was also used as a place of recreation by the mill people. Regarding the growth of eucalyptus, what the team observed was that the coppice growth looked healthier than the originally planted ones. Some of the saplings had changed colour to reddish brown which was said to be due to lack of nutrients in the soil.

On the 17 km run towards the high forests, the land was totally degraded and some calcareus stones were pointed out as a sign of infertility. Also a leguminous plant with bright yellow flowers was seen to be widely prevalent in the area which was said to be an indication that the earth was coal bearing (coal is a product of old forests which implies that the area was once thickly forested).

TABLE 5.19: TIME SERIES DATA FOR ALL THE SPECIES PLANTED BY SPM

Year	No. of seedlings planted	Area covered (in ha)	Survival rate of time of exploitation (in %)	Expected actual quantity (mt.)
1980-1	3,08,600	306	30	not fully harvested (1980-1 to 1984-5)
1981-2	2,40,000	111	30	
1982-3	25,000	101	Failed	
1983-4	8,00,000	284.5	40	
1984-5	3,30,000	208	60	
1985-6	6,12,000	394	65	fully harvested (1985-6 to 1987-8)
1986-7	3,51,000	303	65	
1987-8	9,000	8	50	
1988-9	11,00,000	1,026	65	6,000
1989-90	15,00,000	1,500	45	6,500
1990-1	18,49,500	1,781	40	7,000

Source: SPM, November 1992.

Note: Species planted *E. camaldulensis* and *E. hybrid* percentage share 80 per cent.

TABLE 5.20: QUANTUM OF RAW MATERIALS USED BY SPM FROM VARIOUS SOURCES 1987-8 TO 1991-2

(in mt)

Year	Actual supply of bamboo received from leased areas from APFDC	Purchase of bamboo from APFDC	Procured from outside states (Maharashtra and Assam)
1987-8	60,367	92	11,972
1988-9	45,445	—	7,395
1989-90	58,778	—	54,253
1990-1	61,389	—	44,349
1991-2	90,000 (approx.)	4,412	4,248

Source: SPM, November 1992.

TABLE 5.21: WOOD QUANTITY PURCHASED DIRECTLY BY SPM (EUCALYPTUS, CASUARINA, SUBABUL) 1987-8 TO 1991-2

Year	APFDC	Pvt. sources	Other states	MP	HP	Own plantation	Imported wood pulp
1987-8	95	1,686	5,672	—	—	55	858
1988-9	3,621	8,064	924			35	149
1989-90	992	15,945	2,094	10	72	36	27
1990-1	230	7,201	2,390	14	—	3,073	479
1991-2	363	17,205	6,657	—	—	—	2,012

Source: SPM, November 1992.

Kenaf was tried for a year but was given up with the reason that as a raw material it was 'technically not fit' for the mill and more importantly, because it was very expensive. Also, freight costs are high. As one truck carries only 9 tonnes of *kenaf*, it was uneconomic for the mill to use it.

Imports of Raw Materials

The Director's reports state that raw material was imported by SPM for the years 1980, 1985, 1986, 1987,1989,1990 and 1991. (No information is available regarding the types of raw materials or fromwhere they were imported [see Figure 5.5 on Import of Raw Materials]).

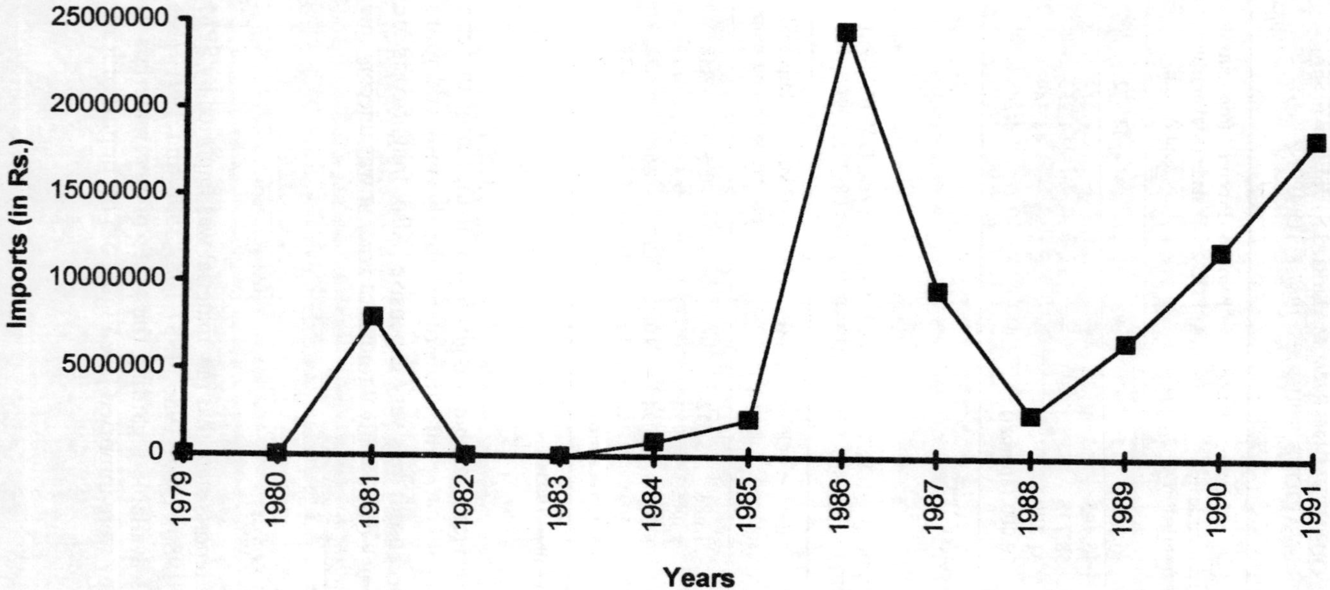

Source: SPM Annual Reports 1979 to 1991

Figure 5.5: Imports of Raw Materials by SPM 1979 to 1991

Research and Development

SPM started a R and D division for developing two new processes and equipment to achieve economy in 1976. During the year 1980, when information first became available, the company reported that a number of R and D jobs were undertaken to improve the process economy and product quality.

R and D is necessary for continuous development. By and large the PPI has in recent years concentrated its R and D on alternate raw material to achieve increased production of paper, to solve various problems mainly shortage of raw material and for better cellulosic material that would yield good paper. This has been discussed in Section 2.

Also a new low pressure, low chemical, process has been developed for pulping rice straw to produce both semi-chemical and bleachable grade pulps. SPM claimed that this would help small paper units to have much simpler and cheaper pulping equipment and better chemical recovery. The company also said that it made efforts to use more and more hardwoods and agricultural residues on the main process line. Experiments in bleaching for attaining greater brightness of pulp were also in progress.

Research was conducted for the use of waste black liquor for de-inking of waste paper, cooking of old rags, reduction in chipping dust loss, improving the fibre recovery system, reduction in fresh consumption, etc., the results to be implemented on a pilot-plant scale.

In addition to this, a new method for making Possolana type cement/mortar out of waste, lime, and sludge from the soda recovery plant and boiler, house coal ash and saw dust has been developed for use in civil construction. The company has filed a patent application to secure patent rights for the process (*Annual Report*, 1980).

Construction of a workshop building at the Research and Development complex at Balanagar, Hyderabad which was taken up in 1978-9 was completed in 1981. It was expected to start functioning in 1981 itself. In the same year an experimental five TPD pilot plant paper machine was designed and installed again, by its own machinery division, in order to undertake trials of special MG and MF quality papers with particular reference to light-weight paper for various types of raw materials. A pilot plant scale impregnator-cum-continuous digester for agricultural residues and one high density waste paper pulper had been designed and installed (*DR*, 1981).

In 1982 work was initiated, to improve the efficiency and operation of the electrolytic caustic soda chlorine plant. The company undertook laboratory studies on the pulping properties of fast-growing species especially *Subabul* (*Leucaena leucocephals*) for adoption as future raw material.

Also, heat-resisting pastel paper has been developed for the manufacture of insulation boards. The industrial filter paper unit was put into actual plant trial in the E.B. Plant, and found to work quite satisfactorily. The company claimed that it saved foreign exchange and reduced dependency on imported asbestos filter paper. Notable equipment design development in high speed lava refiners with 100 per cent indigenous material, a new design of waste paper pulper, and fuel oil water emulsifier have been developed and put into actual plant operation (*AR*, 1982).

In 1983, 1984 and 1986, the company in its annual reports merely stated that it undertook several R and D projects to improve the process economy, efficiency, product quality and new developments. However there was no information provided regarding the 'several' R and D projects (*DR*, 1983, 1984 and 1986). The company did not undertake any R and D activity as is clear from its reports for the years 1985 and 1987.

The Directors' report, 1989-90, said that R and D activities were undertaken in areas of pulping of unconventional raw materials for quality improvement and new process development for the years 1989 and 1990. In 1991, no R and D activity was undertaken.

Environmental Concern of the Industry: Pollution

The most visible solid waste is black coal ash which forms huge mounds—or rather, a long row of squat hillocks on the way to Vampalli catchment area. The effluent treatment plant ordered by the company in consultation with the National Environmental Engineering Research Institute (NEERI), Nagpur, was installed in 1980. The erection of this integrated total effluent treatment plant for the entire mill according to the statutory requirement was partially completed in 1981.

The first phase of the effluent treatment plant was commissioned in 1982. It was expected to reduce the environmental pollution in the surrounding areas as reported in the company's annual report. Offers for the second phase of the effluent treatment plant were also under finalization.

The order for equipment for this second phase of water effluent treatment plant was placed in 1983. The civil work of the secondary clarifier and aerator was commissioned in 1985, to reduce the water pollution in the surrounding areas. There is a tendency today to use the effluents for growing crops. The long-term effects of chemically loaded treated effluents on the soil needs to be understood. This is reported in Chapter 4—Maharashtra and Chapter 8—Tamil Nadu. A recent work by a NEERI scientist shows only the immediate effects. They are:

(1) Pulp and paper mill waste water belongs to the class 'slight to moderate to severe restriction in use' and can be used only for irrigation of salt tolerant crops with appropriate management practices.

(2) The waste water showed beneficial effects on the growth and yield of *jowar*, wheat and *moong* crops grown in the six different types of soils. Moderately tolerant to tolerant crops can be grown with the waste water without affecting the bio-mass yield.

(3) There was marginal decrease in the content of N, P, K, Ca and Mg in stem, leaves and grain of all the three crops irrigated with waste water, but the nutritional quality had not deteriorated. There was an accumulation of sodium in the plant parts of wheat and *moong* plants but *jowar* did not show accumulation in any of the plant parts of all the three crops. Further, zinc and copper in plants remained more or less similar, with respect to all the soils irrigated with waste water.

(4) Waste water application is likely to increase the soluble salts in soil to some extent, but with appropriate management it will not create any salinity problem. Again the question is, will Sirpur Mills care, or its farmers know, how to use the principles of farm management?

    There will not be an appreciable change in the ESP of the soil which will create a severe sodicity problem. Build up of organic matter due to the precipitation of lignin as calcium lignate will help to control deflocculation of soils due to sodium. Soil has a capacity in retaining colour in the waste water and clay, clay and sandy clay loam soils are efficient to retain colour. The nutritional status of soil was not affected due to waste water irrigation as the NPK contents are equally comparable.

(5) Evaluation of percolate quality which joins the ground water, revealed that there would be leaching of some soluble salts, but

contamination of ground water with colour bodies would not be a problem particularly with medium to heavy types of soils with the appropriate loading of waste water. However, coarse-textured soils may show leaching of colours in the long run (Verma 1993).

These are only short-term findings, one needs to understand the long-term effects.

Development

It is no doubt important to know whether or not an industry that is set up in a backward district brings about development. Although this does not happen to be the focus of this study, one cannot but notice certain features. It is important to remember that the mill was set up close to Sirpur town which is well connected by railway. Yet basically it is a tribal Gond area.

No investigation was carried out to find out how much of the local population was absorbed into the industry. However, it is clear that there are a number of Bengalis in the lower cadre of blue and white collar workers. These are the East Bengal refugees who were given shelter in the region.

The mill site has become Kagaznagar but we do not see any element of a 'nagar' (city), except for housing for the mills' white and blue collar workers. Though there is some bazaar or shopping facility a little distance away from Kagaznagar, there is no visible overall development in Sirpur itself.

Welfare Activities and Rural Development Programmes

The SPM brochure lists its activities from 1980 onwards.

The welfare activities for the employees, include aid for technical and educational institutions, clubs, distribution of food grains and other essential commodities through the company's consumer stores, workers educational classes, sports activities, etc. Also, the dairy farm for supply of milk which was started by the company in 1979, is reported to be working satisfactorily, milk being supplied to employees residing in the colony. Schools and hospitals are statutory necessities for an industrial township. An ESIC hospital stands right outside the mill premises; it has evidently been set up by the state, the mill having conveniently forgotten its obligations. Adjacent to the hospital is a school. It is not clear whether it is run by the state or by the mills.

SPM draws water from the upstream of the Peddaganga, and its treated effluent runs into the 'down stream'. This creates hostility with the villagers. In 1978, companies which undertook any rural development work were given income tax concessions under Section 35 CC. SPM's interest was in generating pulpwood trees. It is very likely that SPM, in its rural development programme undertook to provide facilities such as drinking water wells, school buildings and community centres to the rural population in nearby villages. In 1992 the company claimed that the drinking water wells were appreciated by the villagers.

SPM's Rural Development Programme undertook training in improved methods of agriculture, supply of seeds, fertilizers, pesticides, pest control and farm forestry work. Eucalyptus seedlings were distributed even as late as 1992. But SPM's rural development programme was mainly meant to produce raw material for itself. The negative effects of eucalyptus have been discussed in the Mysore Paper Mill section in the chapter on Karnataka.

The company was happy to inform the team, that the programme of 'Rural Development' taken up during 1978 was 'continued, immensely benefiting small and marginal farmers, SC & ST and other weaker sections'. But the real or rather only beneficiaries are the East Bengal refugees who work in the factory and also carry out its farm forestry (Manager of Farm Forestry, personal communication).

SPM claims that every coordinated activity aimed at rural uplift and improvement of the rural economy was taken up with a sense of dedication and social responsibility and that it was a matter of gratification that the Honourable Chief Minister of Andhra Pradesh had occasion to commend their efforts in this field. The Federation of Andhra Pradesh Chambers of Commerce and Industry is said to have been pleased to confer on the company an award for the best effort by an industrial unit in the state to develop the rural economy (*AR* 1982). (The same welfare activities are reported for 1982, 1983, 1984, 1985, 1987, 1991 only, but not for 1986, 1989 and 1990.) The team's observation regarding rural uplift was not as positive as that claimed by the mills. Some aspects of this have been dealt with in earlier sections.

The mills' forest labour is brought in from Balaghat (MP), not only to work in Adilabad but even in distant Kurnool, perhaps because they are more easy to control and manipulate. According to SPM, these labourers are very efficient although their felling practices and the dwindling state of the bamboo forest do not in any way point to this. A Civil Liberties researcher/activist mentioned that in the Kurnool forests, the management had reduced its Balaghat labour almost to the status of serfs, by withholding and manipulating wages. In the Vampalli 'forest', rather 'woods', there were no bamboos. SPM has introduced eucalyptus in an attempt at its mixed forestry. What trend this act portends is far from clear.

SECTION 3 (B)

THE ANDHRA PRADESH PAPER MILLS LTD.

Historical Background

A retired forest officer set up a paper mill at Rajahmundry in 1921. It was finally commissioned in the year 1924 under the name of Carnatic Paper Mills and Company Ltd. Its capacity was 10 tonnes per day. There are no historical records to show when and why, and to whom it was sold. However, what is interesting is that the next known buyer was an Indian and the year was 1937, the year when Congress was elected to power in eleven out of fourteen states, including this area.

In 1937, Dayaram and Sons took over the company, but could not make much headway. The mill remained shut once again from the year 1946. Ultimately, the government took it over in the year 1953 and started running it (APPM's write-up, November 1992).

The state government decided to increase the mill's production to 70 tonnes per day and orders were placed in 1960 for plant and equipment with Empacel of Paris. Due to various factors which have not been made known by the company, the construction of buildings to house the new plant and equipment could not be taken in hand.

Rajahmundry was an ideal location for the integrated pulp and paper mill. The mill is situated on the banks of the largest perennial river of the south, the Godavari. Dense bamboo forests were in close proximity. The town of Rajahmundry had an ample labour force. It was connected to all parts of the country by rail, road and even riverways, with good communication facilities. Since the government could not make any progress to improve the capacity of the mill, in 1964 at the joint suggestion of the central and state government it decided to hand over the mill to G.D. Somani (*Bombay Stock Exchange Directory*, 1978) who already owned a paper mill in Karnataka and had taken good care to replant bamboos in the forests. A formal agreement was entered into between the government and Somani on 13 May 1964, with a specific provision that a separate company be formed to take over the relevant assets, rights and obligations in respect of the project. Pursuant to this agreement, the company was incorporated on 29 June 1964.

The government agreed to transfer, convey and assign to the company all the lands acquired for expansion of AP Paper Mills and to assign to the company the agreements which they had entered into with Empacel, Paris, and other machinery manufacturers and collaborators for the supply of machinery, equipment, and also for engineering and other services to expand this mill to a rated capacity of 70 tonnes per day.

The government also agreed to transfer and hand over to the company the current assets consisting of stores, spares, raw materials, etc., at either cost or market value or book value, whichever was lower. An important question here is why did the government agree to this lowest possible cost?

Pursuant to the agreement between the Government of Andhra Pradesh and G.D. Somani, the Andhra Pradesh Paper Mills Ltd. came into being on 29 June 1964, as one of the first joint sector companies in India, with the participation of the Government of Andhra Pradesh, private promoters and the public. G.D. Somani later 'transferred' his rights and obligations to West Coast Paper Mills Ltd., a company which he also lost to his financier Bangur, a Calcutta-based entrepreneur. Neither the actual date nor the reasons for this are clear (APPM: 'A Brief Resume of the APPM Ltd.', 1992). Bangurs do not figure in India's industrial history. Yet this family holds large assets of over 100 investment companies and 26 manufacturing companies (Roy 1992: 56), stated to have been bought over as sick units from the departing British after the transfer of power. Roy states that Bangurs have property in big cities: 105 buildings in Calcutta alone, and prime property, as well as areas like Bangur Nagar in Goregaon, Bombay. The three generation tangled family has split. AP Paper Mills has gone to Gokulchand (80), and the L.N. branch and the West Coast Paper Mills in Karnataka to RN/SKS branch (Roy 1992: 56, 58). For other details regarding the growth of Bangur family (see WCPM in Karnataka).

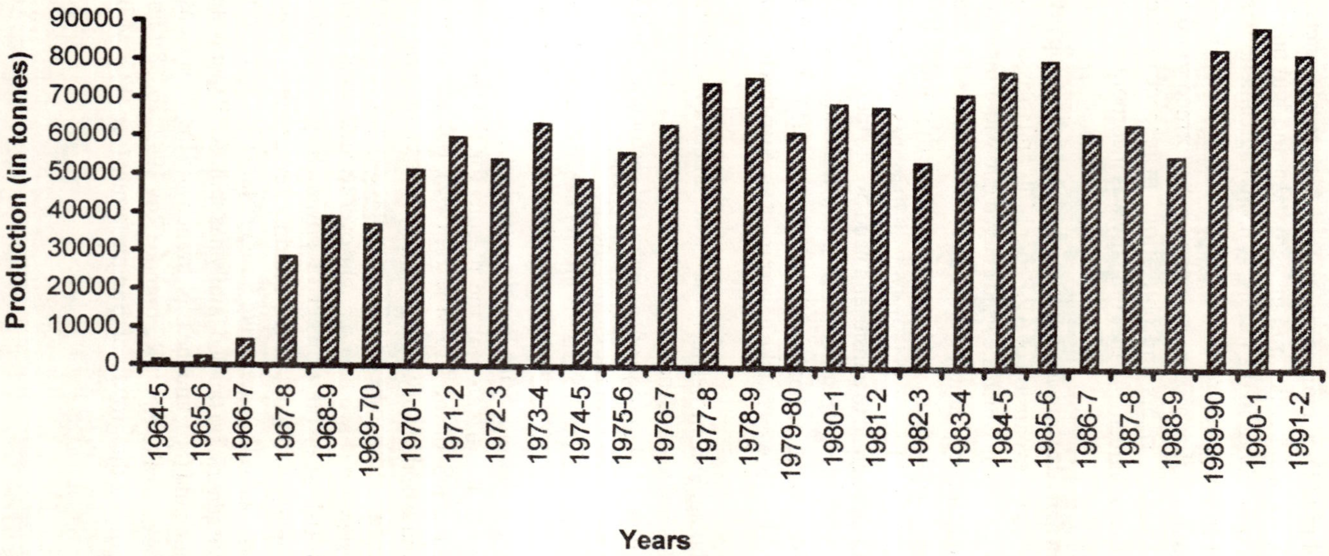

Source: APPM, November 1992

Figure 5.6: Production in tonnes by APPM from 1964-5 to 1991-2

Capital Structure of APPM

APPM was incorporated in June 1964. In 1967, APPM issued 1,34,000 equity shares for cash, while 50,000 preference and 34,000 equity shares without payment in cash were issued to the Andhra Pradesh government. 50,000 preference shares were offered to the public in January 1967.

In January 1970, 75,000 equity shares were offered for public subscription. In October 1972, 75,000 rights equity shares were issued at par in the proportion 1:5. The preference shares were redeemed in 1978-9 and 1,12,500 bonus equity shares issued in the proportion 1:4 in 1980.

Operations

The immediate growth of the mill after 1964 was phenomenal. In keeping with the times, APPM introduced its first (MG) paper machine, a pulp mill and a recovery unit. These were commissioned in October 1966, within 27 months of the company's incorporation. The second paper machine (MF), bleaching plant and other equipment commenced working by May 1967.

DISTRIBUTION OF EQUITY SHARE HOLDINGS
(AS ON 5 SEPT. 1992)

	Category	No. of shares held	% of total
I.	Foreign Buildings	383	0.07
II.	Govt./Govt. Sponsored Financial Institutions	2,71,973	48.35
III.	Bodies Corporate (not covered under I & II)	1,89,824	33.75
IV.	Directors & their relatives	3,319	0.59
V.	Top 50 shareholders (not covered under I, II, III, IV)	21,940	3.90
VI.	Others	75,061	13.34
	Total	5,62,500	100.00

Source: *Stock Exchange Official Directory*, Bombay, 1992.

The installed capacity of APPM was raised to 100 tonnes per day (tdp) in 1970, 250 tdp in 1977 and 275 tdp in 1989. But the actual production in 1965-6 amounted to only 2,077 tonnes as against the rated capacity of 3,000 tonnes per annum (tpa), said to be for want of felts, imported spares and cut in the power supply from March 1966 as claimed by the company in SEOD. From 1966-7, the production of APPM has shown an increase till 1991 except for five years, viz., 1972-3, 1974-5, 1979-80, 1982-3 and 1986-7 when the production was much below the rated capacity. These shortfalls coincide with the other units in the country wherever they may be located. The first shortfall, 1965-6 was a deliberate decision to create a paper famine by PPI. It paid off, and the price control act was lifted in 1968. The early seventy falls were again a concerted protest by the industry against the government's demand for levy paper when the industry as a whole was surging forward to produce value-added speciality and industrial paper. Almost every unit in India had invested in importing MG paper machines. The tantrums of the late seventies were to compel the government to raise the price of levy paper. It succeeded by the mid-1980s in doing so, but meanwhile, it also created a glut so that production had to be reduced to clear this.

The reasons given by APPM for low production during these years are:

1972-3: Power cut ranging up to 100 per cent, non-availability of coal, shortage of railway wagons and agitation in AP.
1974-5 and 1979-80: Prolonged power restrictions and dislocation of operation
1982-3: Labour strike
1986-7: Labour unrest and floods in August 1986.

The actual production figures are represented in Figure 5.7

One of the major complaints of the paper industry is regarding the Paper (Control of Production) Order 1975 which stipulates that 30 per cent of the total output has to be white printing paper and an additional 27 per cent other varieties of writing and printing paper in specified proportions. APPM holds that while the cost of production of white printing paper is higher than that of most other varieties, there is a strict price control on this. APPM seems to forget the heavily subsidized rates at which it gets raw material and infrastructure. In spite of the repeated raising of the controlled price, the mill complained that the production of white printing paper was uneconomic. For instance, during 1976-80, there were three paper price revisions. On 30 June 1979, the price of white printing paper was raised from Rs. 2,750 to Rs. 3,000 per tonne. The second raise was on 11 November

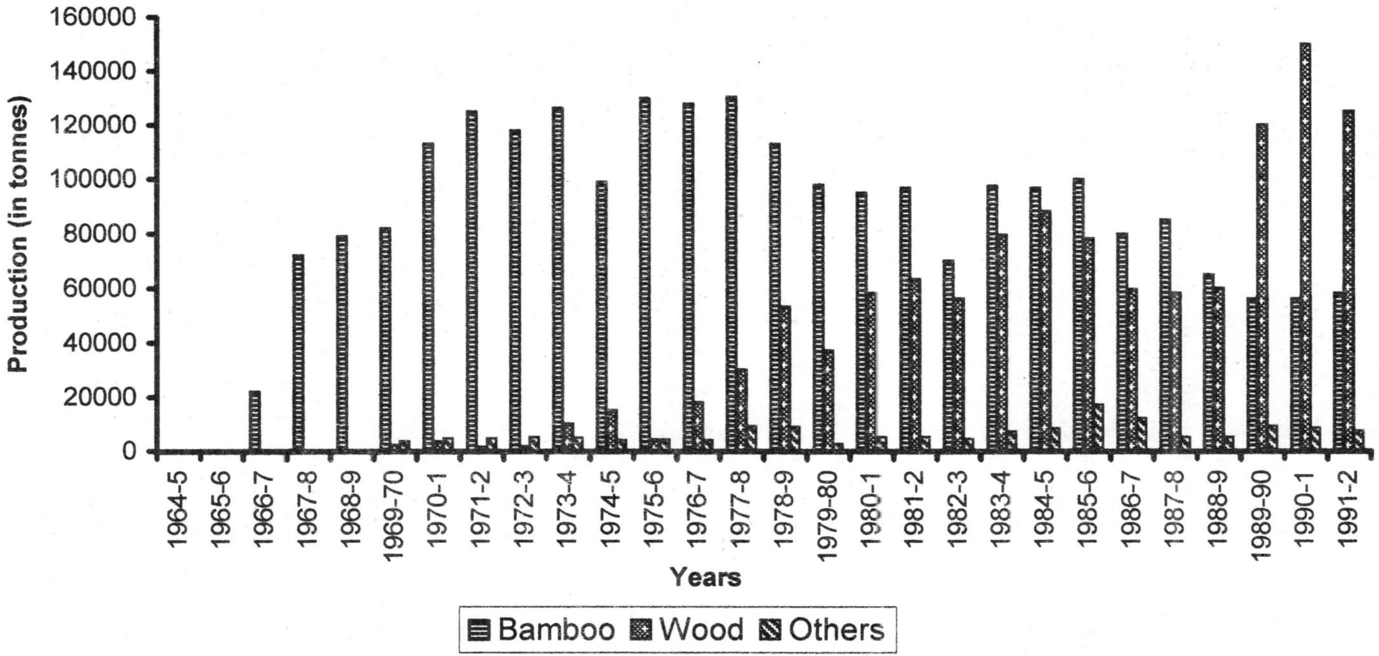

Source: PPM Rajahmundry, Mr Bansali (RM), November 1992

Figure 5.7: Raw Material Consumption by APPM from 1964-5 to 1991-2

1980, to Rs. 3,500 and the third was Rs. 4,200 per tonne on 12 December 1981. Distribution control on cream laid was lifted on 11 November 1980.

The industry claims that the cost of production of its white printing paper is around Rs. 4,500 per tonne, the shortfall sought to be filled by the increased prices of other varieties.

The combined result of all the price increases was a strong resistance from the market and a consequent fall in the sales during the year 1980 and early part of 1981. Imports[3] which were made to meet the shortfall in supply in previous years also added to the discomfiture of the industry. The industry now faced a glut.

Raw Materials

The original lease areas for APPM were Rajahmundry, Eluru and Kurnool. In October 1965, after the present management took over, an agreement was entered into between the company and the Andhra government regarding the supply of raw materials. According to the agreement some forest ranges in Godavari and Kurnool regions were leased to the company for 30 years with effect from 1 October 1965.

In 1970, due to gregarious flowering in some of the ranges allocated to the APPM, the supply of bamboo dwindled. APPM's immediate reaction was to set up its own nursery of fast-growing species. Meanwhile, new forest divisions were allocated in Ankapalli, Rampachodavaram, Narsipatnam, South Khota, Siliru and Vedurumaguram to APPM, by the Andhra Pradesh Forest Development Corporation (APFDC) to raise fast-growing species. Simultaneously, the old 1965 lease was prematurely terminated and a new and larger lease for 1 lakh tonnes of bamboo entered into. This now covered Kakinada, Visakhapatnam and Eluru divisions. It also meant a supply of additional quantity of bamboo.

However, the company found that the forest coupes did not yield the anticipated supply of bamboos. The deficiency was sought to be filled with supplies from farmers and from other states like Maharashtra, 'the Open Market', as it is generally alluded to.

In 1971-2, APPM undertook plantation of fast-growing species for which the state government allocated forest areas in order to combat the problem of ever-increasing cost of production and to achieve higher profits (SEOD). This was before the ban prescribed by the National Commission on Agriculture 1976.

Again in 1978, the government allotted to the company fresh coupes on an *ad hoc* basis. The total forest land that was allotted was capable of yielding about 1 lakh tonnes of bamboo and 1.5 lakh cubic metres of hardwood. Again, for some strange reason, yield from these areas too fell short of the company's requirements. The company therefore turned its efforts to utilize hardwood in larger proportion than bamboo and eventually did succeed.

In April 1980, the state government also increased the rate of royalty on bamboo from Rs. 60 to Rs. 210 per tonne from 1 October 1980, while on hardwood, the increase was from Rs. 30 to only Rs. 100 per tonne. Later, there was a progressive increase of about 11 per cent per annum for the next four years for both hardwood and bamboo. According to the Economic Times Research Bureau (1981), the increased cost of raw materials meant that it worked out to nearly 20 per cent in 1980-1. Whether the Economic Times Research Bureau was correct or not, APPM's R and D efforts were to find non-forest based wood as its raw material.

The company claimed that these rates of royalty were probably the highest in the country and were bound to affect all the mills in the state. It argued that APPM's absorption of paper was limited due to low demand within the state. Hence the competitive position of the company for selling its products outside the state would be badly affected if the royalty rates were to be retained. It further said that the Agricultural Land Ceiling Act did not permit the company to undertake any afforestation, even though it had always offered co-operation (speech by G.C. Bangur, *Times of India*, 1982).

In 1982 APPM made a representation to the Government of Andhra Pradesh for re-consideration of the order which resulted in an abnormal increase in the rates of royalty on bamboo and hardwood, and finally like all other PPI units, APPM too filed a writ petition in the High Court of Andhra Pradesh and again as in other states the Honourable Court of Andhra Pradesh granted an interim order staying the operation of the state government's order, asking the company to pay a royalty higher than the amount the company had been paying in the past, but lower than the rate demanded by the state government (*Economic Times*, 12 January 1983).

The question here is—is the subsidy given for forest materials justifiable? The other side of the coin is that before raising the royalty rates, the Government of Andhra Pradesh constituted an expert committee to advise it on the fixation of royalty both for bamboo and hardwoods. The committee was required to consider various factors while recommending the royalty rates, such as market prices, replacement costs, concessions, if any, and sustainability of supplies etc. (Rangarajan, undated). The government in turn examined and accepted the recommendations of the committee.

Thus it is quite clear that the revised rates of bamboo and hardwoods were fixed after taking into account a variety of important factors including of course the industry's both immediate and long-term needs. It is important to remember that the rate of commercial bamboo and commercial timber fixed by the forest department is much higher than the rate fixed for the paper industry.

But the company is being short-sighted and does not realize what is good for it. Perhaps it is high time that the true value of the precious forest resources be borne by its users.

The annual reports of the Company from 1987-8 to 1990-1 which carry information on raw materials, reveal that during all these years the amount of raw materials supplied was always inadequate and that the shortfall in the requirement had to be supplemented through purchases from neighbouring states and the local market. But when a sceptical Chief Conservator of Forest asked this company, as well as other pulp and paper mill authorities to show him their bills for transportation of wood from other states, they were unable to do so. No bills existed. With growing production the mill also had to import waste paper cuttings from abroad, ending with the same general conclusion that all this has increased the cost of production.

In the annual report for the year 1992, the company boasts, that since the raw material position continued to cause concern, with depleting stocks, lower supply of bamboo and lower allotment by the government, it had to take effective steps to ensure uninterrupted supply of raw materials. This it did, by tying up additional sources of supply, by changing the raw material mix mainly by using casuarina. The company also states that apart from these short-term and medium-term solutions, it has also undertaken schemes of industrial plantations, social forestry, farm forestry and stepped up its R and D efforts for improved seeds and use of unconventional raw materials as long-term solutions. It further says that under the social forestry programme the company has been distributing lakhs of seedlings for plantation (*Economic Times*, 12 January 1983, speech of G.C. Bangur, chairperson of APPM).

Why Casuarina?

The unique feature of APPM operations is that casuarina wood is pulped with its bark as it is very thin. It also implies a saving on debarking. Bamboos yield only 43 per cent bleached pulp. Its supply from the Andhra Pradesh Forest Department is uncertain and the mill receives bamboo in the range from six to 60,000 tonnes. Its earlier lease areas were re-allocated and it has to share its supply with two other units, viz., Rayalaseema and Bhadrachalam Paper boards since 1975. The mill also discovered that casuarina is easily available from the farmer and at cheaper rates than the hardwood auctioned at the state government depots.

The consumption of bamboo, hardwood and other woods by the mill from 1964-5 to 1991-2 is given in Figure 5.7.

Labour

The mills continue to work the coupes and employ tribals of whom 70 per cent to 80 per cent are mainly from Orissa. The rest are local Konda Reddys. They are paid piece-wages for cutting and stacking one tonne of bamboo, the rate being Rs. 160 per tonne. For 'dragging' the felled bamboo to the place of transportation or truck, they are paid Rs. 190 per tonne. This being seasonal labour, it is hired through contractors. The season lasts from October to May. No harvesting is permitted by the Forest Department during the monsoon rains which is an active and vulnerable growth period for the new bamboo shoots. To entice the labour to return to work the next season, the mill bears the conveyance charges to their homes in Orissa (Senior Official, personal communication, 1992).

Profitability

In line with the general trend in the paper industry, APPM also suffered a setback during the year ended June 1976. The gross profit nosedived to Rs. 221.26 lakh from Rs. 487.16 lakh in 1975. However, in 1978 the gross profit shot up by 84 per cent to Rs. 5.89 crore from Rs. 3.20 crore the previous year.

APPM again suffered a setback in its working results in 1982 which was attributed to an increase in the cost of various inputs pushing up the cost of production without any proportionate increase in the sales realization due to sluggish market conditions.

The company also experienced losses in the years 1986, 1987, 1988 and 1989. What is more interesting is that even under such conditions when the company was continuously incurring losses it was still toying with the idea of diversification. The losses were largely due to the inaction of Bangur who was feudal in his outlook and had not cared to modernize or even renovate the mill. He chose to make his monies through diversification. APPM was contemplating a takeover of the Nagarjuna Coated Tubes, promoted by Nagarjuna Steel and AP Industrial Development Corporation (APIDC).

Industry sources aver that APPM would bring in the much-needed funds to put the Nagarjuna Coated Tubes on a recovery trail. One wonders where APPM would acquire the resources for such a purpose. Further, what tilted the APPM in favour of Nagarjuna Coated Tubes? The answer lies in the exclusive technical collaboration of the latter with Allied Tube and Conduit Corporation, USA, the world leader in FLO Coat Technology. However, this takeover made an interesting case of a reviving sick unit (like APPM) attempting to revive another sick unit (*Financial Express*, 1989).

By 1991, the situation improved for APPM, sales jumped by nearly 30 per cent every year to Rs. 148 crore in 1991. Profits quadrupled to Rs. 11 crore. Earning per share rose nearly five times to Rs. 194. The credit for this has been given to M.V.G. Rao, the 'forward looking man', who took over as managing director in 1988. His first job was to solve the workers' problem by giving them a bonus. Since recurrent power shortages crippled production, Rao installed four diesel generation sets of 100 kva and commissioned a 10.5 mw turbo-generator which ensured self-sufficiency in power by 1991. Despite Rao's pleadings, Bangur was unwilling to invest in new machinery or introduce any sort of modernization.

To solve the problem of raw materials where the existing raw material mix at the time was 70 per cent bamboo, and the rest wood, the use of bamboo was progressively reduced to 25 per cent, unheard of in an industry where 50 per cent was typical. The wood chosen was casuarina which had been cultivated over a long period as a shelter belt along the coast. But casuarina can also create problems. This aspect is dealt with in Section 2.

The cost-cutting exercise was extended by Rao and his team to other areas—water, steam and power. Thereafter the company once again became one of the most successful of the country's paper producers. But success does not mean good times forever. Today the company is at the crossroads; this is partly of its own making, confesses its former senior executive. Proud of its recovery it repeatedly raised its price thrice within a year till the paper market crashed. It got caught in the general recession. Its modernization plans announced in 1990, are still on the drawing board. Much of its aging machinery is said to have reached the point of optimum utilization and needs urgent replacement—in fact modernization, to save on the safe but dwindling raw material.

But the Bangurs of Calcutta, who now have a controlling stake of 27 per cent in the company have been involved in their own family disputes, and are showing no interest in ploughing back any profits. There is no doubt that something will have to be done very soon either by the Bangurs or the state government.

It is said by the paper experts that if no modernization takes place immediately, APPM will become a sick mill in five or six years, nullifying all the turnaround efforts (Shenoy 1992: 58-60).

Water Pollution Control Measures by APPM

The company situated on the banks of the river Godavari, is fully conscious of the need to maintain the pristine purity of waters of this great river. It is the constant endeavour of the company to avoid any kind of pollution of this river water and in this connection the company has so far invested an expenditure of Rs. 332 lakh towards pollution control measures.

APPM has successfully adopted a unique system of effluent treatment and disposal, based on the principle of natural evaporation and percolation. The effluent is pumped to the sand shoals on a mid-stream island, Turupulanka, in the middle of the river Godavari at a distance of about 5 km. On these sand shoals a number of 'lagoons' are constructed covering an area of 600 acres taken on lease from the state government. From the lagoons the effluent undergoes evaporation and percolation. This system was suggested in 1969 by the Central Public Health Engineering Research Institute (later known as NEERI), acknowledged leading experts in this field, and approved by the State Government Public Health Authorities. The company has successfully implemented this scheme at a cost of Rs. 60 lakh.

Since then the system has been in operation and the company is on the constant lookout for ways of further improvement. In this context in 1985, the company installed an effluent treatment plant within the mill's premises for primary treatment of effluent before pumping to Turupulanka at a cost of Rs. 112 lakh. However, in order to have a full-fledged treatment system, recently in February 1991, the company successfully commissioned the secondary treatment system at an estimated cost of Rs. 160 lakh. Both the primary and secondary treatment systems are designed to handle an effluent volume of 20 mgd, though the present effluent generation varies between 15-17 mgd. Whenever possible, re-circulation of effluents is done and the valuable fibre and filler material recovered to the maximum possible extent.

In the primary treatment system, effluents from different sections of the mill like the pulp mill, paper machines, recovery, bleaching chambers, etc., are first collected in a sump and the combined effluents after passing through 6a screens and stationary screen chambers are pumped into two primary clarifiers of 150 ft diameter each. The pump house is provided with four 6 mgd capacity pumps of 75 hp each.

The underflow from the clarifiers having a consistency of about 1.5-2.0 per cent is further thickened in a thickener of 95 ft diameter to a consistency of about 6 per cent and is finally de-watered in a specially imported sludge de-watering machine. The cake coming out of this machine is carted out and disposed of as landfill. The primary clarifier overflow which is very clear and with a suspended solids concentration of less than 100 mg/1, is taken into the secondary system.

The secondary treatment is a biological process which involves an activated sludge plant and is designed for an outlet BOD of less than 30 ppm. The clarified effluent from the primary clarifier is taken into the aeration tank where the organic matter (BOD) in the effluent breaks into sludge by bacterial action. The overflow from the aeration tank (an aeration tank of about 5 mgd capacity with eleven mechanical surface aerators of 75 hp) goes into two secondary clarifiers of 160 ft diameter each. The sludge mass gets separated in these clarifiers. The overflow from the secondary clarifiers is taken into a sump from where it is pumped to Turupulanka for final discharge. The secondary clarifier under flow sludge (about 60-70 per cent) is re-circulated back into the aeration tank in order to maintain the bacterial mass. The excess sludge is taken on to the de-watering machine and disposed of as land-fill Nutrients like DAP and urea are added at the inlet of the aeration tank to activate the bacteriological action in the aeration tank. About 1.4 tonnes of urea and 0.6 tonne of DAP are to be added daily.

The secondary clarifier overflow which is very clear, is then pumped to the sand shoals for final disposal. About six pumps of 150 hp each are provided for pumping the effluent to the Turupulanka Island sand shoals. From the pump house, effluent is conveyed through two pipelines of 36 in. and 48 in. in diameter, running all along the left river bank up to a distance of about 5 km. It then crosses the river course under the river bed and drops the effluent into an open channel on the high level Turupulanka Island. From here the effluent flows by gravity up to a distance of about 1.2 km and is distributed into a number of 'lagoons' constructed on the sand shoals. These lagoons are constructed manually with mud and sand. The width of the sand bunds at the bottom is maintained at about 5-7 mts and at the top 1-2 mts. The average depth of the lagoons varies between 1.5–2.0 mts. Normally about 25-30 lagoons are constructed on the sand shoals having a holding capacity of about 25-7 days.

Utmost care is taken to see that sufficient distance of about 2,000 ft is always maintained between the last lagoon and the river course to avoid accidental effluent being mixed directly into the river. Also empty buffer lagoons are provided at the end of the lagoons in all directions to take care of any seepage or accidental breaches from the lagoons.

However, during the flood season as the water level of the river rises, most of the lagoons (sometimes all the lagoons depending on the river water level) get washed off. But since the flow in the river during a flood is of the order of few lakh cusecs of water, the amount of the effluent mixing into the river is negligible, while the quality of river water is not affected because of the tremendous amount of dilution. Once the flood recedes and the water level starts decreasing, the lagoon construction activity on the sand shoals begins immediately. Bulldozers and about 800-1,000 workers from the nearby villages are employed every day for the sandbunds construction work. The bunds construction activity goes on for a period of four to five months. However, the company has employed forty permanent workers to regulate and distribute the effluent into the lagoons round the clock, as also for surveillance, for fear that workers might breach the lagoons. By implementing this unique scheme the company, apart from storing the effluents in the sand shoals, is also providing indirect employment to about 800-1,000 workers from the nearby villages for a period of about six months.

Regular monitoring of river water quality both by the mills as well as the State Pollution Control Board Authorities ensures that the system functions well, by collecting effluent and river water samples at various points every month. Also the Pollution Control Board Authorities have insisted that the mills provide bore wells at a distance of 100 ft away from the river's edge in different directions, in line where the distance between the furthermost lagoon and the river's edge is 2,000 ft. They have also prescribed a BOD value of 15 mg/l for these bore-water samples. Regular monitoring has shown that the BOD value of these bore-water samples have never exceeded 5 mg/l. The analysis reports are regularly submitted to the Pollution Control Board.

The total recurring cost per year for the operation of this entire system works out to about Rs. 225 lakh, which includes construction and maintenance of lagoons, power costs, chemicals, etc. It has been clearly established by the operation of this treatment and disposal method that the river water quality is not affected. The quality of river water 1 km upstream is equivalent to 1 km downstream of the lagoons. This scheme is recognized by experts in the field as novel, ingenious, unique and effective, not adopted anywhere else. In recognition of this unique effluent scheme, the company has bagged the AP State Pollution Control Board award for the most efficient effluent treatments plant for three successive years.

Though the effluent meets the ISI specifications (IS: 2490) for inland surface water discharge (river Godavari) with 30 ppm BOD and 100 ppm of suspended solids after the secondary treatment, the effluents are still being pumped to Turupulanka lagoons for final discharge so that the colour of the effluent is minimized to a great extent through natural evaporation and percolation process. (APPM write-up on water pollution control measures—November 1992.)

SECTION 3 (C)

ITC-Bhadrachalam Paper Boards Ltd. also Known Today as Indian Tobacco Company, Bhadrachalam (ITCB)

Why BPL, and Why in Sarapaka?

Bhadrachalam Paper Boards (BPL) was set up in 1979, a time when the paper industry was in the doldrums. The person who is said to be responsible for this development was Vengal Rao, the then Chief Minister of Andhra Pradesh. He dreamt of turning his birthplace

Khammam into a second Manchester, and invited private industries to venture into the area.

On the other hand, an officer of the Indian Tobacco Company (ITC), stated that the group came forward with plans of setting up a paper board industry to settle its own problems. ITC had been totally dependent on Rohtas' Company Dalima's Paper Mill in Bihar for its paper and paper board requirements for cigarettes. In order to be totally independent and self-sufficient in its raw material needs, it took to production of its own paper board (Raghuveer, personal communication, ITCB).

Secondly, when ITC made a thorough evaluation of the already existing paper mills in Andhra Pradesh—Sirpur and Andhra Pradesh Paper—were producing cultural paper and BPL thought it would be 'difficult to compete' with them in this area while in case of industrial papers, maplitho, craft paper, etc., there was a growing demand but production of these varieties was still low and worth venturing into (Raghuveer: personal communication, November 1992).

Finally the decision was to venture into the production of paper board which it needed for cigarette packing. Paper board, besides being forest-based raw material, also required waste paper and imported chemical pulp as raw material in the ratio of 40:60 respectively. This product would be profitable, and as the industry's official added, it would mean less pressure on forest resources, apart, of course, from feeding its parent concern with paper board, the most required input for cigarette making.

The next important decision was the location for setting up the mill. A survey revealed three available options. One, the village of Eruvandi which required the company to lay a *pucca* (permanent concrete) road. Second was Bhadrachalam which required crossing over the bridge. It was difficult, as machinery could not be transported over this bridge. The third and last option was the village Sarapaka which was very close to a road that the government was building, but prone to floods and seismic activity. Out of these three, the company chose the last area, i.e. village Sarapaka for the mill premises as it required less financial resources for the smooth infrastructure necessary (personal communication, November 1992).

Sarapaka was a tribal village till a few years ago. The outside world knew little of the difficult existence of the people there, though the nearby Bhadrachalam town has attracted pilgrims from all over the country for a long time. The transition of Sarapaka, close to the

Godavari river, from an unknown and uncared for tribal village and its farmlands and forests, to an industrial unit, started in January 1977. Before the commencement of civil construction work of ITCB on the tribal land and forests, a 'deal' was made with the locals that one member of each family who vacated his land would be employed by BPL or ITCB as it is now known. The tribals as well as the forests in which they lived were cleared to make room for the mills. The company was incorporated on 17 July 1979. It was promoted by ITC Limited, Vazir Sultan Tobacco and Andhra Pradesh Industrial Development Corporation.

India Tobacco Company (ITC) which was earlier known as Imperial Tobacco Company was a multinational monopoly company in India's cigarette industry.[4] It is interesting to know, in the first place, why Imperial Tobacoo Company was formed and why it diversified into paper production. ITC and the interconnected Vazir Sultan Tobacco Company (VST) are both its subsidiaries.

Imperial Tobacco Company of India was incorporated in 1910 with headquarters at Calcutta, as a trading company, importing and selling cigarettes manufactured in UK. It remained a cent per cent foreign owned company until 1954, when it was converted into a public limited company. ITC today operates a network of five cigarette factories located in Maharashtra, West Bengal, Uttar Pradesh, Bihar and Karnataka, as also printing and packaging complexes in Bihar and Tamil Nadu.

Ever since ITC's advent in India, it has had a negative impact on the indigenous units prevailing at that time mainly because the ITC cigarette monopoly ruthlessly wiped out some 200 indigenous units. The few surviving indigenous units jog along precariously. Several units had been plagued by idle capacity caused by the monopoly, and forced to take up contract production for the ITC, remaining pitiably servile and silent. The monopoly thus eliminated competition but at the same time, enhanced its production (*Multinationals: A Scourge Proven*, no date).

In the new liberalized economy an attempt has been made by the parent company British American Tobacco Company (BAT) to increase its equity stake from 31 per cent to 51 per cent in India's ITC Ltd. Incidentally, till 1970, BAT held a 70 per cent share which was reduced to 32 per cent in 1982 and recently brought down to 31 per cent. The Chairperson, K.L. Chugh, does not want to give in, and his position has been strengthened by a recent decision of the Finance

Minister to reduce the ceiling on preferential allotments to foreign institutional investors from 24 per cent to 15 per cent. ITC is still lobbying for further reduction of the permitted equity holding in Indian companies to a maximum of 10 per cent, with a proviso that such shares can be sold only to resident Indian Nationals. Barua (1994: 19) reports that Chugh has been able to convince not only the Finance Minister but also the Prime Minister, that ITC is a potential candidate for an Indian multinational. BAT's interest at the moment is cigarette making.

The question that naturally follows is why did ITC diversify and enter into the core sector in 1975, a year when the paper industry was facing problems; and why did it choose the backward Bhadrachalam area for setting up its mills? This study has stated in Section I that ITC was familiar with AP.

The company repeatedly boasts that it chose the most backward area, Bhadrachalam, which is also the heart of the tribal belt, for setting up of their mills. When the company chose Bhadrachalam, what it had in mind was not the backwardness *per se* and its development, but the number of fiscal incentives given by the central government as well as the state government to entrepreneurs, right from 1968, to set up industries in backward areas.

The company was entitled to the following fiscal concessions:

(1) grant of higher development rebate to industries located in backward areas.
(2) grant of exemption from income tax, including corporate tax, for five years after providing for the development rebate.
(3) exemption from payment of import duties on plant and machinery, components, etc., imported by units set up in backward areas.
(4) exemption from excise duties for a period of five years.
(5) exemption from sales tax, both on a variety of raw materials as well as finished products, for units set up in specified backward areas for a period of five years from the date of their going into production.
(6) transport subsidy for taking out the finished products for a period of five years.

In addition to these incentives, the Government of India[5] advocated that it was necessary to offer special incentives to entrepreneurs in backward areas in the matter of fixed assets, i.e. supply of developed plots, built-up accommodation, machinery, equipment, etc.

In addition to these incentives, the company also benefited from the various measures adopted by state governments to attract industries to backward areas. In the case of Andhra Pradesh the measures were:

(1) the State Financial Corporation would provide loans on a concessional rate of seven per cent, and
(2) power would be supplied at concessional rates to power oriented industries varying from Rs. 525 to Rs. 600 per kva.
(3) concessions on sales tax (Chand and Puri 1983: 331-40).

Apart from this, Bhadrachalam proved to be an ideal location for the mill as it is situated on the banks of the mighty perennial river Godavari, barely 25 km from the Thermal Power Station at Paloncha, and 40 km from the coal pitheads of Singareni Collieries for ITCB to generate additional power, and of course, very close to the bamboo and hardwood forests. Finally, it has the necessary infrastructure—a railway station. This proves that even the geographical location was carefully selected to carry on production smoothly without any shortfall of raw materials. The area seems to make up for its backwardness through its ideal location, while the company pinpoints only its 'backwardness'. Backwardness of course gives the company a bonus of abundant subsidies besides cheap labour.

The Managing Director, K.L. Chugh, said in his speech (1977), that a plant was designed for an annual production of 48,000 tonnes of duplex board, kraft and papers and it would occupy 500 acres of virgin forest lands. The layout of the mill was designed in such a manner as to ensure energy conservation, quick dispersal of operational personnel thus saving valuable time, separation of inputs and workers' movement areas to ensure safety, easy access between various sections to facilitate communication and regulation space to provide for expansion and future growth, thus leading to efficient operations and minimum capital investment (*Business Standard*, 11 May 1977).

Another article states that the company had been licensed to manufacture 50,000 tonnes of carton board and writing and printing paper and that it was setting up a plant for manufacture of about 40,000 tonnes per annum of which about 28,000 tonnes was to be paper board and about 12,000 tonnes writing and printing paper (*Business Standard,* 11 May 1977). This however is not quite correct. ITCB did not care to install machine which would produce cultural paper. The question of having to produce 30 per cent levy paper did not bother ITCB.

The capital cost of the project was estimated at Rs. 44 crore, of which plant and machinery would account for Rs. 33 crore. The company decided to make a public issue of Rs. 6.80 crore including Rs. 50 lakh in preference capital.

A noteworthy feature of the project, says one newspaper (*Economic Times*, 18 August 1977), was that 'the process technology was wholly indigenous and there was no foreign collaboration'. This cannot be true. As in all paper mills today, the major equipment—plant and machinery—are fabricated in India and therefore perhaps considered indigenous, but the process technology is held and closely guarded by the parent company. The imported component is generally very small and confined to essentials. Incidentally, the *Economic Times* correspondent forgets that ITC is a mighty American-British multinational.

The board consisted of 12 directors headed by no other than the powerful Ajit Narain Haksar.

The most important feature is that the company sells its products directly to bulk consumers. They are ITC, Vazir Sultan (VST) and WIMCO matches; ITC, of course, being the promoter of ITCB, and VST a subsidiary of ITC. These three consumers between them have the capacity to absorb the total production of carton boards. This is very significant and gives a clue as to how ITC operates. The reason given for bulk setting was rationalized; this was, that it would save the marketing cost of the company. Was it truly so, or is it that ITC diversified into paper production to satisfy its own needs of paper cartonnes for its cigarette production? It had problems with Rohtas which alone produced the speciality paper that ITC cigarette wrappers needed. A paper unit BPL would then be a useful backward linkage for ITC and would be known as ITCB.

Capital Outlay

The total capital outlay of the BPL project was estimated at Rs. 44 crore of which a sum of Rs. 4.40 crorc was for land and buildings, Rs. 21.20 crore for plant and machinery and Rs. 1.10 crore working capital margin. This was being met through share capital of Rs. 11 crore and term loans from Indian financial institutions amounting to Rs. 18.65 crore (including a central subsidy of Rs. 0.15 crore). Subsequently, the cost of the project went up by Rs. 5.90 crore to Rs. 49.90 crore. This was partly financed by additional term loans of Rs. 3.76 crore from financial institutions. An application for an additional term loan of Rs. 3.20 crore was made to the IDBI. In percentage terms the equity works out to 35 per cent—ITC, 5 per cent by the AP government and 40 per cent from various financial institutions such as IDBI, IFCI, ICICI and LIC which sanctioned term loans of Rs. 10.50 crore (to be reduced to the extent of a central subsidy of Rs. 0.15 crore when received), Rs. 3.10 crore, Rs. 3.00 crore and Rs. 1.50 crore respectively. A loan of Rs. 25 lakh was raised from UTI through private placement of debentures. The equity in detail from 1975 is given in the ITCB capital structure.

ITCB Capital Structure

The ITCB was incorporated in July 1975. The 7 equity shares were subscribed for signatories to the Memorandum of Association and 23 equity shares allotted to others. 50,000 preference and 1,04,99,970 equity shares were offered at par through a prospectus in September 1977. Of the equity shares reserved for allotment (a) 31,99,970 were given to ITC Ltd., (b) 10,00,000 to Vazir Sultan Tobacco Co. Ltd., and (c) 10,00,000 to APIDC, 63,00,000 equity shares and 50,000 preference shares were offered to the public.

10,650 equity shares were forfeited and reissued at par in 1979-80 (arrears Rs. 875). 50 shares were forfeited and reissued at par in 1982-3, and 68,259 equity shares issued at par to financial institutions on conversion of loans in 1984-5. Preference shares were redeemed

DISTRIBUTION OF EQUITY SHARE HOLDINGS OF ITCB
(AS ON 7 AUGUST 1993)

Category		No. of shares held	% of total
I.	Foreign Buildings	7,523	0.04
II.	Govt./Govt. Sponsored Financial Institutions	79,00,089	45.52
III.	Bodies Corporate (not covered under I & II)	70,02,485	40.35
IV.	Directors & their relatives	14,225	0.08
V.	Top 50 shareholders (not covered under I, II, III, IV)	3,17,640	1.83
VI.	Others	21,13,923	12.18
	Total	173,55,885	100.00

Source: Stock Exchange Official Directory, Bombay, 1993.

in 1990-1, while in 1991-2 23,53,220 equity shares (Premium Rs. 25) and 14,17,830 equity (Premium Rs. 30) were allotted on part conversion of zero-interest debentures and 12.5 per cent debentures respectively. 18,82,448 equity shares (Premium Rs. 52.50 per share) and 11,34,128 equity shares (Premium Rs. 65 per share) were allotted on 15 December 1992, in conversion of zero-interest debentures and 12.5 per cent debentures respectively in 1992-3.

Mills Operations

The mill went into commerical production on 1 October 1979, in a record time of 27 months with an initial investment of Rs. 50 crore on an annual installed capacity of 40,200 tonnes.

From a mere single product, paper board, and for official purposes a three-product range was declared in 1979, these being duplex board, cream wove and kraft papers. The company has now a product range of more than twenty varieties. They are match box board, gypsum base board, liquid packaging board, absorbent kraft, maplitho, plain paper copier, diazo paper, super graphic paper, poster, chromo paper, art board, florescent paper and paper boards. The market share of duplex board and absorbent kraft exceeds 25 per cent while that of maplitho papers is over 10 per cent.

Starting out in 1979-80 with a production of 26,906 tonnes and capacity utilization of 70 per cent Bhadrachalam's output increased continuously except for the years 1986-7, 1989-90 and 1990-1.

The production is officially said to have consisted of duplex board, cream wove and kraft papers. This of course does not conform with what a former employer stated. In the year 1980-1 the company claimed that maplitho was launched, the production increased to 42,013 tonnes and sales almost doubled to Rs. 33.54 crore. This was attributed to overall operating efficiency and better capacity utilization.

In 1982-3, sales increased by 12.6 per cent while the production increased by 8.7 per cent. This year too, new higher value-added products were developed and introduced into the market.

In the year 1986-7, the company first experienced a marginal decrease in production and sales, due to sluggish market conditions accompanied by labour unrest and massive power cuts imposed between April-July 1987 (*Stock Exchange Directory*, 1987). The sluggish market conditions were due to a combination of factors created basically by PPI and over-production by small-scale units that the government was compelled to set up. Under pressure from PPI with BPL as the leader, the paper control order was lifted in 1987 (SEOD).

In 1989-90, though production declined, the position of sales was very favourable. In its *Annual Report* of 1990, the company states 'The buoyancy in the market, noticed towards the end of 1988, prevailed during the year as well. With increased volume of production and product range catering primarily to the premium segment of the market your company was able to improve its market share in a growing market.'

Having achieved its major purposes of getting rid of social obligation and scrapping of the paper control order, the company openly admitted to us in November 1992, that its aim was to produce only speciality papers which would serve only a particular upper segment of people, and those addicted to cigarettes of particular brands. The over-confident PPI began to relentlesly push up the price of paper. The paper market crashed, and production fell by 3 per cent in the year 1991.

To complicate matters further, market societies were faced with a recession which hit the paper industry. But ITCB would like us to believe that its fall in production was due to floods in August 1990 which brought operations in the mill to a complete halt for three days, accompanied by an illegal strike by workers for four weeks. BPL goes on to say that operations were further affected by transportation bottlenecks, as well as shortage of diesel due to the hostilities in the Gulf. In the same year input costs escalated due to a steep rise in petroleum prices from October 1990, as well as 100 per cent increases in customs duty on waste paper and pulp. The capacity utilization in 1989 was 134 per cent which declined by 4 per cent in the year 1990 (130 per cent). The production figures for 1979-80 to 1992-3 are represented in Figure 5.8.

Paper Control Orders and BPL

The commerical production of Bhadrachalam Paper Boards Limited started on 1 October 1979. The Paper (Regulation of Production) Control Order in existence from 1 April 1978, stated that 63 per cent of the Company's production was to be of specified varieties. Out of this, 30 per cent was to be of white printing paper to be supplied against government allocations at a price of Rs. 2,750 per tonne.

In the case of Bhadrachalam there is no indication that it produced

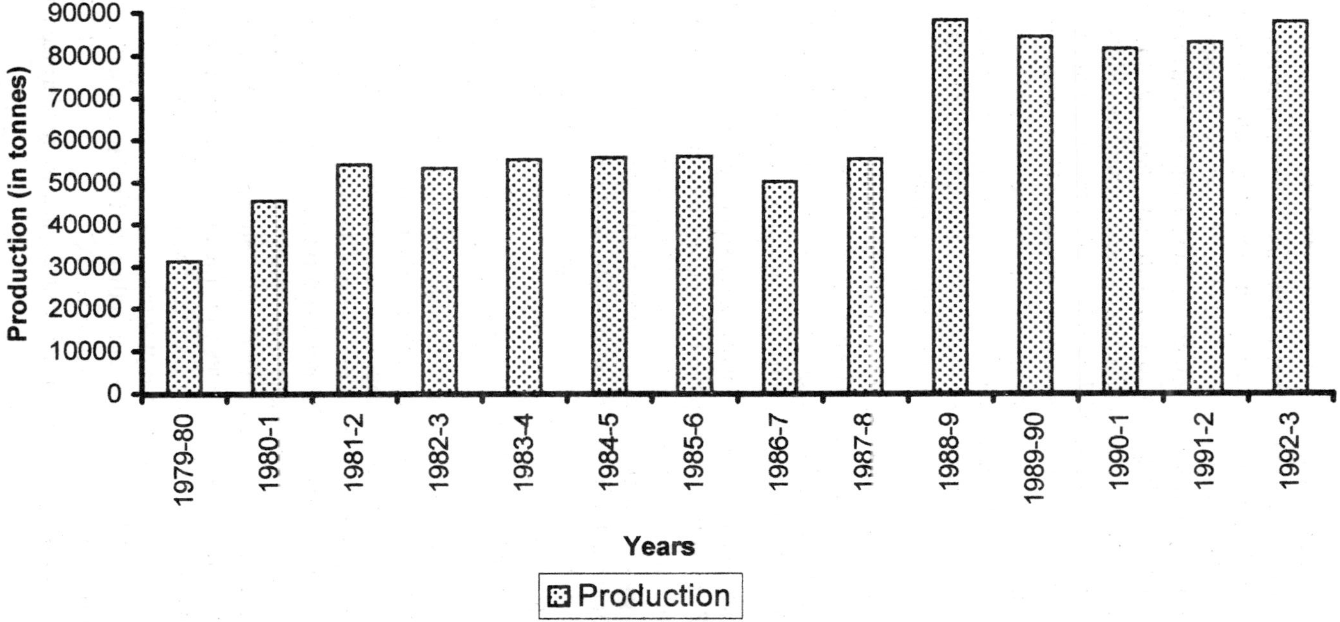

Source: ITCB Mills 1992 and SEOD

Figure 5.8: Production Figures of BPL from 1979-80 to 1992-3

white printing paper at all, since the first machine installed in the mill was an MG machine which is a combination of paper and board machine, and can only produce paper meant for cigarettes, matchboxes and duplex boards, i.e. the so-called industrial speciality papers.

A former employee inadvertently let slip the information that ITCB 'did not break its obligation' to supply the levy paper but buys white printing paper from a small paper mill—Telengana Paper Mills/Delta Paper Mills and requests this mill to give the levy paper to the government on its behalf. This act not only saved ITCB the trouble and cost of producing levy paper apart from the capital cost of installing suitable machinery for it, but it is also alleged to have turned profitable. Asking the small paper mills to deposit the levy paper directly to the government on 'its behalf' had two major advantages:

1. firstly, ITCB saved itself the payment of excise duty.
2. secondly, since ITCB is a bulk buyer of all the output of a small paper mill, it further got an extra discount for two months (A Research Scholar, personal communication, 1992).

The second act of the company raises still other serious questions. The AP Industrial Development Corporation is investing money in crores of rupees (see section of Capital Structure) which is actually the tax payers' money collected through taxation. Apart from this, the company gets concessions on valuable forest resources which again belong to the local people of this region who have lived there for hundreds of years. Are these concessions justifiable? Is the company producing paper which our country desperately needs? 60 to 70 per cent of its production is going to the parent concern as cigarette-wrap. The remaining 30 per cent goes to specialized markets like Terapacks for Dhara, Amul, Frooti, etc., to serve the needs of the elite.

The mill takes away precious resources—money as well as material with absolutely no return to the country. In such cases why should the government and its financial institutions step in and support large concerns which produce only value-added products and even carcinogenous products like cigarettes? One cannot overlook the fact that next to fossil fuel, cigarettes yield the highest tax revenue for the government. The company is also fleecing the government, using its resources for an unproductive purpose.

BPL, or ITCB as it is more correctly called today, is itself an outgrowth of a powerful multinational ITC group. Would ITCB be frightened of liberalization? Yes, because it has not modernized itself to international standards. A senior executive stated, 'liberalization has positive advantages to the consumer who will now get wider choice, but to the producer, opening out the flood gates is a shock'. India with its cheap labour, ready markets and skilled personnel, is a boon to new entrants like Sinar Mas. ITCB is gearing up for competition by becoming more energy-efficient to cut down costs, and will resort to oxygen bleaching 'for high brightness' in order to meet global challenges. The same executive complained that the government has liberalized the licensing system, but not the system which will permit the industry to cultivate its own plantations—at the rate of 2,000 ha per year even though it has adequate quantity of treated effluents to water it (Senior Executive, personal communication, March 1994, New Delhi).

ITC has set up a trading 'outpost' in Singapore and is opening another in Dubai. It wants to lease small-scale units for making handicrafts and procuring gems and jewellery and spices; it plans to acquire 24 per cent stakes in companies exporting granite, leather, garments, and to set up its own export units for processed foods, electronics, fatty chemicals, etc. It is also targeting imports of fertilizers, seeds, pulses, edible oil, scrap and newsprint (*Business Today*, 7 September 1992).

Product Development through R&D

The company claimed that right from its inception, it has believed in value-added products and has worked relentlessly to develop products for newly-emerging demands. It first succeeded in 1981 with the development of duplex board for 'matchbox outers' to substitute wood. Another product developed was 'topak board' for Corrugated Fibreboard Containers (CFC) for export packaging of tobacco. However, the most significant breakthrough, as stated by the mill, was the development of board for packaging of liquids in a Tetrapak system which has earned the company, clients such as Amul, Vijaya, and manufacturers of juices such as Frooti, Volfruit, Treetop, refined vegetable oils such as Dhara, etc. The credit for this development is given to the innovative in-house Research and Development Department which has led to a wide range of customer-oriented, super quality products tailor-made for specific applications (BPL, 19 March 1990 and *Money*, 4 March 1991)

Types of paper produced, their uses and prices are given in Table 5.22.

TABLE 5.22: TYPES OF PAPER PRODUCED BY BPL—PRICES AND USES

Types of paper	Price	Uses
Duplex Board (70 % production) 3 types (a) Vanishable (b) Non vanishable (c) Matchbox shell	Rs. 1,800 per tonne	(a) and (b) packaging of consumer items like tea, cigarettes, cerelac packing, everyday milk whitener, toothpaste (c) matchbox industry
Tetrapack Board	Rs. 800 per tonne	Packing of fruit juice, milk
Topak Board	Rs. 2,400 to Rs. 2,500 per tonne	Corrugated boxes, packing of tobacco and fruit for export
Absorbent Kraft	Rs. 17,000 per tonne	Lamination
Maplitho varieties	Rs. 21,000 per tonne	Writing and Printing
Super High Bright Paper (made up of 100 % imported pulp)	Rs. 24,000 per tonne	Balance sheets, mainly exported to Sri Lanka, Bangladesh and Singapore
Unconventional variety of Recycled paper	Rs. 20,000 per tonne	Computer stationery
Machine Glazed Poster Paper	Rs. 25,000 per tonne	Printing of posters, soap wrappers, matchbox
Chromo Paper, Chromo Board, Flourescent Paper Board	Rs. 25,000 per tonne	Printing
Diazobase	Rs. 22,000 per tonne	Ammonia painting, engineering drawing

Source: BPL, November 1992.

Technology

Advanced comprehensive technology for 'every aspect' of paper making was seen to be incorporated by the mills. BPL's paper technologist claimed that the mill has a highly versatile Paper and Board combination machine, the first of its kind in India, which produces a multiple range of cost-effective superior quality products. But BPL has not cared to use pulping technology that now saves precious cellulosic forest raw material or safe bleaching agents that public sector units in the south have. The mill continues to use obsolete chlorine dioxide technology for bleaching, even for the pulp it needs for its value-added papers and boards. The company official justifies the use of the dangerous chlorine dioxide on the grounds that it gives 'higher brightness which provides a unique marketing thrust and export opportunities'. He also claimed that the advanced electronic process instrumentation ensured a high degree of accuracy and consistency throughout the manufacturing process. The study team's observation is that automation is more the rule than the exception. The official was nearer the truth when he added that 'the electronic equipment helped in bringing down the labour power usually required for operating machines'. So much for the development of a backward area by a modern industry (personal communication, *Field Work*, November 1992).

Raw Material—Royalty Concessions

At the point of inception, the requirement of the raw material at full capacity (50,000 tonnes of paper per year) was 1 lakh tonnes of cellulosic raw materials, viz., bamboo and hardwoods annually, 60,000 tonnes of coal, 64 million units of power per year, and 10 million gallons of water daily.

The Government of Andhra Pradesh entered into a 20-year lease agreement with the BPL where it agreed to supply 60,000 tonnes of bamboo annually, against the requirement of 74,000 tonnes at full capacity production and 26,000 tonnes of hardwood which would be more than the requirement of 33,000 tonnes (*Economic Times*, 9 September 1978).

In 1979, the Government of Andhra Pradesh granted a package of incentives to the company which included a 50 per cent concession on royalty for the principal raw material, bamboo. However in November 1981, it announced a hike in royalty rates with effect from April 1980, which meant an immediate additional cost to the company of Rs. 1 crore per year. It was IDBI which intervened in this matter (being BPL's principal financier) and recommended to the government that it give a 10-year royalty holiday on bamboo. Since this was asking too much, the government agreed only to 5 years (50 per cent) concession with a provision for review at the end of 3 years. The company is so used to concessions and subsidies from the government that even a negligible hike in any rates is termed as an 'extraordinary hike' (*Business India*, 5 July 1982).

The 50 per cent concession which the government agreed on, for a period of five years was terminated midway at the end of 3 years while the 25 per cent concession in power tariff which was given for 10 years was discontinued at the end of 5 years. The Chief MD, K.L. Chugh was very critical regarding these actions of the government.

He claimed that the company was severely discriminated against and said that because of this premature withdrawal of committed concessions, the company was on the verge of sinking (*Economic Times*, 20 April 1985).

The company needs to be reminded that these concessions were subject to review midway. It should have been aware of this clause when it was accepting the concessions from the year 1981. The concessions given were accepted coolly, but once they were terminated, the company made a big issue out of it.

The mills' Chief Forest Manager while giving his views on the raw material question said that in reality the quantum of bamboo supplied by APFD was much less, as the bamboo supplied was 60,000 mt with moisture while the mills used bone-dry wood by removing its moisture, which then weighed 46,200 mt (bamboo contains 25 per cent moisture). At this point of time the actual requirement of the mill was 58,000 BDT (bone-dry tonnes). This shortfall was met by sending tenders to Andhra Pradesh Forest Development Corporation or through purchase from other states like MP, Maharashtra, Orissa, Assam. Since this turned out to be a costly affair, the mill started purchasing from middlemen or contractors who supplied the factory with bamboo according to its exact demand. Most of the time, the mill is allotted coupes by the Forest Department for bamboo working, which is carried on from October to June. In 1992, the allotment was made at the end of May instead of in September. Because of this delay, the company was unable to recruit a workforce, and when it did, the monsoons commenced. So the mill had to purchase huge quantities from Maharashtra, as claimed by BPL's Chief Forest Manager (personal communication, November 1992).

Regarding hardwoods, a mill employee informed us that there was no specific agreement with the Forest Department (FD) regarding its supply. In 1978, there was only an understanding that the FD would supply the mills lops and tops of miscellaneous wood, and the trunk would be used timber processing industry. Much wood used to be felled for timber.

In 1980-1, the mill was supplied with 74,000 mt of hardwood with bark, which had to be debarked. It complained that debarking meant a loss of 20 per cent. Removal of moisture meant a further loss of 35 per cent. In other words 74,000 mt yielded 38,877 bone-dry metric tonnes for actual use (BDT). Wood and bamboo are not actually weighed but calculated in terms of notional tonnes. The Forest

Department considers 2 cubic metres of wood = 1 mt, while the mills protested that in actuality it should be 2.8 cubic metres = 1 mt.

The statement of bamboo and wood receipts from 1979-80 to 1991-2 is given in Tables 5.23 and 5.24.

Import of Waste Paper

PPI units recycle old paper as cellulosic raw material. Most units including BPL till recently preferred waste paper from abroad. ITCB requires 40 per cent of waste paper which it imports from the US, Canada and Singapore, which it then converts into pulp. It was claimed that imported waste paper from Singapore was of much better quality than the paper produced in India (personal communication, November 1992).

The company imported 12,000 tonnes of waste paper from the US and Canada. In 1988, it also purchased 9,000 tonnes in the domestic market (*Business Standard*, 17 April 1989).

R.K. Kutty, Managing Director (MD) of ITCB, was said to be unhappy in 1991 on the import issue because according to him, the mills resorted to imports of fibres and waste paper to reduce dependence on farm and forest based raw materials. But the truth of the matter is that waste used to be imported at a zero rate of customs duty. Suddenly ITCB found itself slapped with a 10 per cent duty, which later was increased to 15 per cent, then further to 20 per cent and then stood pegged at 40 per cent in 1991. A depreciating rupee, and increasing freight rates further contributed to the company's woes. Together they caused a hike in the cost of raw materials and unhappiness to the MD. Most PPI including BPL recycle local paper and rags. With a number of good adhesives available, the prejudice for the inferior local paper is not voiced (*Express Investment Week*, 1991).

Social Forestry and Farm Forestry Activities of BPL

In 1979, BPL informed the state government of its willingness to make joint bamboo plantations with AP Forest Development Corporation over an area of 18,000 ha. This plan of the company did not materialize due to the Land Ceiling Act which does not allow corporate land-holdings for pulpwood plantations (*Economic Times*, 21 May 1979). It is also true that the Forest Department is afraid of such deals and of the pressures that the powerful PPI can exert on state government's

TABLE 5.23: RAW MATERIALS—STATEMENT OF BAMBOO RECEIPTS 1979-80 TO 1991-2

Year	AP Lease B.	AP Auction	APFDC	AP Pvt	MP	Orissa	Maharashtra	Assam	W. Bengal	Total
1979-80	45,461	—	—	—	875	1,904	—	—	—	48,240
1980-1	50,031	47	—	—	3,427	170	—	—	—	53,675
1981-2	56,643	45	—	—	6,206	287	—	665	—	63,846
1982-3	61,799	—	—	—	—	139	—	—	—	61,938
1983-4	55,990	—	—	—	9,304	—	—	—	—	65,294
1984-5	50,284	—	—	—	—	—	5,608	3,571	—	59,463
1985-6	42,975	264	—	—	1,677	—	10,395	1,661	—	56,972
1986-7	51,173	—	—	—	—	—	—	—	—	51,173
1987-8	77,903	131	—	46	255	40	—	—	—	78,375
1988-9										
1989-90	33,815	405	2,240	54	1,357	8,698	10,584	497	1,440	59,090
1990-1	29,147	375	5,966	108	846	6,695	7,264	558	43	51,002
1991-2	47,057	—	4,723	447	1,986	2,134	—	3,509	—	59,856

Source: BPL, October 1992.

TABLE 5.24: RAW MATERIALS—STATEMENT OF HARDWOOD RECEIPTS 1979–80 TO 1991–2

Year	AP Govt. Mixed Wood	AP Pvt. Mixed Wood	Eucalyptus Pvt.	Casuarina Pvt.	Subabul Pvt.	AP Timber Mixed	MP Forests	AP Auction	Total
1979-80	22,037	—	—	—	—	916	—	—	22,953
1980-1	38,877	2,695	—	—	—	15	1,499	—	43,086
1981-2	31,783	15,792	—	—	—	—	160	2,918	50,653
1982-3	29,276	1,088	—	—	—	—	—	—	30,364
1983-4	29,175	1,069	—	—	—	—	—	—	30,244
1984-5	6,688	19,402	3,550	1,532	—	—	—	—	31,172
1985-6	6,466	16,117	5,446	803	—	—	—	—	28,832
1986-7	—	11,147	2,407	866	—	—	—	—	14,420
1987-9 (18 months)	—	20,964	17,895	308	—	—	—	658	39,825
1989-90	—	23,351	28,580	13,702	—	—	—	—	65,633
1990-1	—	12,002	21,987	16,040	—	—	—	—	50,629
1991-2	—	14,940	21,858	13,962	2,879	—	—	—	53,639

Source: BPL, October 1992.

forest officials. The National Commission of Agriculture (1976) had for-bidden plantation activity by the PPI.

The company distributed 4 lakh seedlings in 1981-2 as a part of its social forestry programme (*Field Work,* November 1992). These operations were taken to relatively inaccessible areas by providing infrastructural facilities.

The company is also reported to have undertaken bamboo rejuvenation schemes in limited areas in a phased manner at its own expense in 1983 (*Economic Times,* 30 March 1983). No mention was made of this important activity in 1992 when the team visited the mills, its catchment area and farm forestry location.

In 1984-5, the company in association with NABARD undertook planting 10,500 ha of fallow and marginal agricultural land over 7 years. During the year 235 ha said to have been planted and also 31 lakh saplings distributed.

In 1990-1, about 1,400 ha of marginal land were claimed to have been brought under farm forestry. Also, the company received approval for planting 1,500 ha annually on marginal farmland from NABARD. NABARD in principle agreed to finance loans at a concessional rate of interest of 12.5 per cent to the farmers as well as to the company. In addition to this, the Company responded positively to a proposal from the APFDC to raise pulpwood plantations in the vicinity of the mill spread over 12,000 ha in an 8-year cycle (*Annual Report*, 1990-l).

The company has sought to alleviate its pulp problems by encouraging 'farmer plantations' of fast-growing varieties of trees like *subabul* and eucalyptus. For this it followed its social forestry programme (embarked upon in 1982) with a farm forestry programme from 1987 in Khammam and Guntur districts. The programme, as noted by the company, was targeted towards the raising of superior quality pulpwood on a sustained basis and increasing the productivity of marginal lands. It was being implemented under the refinancing scheme of NABARD, under which farmers raised plantations of fast-growing species on their marginal lands. High quality seedlings were provided from the company's nurseries.

The R and D Department played a major role in the production of raw material. It adopted the clonal selection and multiplication methods of propagation of improved planting stock, a technology successfully developed and implemented in Brazil. By this method, the best phenotype specimens Candidate Plus Trees (CPTs) were of *E. tereticornis* identified for testing and cloning. The best clones from these areas were then vegetatively propagated on a large scale, which involved the following process.

The clones are first planted in cubets which have a special type of soil (or sand). They are then injected with a hormone and kept for 35 days. In this period they develop roots. They are then placed in mist chambers made of polythene. During the day the chamber gets heated very fast. The sprinklers in the chamber are switched on to maintain humidity. The plants are kept in this state for 70 days. After this they are exposed to some light for 10 days where the rooted plants stabilize while the non-rooted ones die. Finally, they are brought into the open sunlight where further growth takes place.

The eucalyptus plants grown in this fashion are taller and sturdier. The company was sure that its own plantations developed on these lines would solve the raw material problem. It is not aware that the joint Indian and FAO scientists have found the cellulosic quality of *E. Tereticornis* grown in India is not quite up to the standards of paper pulp. GOI has not cared to convey this 1980 finding to PPI units. The company claimed that it made considerable investment in acquiring this knowledge and technology and had commenced a pilot clonal multiplication research centre at Bhadrachalam. In 1991, the mist chamber capacity for vegetative propagation of genetically superior clones was raised from 25,000 rooted plants to 1,50,000 rooted plants a year.

The company brought under plantation 1,800 ha in social forestry in AP by 1989, while under farm forestry 1,500 ha were sought to be brought under plantation. Since the plantation required a 7-year gestation period before the trees were large enough for industrial felling, the company provided the farmers seeds and biotechnology while NABARD provided them with loans. Under this scheme (FF) the farmer-entrepreneurs were guaranteed a buy back of the pulpwood at harvest at the maximum support price or the market price, whichever was higher.

One of the plantations that we visited under this scheme was undertaken on large farm land. Eucalyptus seedlings of about 2 in. growth were combined with intercrop such as green gram, black gram, groundnut, horsegram, etc. These are leguminous crops which help by adding atmospheric nitrogen to the soil. The plot was well protected from damage by cattle, when the leguminous plants were young. No judgement could be made on the effect of eucalyptus on the intercrops as they had still not shed their leaves (*Field Observation*, November 1992).

Effluents

BPL has a modern effluent treatment system installed at a cost of about Rs.70 lakh designed by NEERI, said to be in accordance with the parameters of ISI and the AP State Pollution Control Board. Recovery of chemicals being 92 per cent, the effluents containing the remaining 8 per cent chemicals were let out into the downstream of river Eruvandi. An ITCB employee claims that now the scene has changed, that 80 per cent of treated effluent water is illegally used for irrigation by farmers. Its chemicals have no doubt given a boost to growth. The mill has plans to form a cooperative and then supply the effluent water to the farmers free of cost. It hopes to increase the acreage under effluent irrigation. The ill effects of the chemical load will begin to show after about 10 years. The lignin in the effluent, the company official said, acts as humus. This is not correct. In fact, it is a non-bio-degradable organic matter used as landfill along with lime sludge by other pulp and paper industries. The salinity of soil increases as the effluent contains sodium. But according to a company executive, there was already an answer to this problem of salinity as the area is prone to floods. The flood water dissolves the salt, making the soil more fertile (Raghuveer, personal communication, November 1992).

Solid Waste

BPL uses coal ash for making bricks. These can be used for building purposes as they have enough strength. The credit for this goes to BPL's R and D wing. If this venture is a success BPL will soon be looking out for a market for its products (personal communication *Field Work*, November 1992).

Area Development

A core industry should initiate some sort of development, and should give rise to ancillary industries, trade and commerce, schools, hospitals, transport, etc., to say the least. If no immediate development takes place in a totally backward area it is understandable. But if even after two decades, no sign of any development is visible, the outlook is negative. The walled township contains neat housing, said to be built in 'Canadian village' style, for the executives and a section of blue and white collar workers. Others commute from Sarpaka village 10 km away, by company bus. The township has a small cooperative shop, but no hospital. It has two clubs, one for the officers and the other for lower-level employees. The company did not boast of any other welfare services.

Summary and Conclusion

Andhra remained a backwater as long as it was hitched to Madras Presidency. Telengana had proved to be a haven for Birla to enter into production of paper. The ancient AP Mill was handed over on a platter to the Bombay-based Somani as there was no local industrialist to take it over. In order to rectify the highly uneven development in AP, a local merchant of Rayalaseema was assisted by the state to set up a paper mill in that region. It appears that the giant cigarette company ITC was also interested in setting up its own paper unit. Its decision coincided with the interest of a powerful minister who hails from Bhadrachalam. ITC, therefore, was supported by state funds in setting up its mill in that area. In fact all the three mills are financially supported by AP state resources. Thapar's AP Rayon is outside the scope of this investigation. Three out of the four forest-based units draw their raw material from the northern region, mainly Telengana. Sirpur Mills has totally devastated its catchment area and the rest of the forest region is under strain. The Forest Department continues to supply bamboo to the paper mills but the mills are 'defiant' and claim that they produce their own raw material through farm forestry. Two species of eucalyptus and one of casuarina have been chosen; the latter is in the delta region, the rice bowl of Andhra Pradesh. Farm forestry is likely to create a negative impact mainly on the poor peasant due to warped agrarian relations in a country where agrarian reform did not take off at all.

Recommendations

The tribal, forest and paper belt are one and the same. If human rights are to be given priority, then the government of India must give the tribals the rights to their land. Developmental process in the region must begin and PPI must contribute to it.

The AP Forest Department has done commendable research and its FDC has raised a range of plantations and NABARD has released loans in 1994-5 for bamboo and eucalyptus plantations (Haque, personal communication). Government of India should forthwith stop giving wrong mandates to these departments. On the contrary it should strengthen the forests by reclothing them with mixed native species and eco-friendly pulpable species.

The status of the small and marginal farmer in Telengana and the Delta area and their relationship with the large landowners should be properly investigated by the Government of India and the rights of the farmer restored and protected. They should be further helped to grow eco-friendly pulpable crops that match with the soil and climatic conditions, be it bamboo, *kenaf*, hemp, or agave if they wish. This is an alternate in development for the lowest range of the farmers and for PPI too. The AP government has made heavy investment in all the three PPI units. Government of India should reciprocate by indicating the negative features of *E. Tereticornis* as pulpable raw material and the futility of cultivating this as well as *E. Camaldulensis* in regions unsuitable to it. This information must also be conveyed to NABARD.

All the three mills should not only follow bamboo felling practices, but should also incorporate the technique suggested by FD for preservation of moisture and growth of the rhizome.

NOTES

1. Rayalaseema Paper Mill was rated highly by a public sector financial institution. Yet it folded up. The question is why? Mismanagement is an over-simplified answer. A technologist, an insider, spoke about the split within the family where each son chose to utilize the company's resources to start afresh in new fields of production. The financial institution's assessment and the behaviour of the second generation of the Guptas do not match. As one perceptive scholar observed, 'the sickness' of the mill seems to have been engineered. Raw material from Kurnool is shared by the other three, but in reality diverted to Sirpur. All the three corporations were trying to get control of the sick unit. Modernization of a sick unit qualifies for tax concessions while setting up of a brand new unit does not, says L.M. Thapar, the MD and moving spirit of BILT. He should know. Over the last couple of years he has obtained control of the Titaghur Paper Mill and Seva Paper Mills, both in Orissa, and now also the Rayalaseema Paper Mills (RPM). Since RPM is not under production, it is outside the scope of the present study.
2. In the late 1920s, subsistence farmers in Guntur took to a radically different capital intensive crop, viz., Flue-Cured Virginia Tobacco (FCVT). This was possible because Indian Leaf Tobacco Development Ltd. (ILTD)

entered into a formal contractual relationship with the peasants providing finance, seedlings and fertilizers, in return ensuring purchase of FCV tobacco at specified prices. The value of inputs provided were deducted at the time of transfer of the produce (Nata Duvvury 1987: 9). It is significant that in Guntur and other areas, FCVT was cultivated in coastal Andhra Pradesh.

Around this time two categories of landowners were selling away their land. These were the impoverished peasants as well as the Brahmins, the non-cultivating landowners were migrating to urban areas. Such lands were bought by the rich peasants (Nata Duvvury: 18). For cultivation of tobacco the cultivators preferred to lease rather than purchase the land for cultivation, as they had perceived that tobacco cultivation led to exhaustion of the soil (Sayana 1949: 191 quoted in Duvvury 1987: 20). Richer peasants who had amassed surplus from commercial cultivation were increasingly displacing the moneylender castes from the rural credit scene (Duvvury 1987: 17). This is the second reason to show that backward linkages with the primary sector can be detrimental to the farmer. Farm and social forestry projects have now spread all over the state. The state had taken the initiative to set up PPI units in all the three regions and has now possibly invited trouble both in terms of loss of fertility of even prime land, as well as pauperizing the poor peasants. This is a question that the state has to have a close look at and bring about necessary changes to prevent disaster. Meanwhile, we take a closer look at it in the section which deals with the question of augmenting raw material.

3. Imports in 1981-2 were estimated at a little above 26,000 tonnes against 32,000 tonnes in 1980-1 and 21,700 tonnes in 1979-80 (Vikram Nair 1981).
4. A brief early history of ITC: At the beginning of this century there were about 500 cigarette/tobacco manufacturers in Britain, the oldest and by far the largest one being W.D & H.O Wills Ltd. of Bristol founded in the eighteenth century.

   In May 1901, the dominant American Tobacco Company (American Tobacco) from the United States entered the United Kingdom and purchased Ogden's Ltd., with the avowed purpose of capturing the British and European markets.

   In the face of this aggressive competition 13 United Kingdom manufacturers, including Wills, decided to amalgamate to form the Imperial Tobacco Company in 1901. Imperial Tobacco explored the possibilities of entering the United States market, but came to terms with American Tobacco. Each agreed to keep away from the other's home market and not to export to other markets except through the British American Tobacco Company Limited (BAT), a newly-formed company with two-third and one-third shares owned by American Tobacco and Imperial Tobacco respectively. Today BAT, as a multinational, monopolizes the cigarette industry in 50 countries through 150 dominant factories, mostly located in the erstwhile British Empire.

5. The National Development Council (NDC) in its meeting held on 13 September 1968 decided that two working groups be set up for studying the question of regional imbalances. The Planning Commission set up two working groups: (a) Pande Committee: For specifying the criteria for identification of backward areas, (b) Wanchoo Committee: For recommending, granting fiscal and financial incentives in the backward areas.

REFERENCES

Andhra Pradesh District Gazetteer, Krishna, Hyderabad.

Anon., 'Andhra Pradesh Forest Department R and D Circle, 1992', *Forestry Research in Andhra Pradesh.*

——— 1908, *Imperial Gazetteer of India,* Provincial Series, Rajahmundry, Madras, Calcutta.

——— 1977, 'Bhadrachalam Paperboards Limited', *Business Standard,* 11 May.

——— 1978, 'BPL to Enter Market', *Economic Times*, 18 August.

——— 1978, 'BPL—Largest Paper Making Unit, Taking Shape', *Economic Times*, 9 September.

——— 1982, 'BPL's Royal Ditch', *Business India*, 5 July.

——— 1983, 'BPL Sets Record', *Economic Times*, 30 March.

——— 1985, 'Withdrawal of Relief Hits BPL', *Economic Times*, 20 April.

——— 1989, 'Bhadrachalam to Expand Capacity', *Business Standard*, 17 April.

——— 1989, 'Planning Nagarjuna Tubes Takeover Bitten by "Diversification Bug"', *Financial Express*, 26 December.

——— 1990, 'BPL Poised for Faster Growth', *Economic Times*, 19 March.

——— 1991, 'BPL—How They Did It?', *Express Investment Week*, 6-12 May.

——— 1991, 'BPL Turnaround Through Technology', *Money*, 4 March.

——— 1992, 'Bhadrachalam Paper Board, 1992', *Business Today*, 7 September.

——— 1992, *Farm Forestry Project*, SPM's handout.

——— 1992, *Social Forestry Activities of SPM*, SPM's handout.

Andhra Pradesh Forest Development Corporation, 1991, *Annual Report 1987-8*, Hyderabad.

Andhra Pradesh Paper Mills, *Annual Reports*, 1979-80 to 1990-1.

———, 'A Brief Resume of APPM Raj', November 1992.

———, 1992, *Water Pollution Control Measures.*

Sirpur Paper Mills, *Annual Reports*, 1979-80 to 1990-1.

Bangur, G.C. 1982, Chairperson of APPM, Speech, *The Hindu,* 21 January.

——— Chairperson's Speech, APPM, *Economic Times,* 12 January.

Barua, Sanjay 1994, 'ITC May Tell BAT to Back Off', *Times of India*, 4 May.

Bhadrachalam Paper Boards (ITCB), *Annual Reports*, 1979-80 to 1990-3.

Chand, M. and V.K. Puri 1983, *Regional Planning in India*, New Delhi, Allied Publishers.

CMIE (n.d.), 'Multinationals—A Scourge Proven', CMIE File.

Dastagir, M.G., K.K. Suresh, and C. Surendran 1989, 'Inter-cropping with Different Aged Trees', *National Seminar on Casuarina*, Coimbatore.

Duvvury, Nata 1987, 'Imperialism and the Social Framework of Agriculture in Dryland South India: A Study of Tobacco in Guntur', *Economic History of Andhra Pradesh,* vol. I, Conference Paper, Sri Krishnadevaraya University, Anantpur.

Francis, W. 1905, *Madras District Gazetteer, Visakhapatnam*, Government Press, Madras.

——— 1907, *Madras District Gazetteer, East Godavari*, Government Press, Madras.

Government of Andhra Pradesh 1993, *Facts and Figures*, Forest Department, Hyderabad.

Hemingway, F.R. 1907, *Madras District Gazetteer, Godavari*, Government Press, Madras.

ICICI 1992, *A Sectoral Study of Paper Industry in India.*

Kholi, R.K. 1990, 'Allelopathic Properties of Eucalyptus', *Andhra Pradesh Forest Development Corporation Annual Report, 1987-8.* (Forestry Research in Andhra Pradesh 1992—Marching Towards 21st Century.)

Nair, Vikram 1981, 'APPM Poised for Growth', *Economic Times.*

Narayana Rao, B.K. 1960, *Agricultural Development in Hyderabad State 1900-56,* Keshav Prakashan.

Narayana Rao, J.S. 1987, *Fifth Annual Conference, AP Economic Association*, Osmania University, 17/18 January, Hyderabad.

Narayan Rao, K.V. 1973, *The Emergence of Andhra Pradesh*, Popular Prakashan, Bombay.

Prakash Rao, K.V. *Working Plan for the Forests of East Godavari District*, 1991-2 to 2005-6, Part II, GOAP, Forest Department.

——— *Working Plan for the Forests of East Godavari* 1991-2 to 2005-6, Part I, GOAP, Forest Department, Rajahmundry.

Puri, G.G. 1958, 'Contributions to the Ecology of Casuarina equisetifolia', *Indian Forester*, vol. LXXXIV, no. 2.

Rajgopal, M.V. 1976, *Andhra Pradesh District Gazetteer, Adilabad,* Hyderabad.

——— 1976, *Andhra Pradesh District Gazetteer, Warangal,* Hyderabad.

——— 1977, *Andhra Pradesh District Gazetteer, Khammam,* Hyderabad.

Ramensan, N. 1979, *Andhra Pradesh District Gazetteer, Visakhapatnam*, Hyderabad.

Ramesen, R. 1979, *Andhra Pradesh District Gazetteer, East Godhavari,* Hyderabad.

Rangarajan (n.d.), *Bamboo Resources in AP.*

Rao, A.L., *A General Purpose and Training Manual-cum-Field Guide for Raising and Maintaining Nurseries and Plantation of Casuarina equisetifolia*, APPM Ltd., Rajahmundry.

Rao, G.N. 1978, 'Stagnation and Decay in the Agricultural Economy of Coastal Andhra', *Artha Vijnana*, vol. XX, no. 3.

——— 1988 *Transition in the Agrarian Economy of Andhra Pradesh*, Presidential address to Andhra Pradesh History Congress, 4-8 January.

Reddy, B.K. Sudhakar, 'Regional Industrialization, Partial Transformation of a Late Starter', *Rural Industrialization.*

Roy, Abijit, 1992, 'The Bangur group splintering for the big Break', *Business Today*, September.

Sayana, V.V. 1949, *The Agrarian Problem of Madras Province,* Business Week Press, Madras.

Shankaran, K.V. 1993, 'Decomposition of Leaf Litter of Albizzia (Paraserianthes falcataria), Eucalypt (Eucalyptus teriticornis) and teak (Tectona Grandis) in Kerala, India', in *Forest Ecology and Management*, Amsterdam.

Shenoy, Meera 1992, 'APPM's Turnaround is not Enough', *Business India*, 30 March-12 April.

Sirpur Paper Mills, *Annual Reports*, 1979-80 to 1992-3, Sirpur.

Stock Exchange Official Directory (SEOD) Papers, Bombay, 1964-5 to 1993-4.

Stock Exchange Official Papers, 1979, Bombay, 1979 to 1993.

Subramaniam, K.N. et al. 1992, *Casuarina Tree of Multiple Utility,* Institute of Forest Genetics and Tree Breeding, Coimbatore.

Thirwat, Sukam 1955, 'The Eucalyptus for Tropical Climate', *Indian Forester*, vol. LXXXI, no. 4, Dehradun.

Venkatachalam, A.V., R.S. Vinaya Rai and G. Kumaravelu (n.d.), *Effect of Casuarina Equisetifolia; Forest on Soil Properties* (n.d.) (Mimeo).

Verma, Mamta, 1993, 'Effect of Pulp and Paper Mill Waste Water on the Mineral Nutrition and Growth of Crop-Plants', unpublished Ph.D. thesis, Nagpur University.

Vinaya, Rai R.S., V.M. Srinivasan, C. Surendran and G. Kumaravelu 1989, 'Comparative Efficacy of Casuarina Equisetifolia, F.R.G. Forest as an Agro Forestry Species', *National Seminar on Casuarina, Neyveli.*